Fundamentals of

Sustainable Chemical Science

Fundamentals of Sustainable Chemical Science

Stanley E. Manahan

CRC Press is an imprint of the
Taylor & Francis Group, an **informa** business

CRC Press
Taylor & Francis Group
6000 Broken Sound Parkway NW, Suite 300
Boca Raton, FL 33487-2742

Printed in the United States of America on acid-free paper
10 9 8 7 6 5 4 3 2 1

International Standard Book Number-13: 978-1-4398-0239-7 (Softcover)

Visit the Taylor & Francis Web site at
http://www.taylorandfrancis.com

and the CRC Press Web site at
http://www.crcpress.com

CONTENTS

PREFACE

Fundamentals of Environmental Chemistry, 3rd edition, is designed to build upon the approach successfully employed in the first two editions while also expanding the scope of the book into the strongly emerging area of Sustainability Science and Technology, which includes green chemistry and industrial ecology. The book takes into account the needs of those who have little or no knowledge of chemistry, but who require the basics of chemical science for their trade, profession, or study curriculum, as well as for readers who want to have an understanding of the fundamentals of sustainable chemistry and its crucial role in maintaining a livable planet.

The book are a basic course in chemical science that includes the fundamentals of organic chemistry and biochemistry. In presenting the fundamentals, every effort is made to relate them to real-world examples from environmental chemistry, green chemistry, and related areas while still maintaining brevity and simplicity.

One of the unique features of the book is a "mini-course" in chemistry presented in the first few pages of Chapter 1 and consisting of the most basic concepts and terms needed by the reader to really begin to understand chemistry. To study chemistry, it is necessary to know a few of the essentials, i.e., what an atom is and what is meant by elements, chemical formulas, chemical bonds, molecular mass, and chemical reactions. With these terms defined in very basic ways, it is then possible to go into greater detail concerning chemical concepts without having to assume—as many introductory chemistry books do somewhat awkwardly—that the reader knows the meaning of these terms.

Chapter 2 discusses matter largely on the basis of its physical nature and behavior, and introduces physical and chemical properties, states of matter, the mole as a quantity of matter, and other ideas required to visualize chemical substances as physical entities. Chapters 3–5 cover the core of chemical knowledge constructed as a language in which the elements and the atoms that form them (Chapter 3) are presented as letters of an alphabet, the compounds made up from the elements (Chapter 4) are analogous to words, the reactions by which compounds are synthesized and changed (Chapter 5) are like sentences in the chemical language, and the mathematical aspects hold it all together quantitatively. Chapters 6–8 constitute the remainder of the material that is usually regarded as essential in general chemistry. A basic coverage of organic chemistry is presented in Chapter 9. Although this topic is often omitted at the beginning chemistry level, those who deal with the real world of environmental pollution, hazardous wastes, agricultural science, and other applied areas quickly realize that a rudimentary understanding of organic chemistry is required. Chapter 10 covers biological chemistry, an area essential to understanding material presented in later chapters that deal with environmental and toxicological chemistry.

The author welcomes input from readers. Comments and questions may be sent to the author at the following e-mail address: manahans@missouri.edu

AUTHOR

Stanley E. Manahan is a professor of chemistry at the University of Missouri (Columbia) where he has been on the faculty since 1965. He received his A.B. in chemistry from the Emporia State University in 1960 and his PhD in analytical chemistry from the University of Kansas in 1965. Since 1968, his primary research and professional activities have been in environmental chemistry, toxicological chemistry, waste gasification, and gasification of biomass for energy production. His classic textbook, *Environmental Chemistry*, 8th edition (CRC Press, Boca Raton, Florida, 2004), has been in print continuously in various editions since 1972 and is the longest standing title on this subject in the world. Other books that he has written are *Green Chemistry and the Ten Commandments of Sustainability*, 2nd edition (ChemChar Research, Inc., 2006), *Green Science and Technology: The Path to a Sustainable Future*, 2nd edition (CRC Press/Taylor & Francis, 2006), *Toxicological Chemistry and Biochemistry*, 3rd edition (CRC Press/Lewis Publishers, 2001), *Industrial Ecology: Environmental Chemistry and Hazardous Waste* (CRC Press/Lewis Publishers, 1999), *Environmental Science and Technology* (CRC Press/Lewis Publishers, 1997), *Hazardous Waste Chemistry, Toxicology and Treatment* (Lewis Publishers, 1992), *Quantitative Chemical Analysis*, (Brooks/Cole, 1986), and *General Applied Chemistry*, 2nd edition (Willard Grant Press, 1982).

Dr. Manahan has lectured on the topics of environmental chemistry, toxicological chemistry, waste treatment, and green chemistry throughout the U.S. as an American Chemical Society Local Section Tour Speaker. He has also presented plenary lectures on these topics at international meetings in Puerto Rico, the University of the Andes in Mérida in Venezuela, Hokkaido University in Japan, the National Autonomous University in Mexico City, and in Italy and France. Since 1998, he has taught a short course annually at the National Autonomous University of Mexico. He was the recipient of the Year 2000 Award of the Environmental Chemistry Division of the Italian Chemical Society.

1. INTRODUCTION TO CHEMISTRY AND GREEN CHEMISTRY

1.1. IF WE DO NOT CHANGE DIRECTION

An old Chinese proverb states, "If we do not change direction, we are likely to end up where we are headed." At no time has this statement been more appropriate than as it applies to humankind at the present time. Blessed with a miniscule, but unique and remarkable, speck of the universe that has conditions conducive to life, humans are on a course that, if not altered, will result in destruction of the only home that they have or ever can have. About 2 billion years ago, one other type of organism, photosynthetic cyanobacteria, used captured solar energy to produce biomass and liberate atmospheric elemental oxygen to the atmosphere, making possible all of the life forms that require this gas for their metabolic processes. This was a planet-altering process that determined the life forms that have existed on Earth in all the eons since it occurred and that caused massive chemical change, such as the formation of iron oxide deposits from the oxidation of soluble iron in water. Until now, no type of organism has caused such a drastic change on Earth, particularly in its atmosphere. However, by burning enormous quantities of fossil fuels at an ever-increasing rate, humans are well on the way to doubling pre-industrial levels of atmospheric carbon dioxide. Virtually all reputable authorities agree that this will have a significant warming effect on the global climate. Whereas the atmosphere created by the humble single-celled cyanobacteria made possible the development of millions of kinds of new species, what humans are doing to the climate now will almost certainly result in the extinction of hundreds of thousands—perhaps millions—of species.

The challenge facing humankind today is **sustainability**, the maintenance and enhancement of conditions that will enable humans and other organisms to exist on Planet Earth. This means living within the limits of materials extracted from Earth or taken from its atmosphere or oceans. It means, especially, dealing sustainably with energy, essential to modern civilizations, but, with the present reliance on fossil fuels that pump global-warming carbon dioxide into the atmosphere, unsustainable with present patterns of acquisition and use.

So, collectively, humankind faces a monumental challenge. It is relatively easy to see the enormous problems that face us. It is also easy for some to debunk what they consider to be alarmist rhetoric and to contend that doubling the small fraction of a percent of carbon dioxide in the atmosphere within a few decades cannot possibly seriously damage Earth's carrying capacity or that cropland is so abundant and can be made so productive that diverting a significant portion of it to the intensive cultivation of corn-based ethanol fuel cannot have a significant adverse effect on world food production. Others contend that the situation is hopeless and that, in a world in which people of similar heritage who differ mainly in details of their creeds persist in trying to kill each other, it is futile to even try to get Earth's diverse peoples to cooperate on a global scale to ensure sustainability. Evidence is mounting against the deniers. It is a fact that glaciers are melting at a pace never before seen in the time over which historical records exist. It is a fact that people in cities in the far Southern Hemisphere over which the Antarctic ozone hole spreads during the early spring months of September and October find it necessary to take protective measures to prevent debilitating exposure to ultraviolet radiation that penetrates the stratospheric ozone layer thinned by the effects of chlorofluorocarbon (Freon) compounds. It is a fact that fluctuating prices of gasoline show that limits are being reached on available resources of petroleum.

As a result of the challenges facing humankind and its relationship with Planet Earth, thoughtful people have begun to take action to try to ensure their survival and that of their descendents. Measures to prevent loss of essential Earth support systems have been ongoing in some sectors for quite some time. By 1900, it was recognized that the cultivation of soil in the U.S. that had lain undisturbed until the arrival of European settlers was causing unacceptable soil erosion. The great damage caused to former prairie lands of the U.S. Great Plains by drought and wind erosion during the dustbowl days of the 1930s led to intensive government-funded initiatives in soil conservation. Rachel Carson's classic 1962 book *Silent Spring* brought to public attention the damage caused to wildlife by the indiscriminate use of persistent pesticides and helped bring about the massive environmental improvement efforts of much of the world in the latter part of the 20th century. By around 2000, the environmental movement had developed an emphasis upon sustainability and self-regulating systems for environmental preservation. Industrial ecology, which treats industrial enterprises as mutually beneficial interacting systems analogous to species in natural ecosystems, had emerged as a dynamic discipline. Since the mid-1990s, there has been a strong "green" movement in which human enterprises are dedicated to environmental protection, efficient uses of materials and energy, maximum recycling, and minimum generation of waste. Now many individuals and enterprises are dedicated to green engineering and, specifically in chemistry, to green chemistry.

1.2. THE ESSENTIAL ROLE OF CHEMISTRY

Chemistry is unavoidable. We eat chemicals. We are made of chemicals. We are surrounded by chemicals. All of these things are true because chemistry is the science of matter; all things are chemical. Therefore, chemistry deals with the air we breathe, the water we drink, the soil that grows our food, and vital life substances and processes. Our own bodies contain a vast variety of chemical substances and are tremendously sophisticated chemical factories that carry out an incredible number of complex chemical processes.

There is a tremendous concern today about the uses—and particularly the misuses—of chemistry as it relates to the environment. Ongoing events serve as constant reminders of threats to the environment ranging from individual exposures to toxicants to phenomena on a global scale that may cause massive, perhaps catastrophic, alterations in climate as discussed above. These include, as examples, air quality in Beijing so bad that it raised concerns over the health of athletes in the 2008 Summer Olympics, pets dying from consumption of food illegally laced with toxic melamine put into animal food to artificially raise its nitrogen content (used as a measure of protein), and tanks of toxic chlorine gas combined with terrorist explosives to add to the toll and misery caused by these diabolical devices in Iraq. Furthermore, large numbers of employees must deal with hazardous substances and wastes in laboratories and the workplace. All such matters involve chemistry for understanding of the problems and for arriving at solutions to them.

The central role that chemistry must play in dealing with the challenges facing humankind and its stewardship of Planet Earth is undeniable. People in a large variety of areas and enlightened citizens need to have some basic knowledge of chemistry. The purpose of this book is to provide such knowledge for a broad range of people who want and need it. The book presents an overview of chemistry, including organic chemistry (Chapter 9) and biological chemistry (Chapter 10), at a fundamental level. The book relates to sustainability science, environmental chemistry, and green chemistry.

1.3. ENVIRONMENTAL CHEMISTRY AND GREEN CHEMISTRY

Environmental chemistry is that branch of chemistry that deals with the origins, transport, reactions, effects, and fates of chemical species in the water, air, earth, and living environments and the influence of human activities thereon.[1] A related discipline, toxicological chemistry, is the chemistry of toxic substances with emphasis upon their interaction with biologic tissue and living systems.[2] Besides being an essential, vital discipline in its own right, environmental chemistry provides an excellent framework for the study of chemistry, dealing with "general chemistry," organic chemistry, chemical analysis, physical chemistry, photochemistry, geochemistry, and biological chemistry.

In its earlier stages of development dating from around the 1960s, the emphasis in environmental chemistry was upon detection of pollution, cataloging its adverse effects, and controlling pollutants once they were produced. This resulted in an

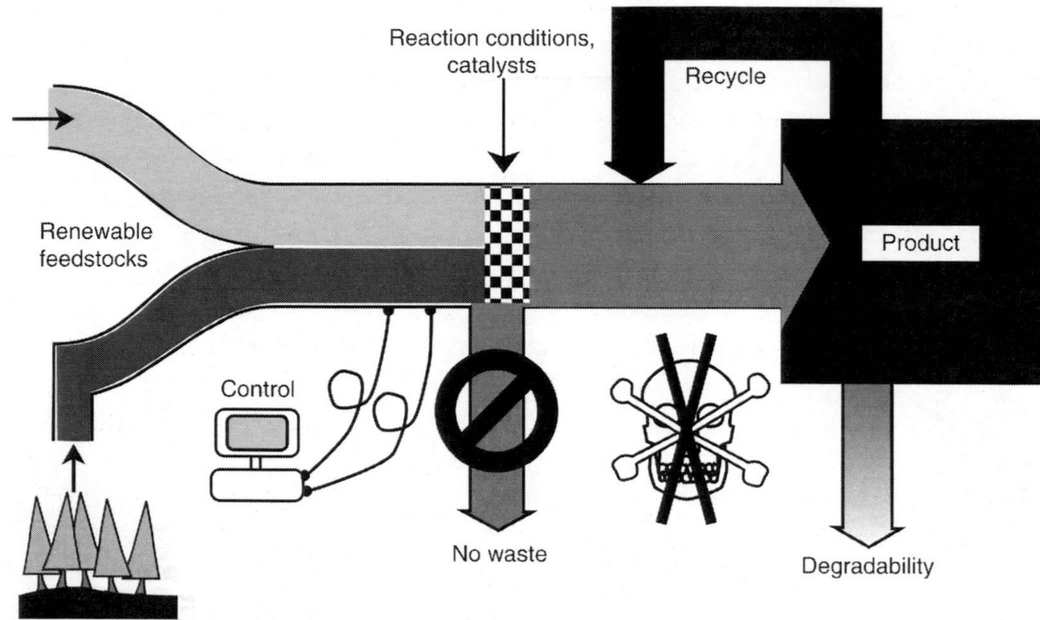

Figure 1.1. Illustration of the definition of green chemistry.

abundance of laws and regulations in countries throughout the world that, in many areas, prevented pollution from becoming worse and in a number of cases improved air and water quality and controlled toxic and hazardous substances improperly disposed on or buried beneath Earth's surface. Even as these efforts gained success, it became obvious that such measures were increasingly expensive to implement as allowable pollutant levels became lower. The "command-and-control" approach was hard to monitor and potential polluters were always tempted to avoid controls to save costs. A better way was needed. Insofar as chemistry was concerned, a better way emerged in the practice of green chemistry.

Formulated in the latter 1990s by Paul Anastas and others, green chemistry is the practice of chemical science and engineering in ways that minimize consumption of materials and energy, produce the least possible amount of waste, and is inherently safe.[3] Figure 1.1 illustrates this definition of green chemistry. Ideally, green chemistry is practiced within a framework of **industrial ecology**, defined as a comprehensive approach to production, distribution, utilization, and termination of goods and services in a manner that maximizes mutually beneficial utilization of materials and energy while preventing the production of wastes and pollutants.[4] Within the short time that it has been recognized as a distinct area of endeavor, green chemistry has become an integral, important part of the practice of chemical science.

1.4. A MINI-COURSE IN CHEMISTRY

It is much easier to learn chemistry if one already knows some chemistry! That is, in order to go into any detail on any chemical topic, it is extremely helpful to have some very rudimentary knowledge of chemistry as a whole. For example, a crucial part of chemistry is an understanding of the nature of chemical compounds, the chemical formulas used to describe them, and the chemical bonds that hold them together; these are topics addressed in Chapter 4 of this book. However, to understand these concepts, it is very helpful to know some things about the chemical reactions by which chemical compounds are formed, as addressed in Chapter 5. To work around this problem, Chapter 1 provides a highly condensed, simplified, but meaningful overview of chemistry to give the reader the essential concepts and terms required to understand more advanced chemical material.

1.5. THE BUILDING BLOCKS OF MATTER

All matter is composed of only about a hundred fundamental kinds of matter called **elements**. Each element is made up of very small entities called **atoms**; all atoms of the same element behave identically chemically. The study of chemistry, therefore, can logically begin with elements and the atoms of which they are composed.

Subatomic Particles and Atoms

Figure 1.2 represents an atom of deuterium, a form of the element hydrogen. It is seen that such an atom is made up of even smaller **subatomic particles**—positively charged **protons**, negatively charged **electrons**, and uncharged (neutral)

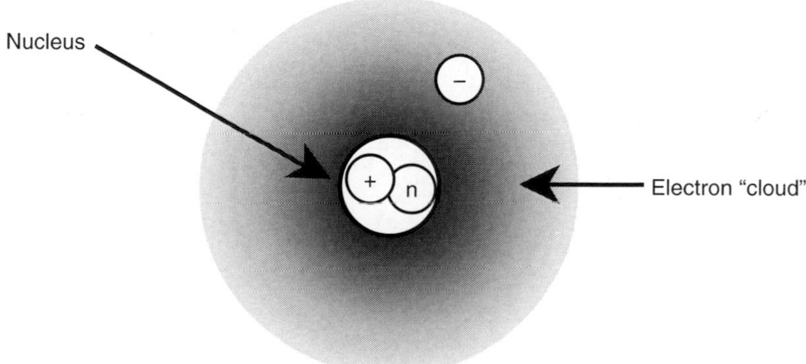

Figure 1.2. Representation of a deuterium atom. The nucleus contains one proton (+) and one neutron (n). The electron (–) is in constant, rapid motion around the nucleus, forming a cloud of negative electrical charge, the density of which drops off with increasing distance from the nucleus.

neutrons. Protons and neutrons have relatively high masses compared with electrons and are contained in the positively charged **nucleus** of the atom. The nucleus has essentially all the mass, but occupies virtually none of the volume, of the atom. An uncharged atom has the same number of electrons as protons. The electrons in an atom are contained in a cloud of negative charge around the nucleus that occupies most of the volume of the atom.

Atoms and Elements

All of the literally millions of different substances are composed of only around 100 elements. Each atom of a particular element is chemically identical to every other atom and contains the same number of protons in its nucleus. This number of protons in the nucleus of each atom of an element is the **atomic number** of the element. Atomic numbers are integers ranging from 1 to more than 100, each of which denotes a particular element. In addition to atomic numbers, each element has a name and a **chemical symbol**, such as carbon, C; potassium, K (for its Latin name kalium); or cadmium, Cd. In addition to atomic number, name, and chemical symbol, each element has an **atomic mass** (atomic weight). The atomic mass of each element is the average mass of all atoms of the element, including the various isotopes of which it consists. The **atomic mass unit, u** (also called the **dalton**), is used to express masses of individual atoms and molecules (aggregates of atoms). These terms are illustrated for carbon and nitrogen in Figure 1.3.

Although atoms of the same element are *chemically* identical, atoms of most elements consist of two or more **isotopes** that have different numbers of neutrons in

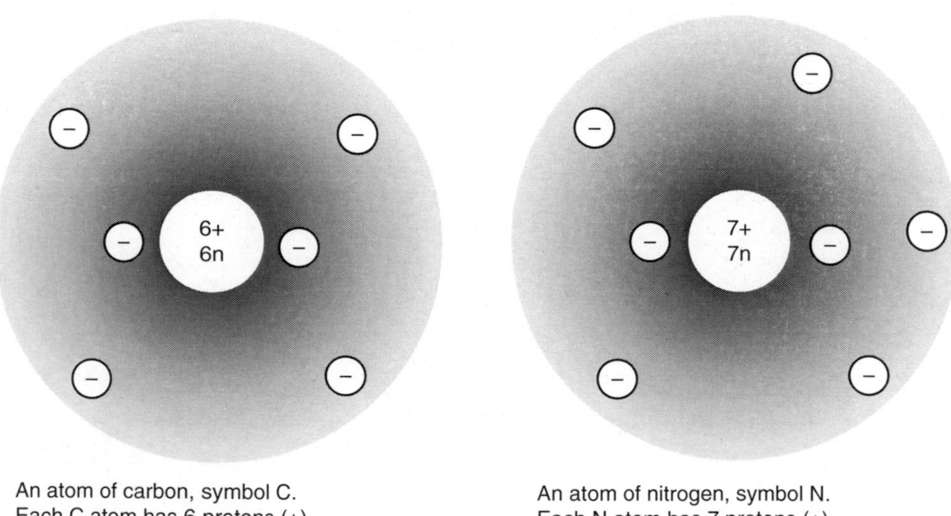

Figure 1.3. Atoms of carbon and nitrogen.

Table 1.1. Some of the More Important Common Elements

Element	Symbol	Atomic Number	Atomic Mass (Relative to Carbon-12)
Argon	Ar	18	39.948
Bromine	Br	35	79.904
Calcium	Ca	20	40.08
Carbon	C	6	12.01115
Chlorine	Cl	17	35.453
Copper	Cu	29	63.546
Fluorine	F	9	18.998403
Helium	He	2	4.00260
Hydrogen	H	1	1.0080
Iron	Fe	26	55.847
Magnesium	Mg	12	24.305
Mercury	Hg	80	200.59
Neon	Ne	10	20.179
Nitrogen	N	7	14.0067
Oxygen	O	8	15.9994
Potassium	K	19	39.0983
Silicon	Si	14	28.0855
Sodium	Na	11	22.9898
Sulfur	S	16	32.06

their nuclei. Some isotopes are **radioactive isotopes** or **radionuclides**, which have unstable nuclei that give off charged particles and gamma rays in the form of **radioactivity**. This process of **radioactive decay** changes atoms of a particular element into atoms of another element.

Throughout this book, reference is made to various elements. A list of the known elements is at the end of Chapter 3. Fortunately, most of the chemistry covered in this book requires familiarity with only about 25–30 elements. An abbreviated list of a few of the most important elements that the reader should learn at this point is given in Table 1.1.

The Periodic Table

When elements are considered in order of increasing atomic number, it is observed that their properties are repeated in a periodic manner. For example, elements with atomic numbers 2, 10, and 18 are gases that do not undergo chemical

Transition Elements

Period	IA 1	IIA 2	IIIB 3	IVB 4	VB 5	VIB 6	VIIB 7	VIII 8	VIII 9	VIII 10	IB 11	IIB 12	IIIA 13	IVA 14	VA 15	VIA 16	VIIA 17	Noble Gases 18
1	1 H 1.008																	2 He 4.003
2	3 Li 6.941	4 Be 9.012											5 B 10.81	6 C 12.01	7 N 14.01	8 O 16.00	9 F 19.00	10 Ne 20.18
3	11 Na 22.99	12 Mg 24.3											13 Al 26.98	14 Si 28.09	15 P 30.97	16 S 32.07	17 Cl 35.45	18 Ar 39.95
4	19 K 39.10	20 Ca 40.08	21 Sc 44.96	22 Ti 47.88	23 V 50.94	24 Cr 52.00	25 Mn 54.94	26 Fe 55.85	27 Co 58.93	28 Ni 58.69	29 Cu 63.55	30 Zn 65.39	31 Ga 69.72	32 Ge 72.59	33 As 74.92	34 Se 78.96	35 Br 79.9	36 Kr 83.8
5	37 Rb 85.47	38 Sr 87.62	39 Y 88.91	40 Zr 91.22	41 Nb 92.91	42 Mo 95.94	43 Tc 98.91	44 Ru 101.1	45 Rh 102.9	46 Pd 106.4	47 Ag 107.9	48 Cd 112.4	49 In 114.8	50 Sn 118.7	51 Sb 121.8	52 Te 127.6	53 I 126.9	54 Xe 131.3
6	55 Cs 132.9	56 Ba 137.3	57 La 138.9 *	72 Hf 178.5	73 Ta 180.9	74 W 183.8	75 Re 186.2	76 Os 190.2	77 Ir 192.2	78 Pt 195.1	79 Au 197.0	80 Hg 200.6	81 Tl 204.4	82 Pb 207.2	83 Bi 209.0	84 Po (210)	85 At (210)	86 Rn (222)
7	87 Fr (223)	88 Ra (226)	89 Ac (227) *	104 Rf (261)	105 Ha (262)	106 Sg (263)	107 Ns (262)	108 Ha (265)	109 Mt (266)									

Inner Transition Elements

Series	Period														
Lanthanide series *	6	58 Ce 140.1	59 Pr 140.9	60 Nd 144.2	61 Pm 144.9	62 Sm 150.4	63 Eu 152.0	64 Gd 157.2	65 Tb 158.9	66 Dy 162.5	67 Ho 164.9	68 Er 167.3	69 Tm 168.9	70 Yb 173.0	71 Lu 175.0
Actinide series *	7	90 Th 232.0	91 Pa 231.0	92 U 238.0	93 Np 237.0	94 Pu 239.1	95 Am 243.1	96 Cm 247.1	97 Bk 247.1	98 Cf 252.1	99 Es 252.1	100 Fm 257.1	101 Md 256.1	102 No 259.1	103 Lr 260.1

Figure 1.4. The periodic table of the elements.

reactions and consist of individual molecules, whereas those with atomic numbers larger by one—3, 11, and 19—are unstable, highly reactive metals. An arrangement of the elements in a manner that reflects this recurring behavior is known as the periodic table (Figure 1.4). The periodic table is extremely useful in understanding chemistry and predicting chemical behavior. The entry for each element in the periodic table gives the element's atomic number, symbol, and atomic mass. More-detailed versions of the table include each element name and other information as well.

Features of the Periodic Table

The periodic table gets its name from the fact that the properties of elements are repeated periodically in going from left to right across a horizontal row of elements. The table is arranged such that an element has properties similar to those of other elements above or below it in the table. Elements with similar chemical properties are called **groups** of elements and are contained in vertical columns in the periodic table.

1.6. CHEMICAL BONDS AND COMPOUNDS

Only a few elements, particularly the noble gases, exist as individual atoms; most atoms are joined by chemical bonds to other atoms. This can be illustrated very simply by elemental hydrogen, which exists as **molecules**, each consisting of 2 H atoms linked by a **chemical bond** as shown in Figure 1.5. Because hydrogen molecules contain 2 H atoms, they are said to be diatomic and are denoted by the **chemical formula** H_2. The H atoms in the H_2 molecule are held together by a **covalent bond**

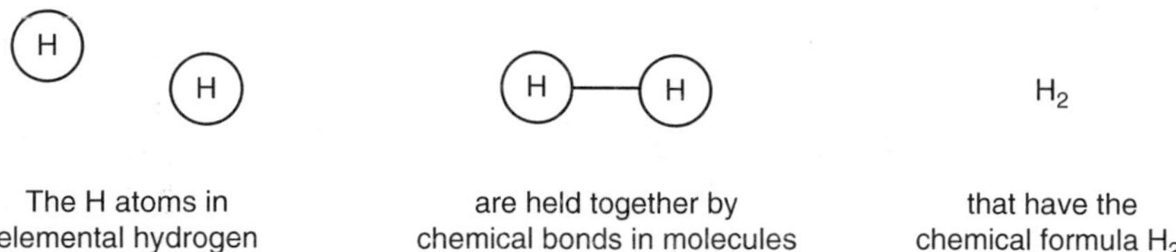

Figure 1.5. Molecule of H_2.

made up of 2 electrons, each contributed by one of the H atoms, and shared between the atoms.

As discussed in latter parts of this book, chemical bonds are extremely important in environmental chemistry and green chemistry. For example, the very strong bonds between atoms of carbon and those of oxygen in the compound carbon dioxide are responsible for the high stability of this gas in Earth's atmosphere, enabling it to persist and contribute to "greenhouse" warming of the atmosphere. The relatively weak bonds between nitrogen and oxygen in air-pollutant nitrogen dioxide make this compound susceptible to breaking apart when it absorbs radiation from the Sun, leading to reactive free oxygen atoms that start the process by which photochemical smog is formed.

Chemical Compounds

Most substances consist of two or more elements joined by chemical bonds. As an example, consider the chemical combination of the elements hydrogen and oxygen shown in Figure 1.6. Oxygen, chemical symbol O, has an atomic number of 8 and an atomic mass of 16.00 and exists in the elemental form comprising about 20% of air as diatomic molecules of O_2, each composed of two atoms of oxygen joined by a chemical bond. Hydrogen atoms combine with oxygen atoms to form molecules in which 2 H atoms are bonded to 1 O atom in a substance with the chemical formula H_2O (water). A substance such as H_2O that consists of a chemically bonded combination of two or more elements is called a **chemical compound.** (A chemical compound is a substance that consists of atoms of two or more different elements bonded together.) In the chemical formula for water, the letters H and O

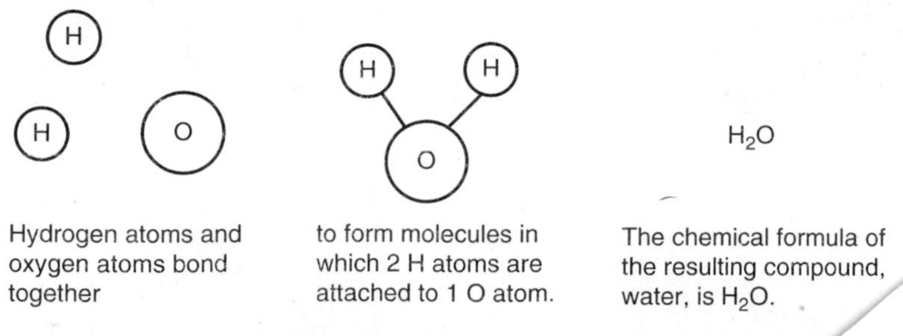

Figure 1.6. A molecule of water, H_2O, formed from 2 H atoms and 1 O at[illegible] chemical bonds.

are the chemical symbols of the two elements in the compound and the subscript 2 indicates that there are 2 H atoms per O atom. (The absence of a subscript after the O denotes the presence of just 1 O atom in the molecule.) Each of the chemical bonds holding a hydrogen atom to the oxygen atom in the water molecule is composed of two electrons shared between the hydrogen and oxygen atoms.

Ionic Bonds

As shown in Figure 1.7, the transfer of electrons from one atom to another produces charged species called **ions**. Positively charged ions are called **cations** and negatively charged ions are called **anions**. Ions that make up a solid compound are held together by **ionic bonds** in a **crystalline lattice** consisting of an ordered arrangement of the ions in which each cation is largely surrounded by anions and each anion by cations. The attracting forces of the oppositely charged ions in the crystalline lattice constitute the ionic bonds in the compound.

The formation of the ionic compound magnesium oxide is shown in Figure 1.7. In naming this compound, the cation is simply given the name of the element from which it was formed, magnesium. However, the ending of the name of the anion, ox*ide*, is different from that of the element from which it was formed, ox*ygen*.

Rather than individual atoms that have lost or gained electrons, many ions are groups of atoms bonded together covalently and having a net charge. A common example of such an ion is the ammonium ion, NH_4^+

$$\begin{array}{ccc} & H & + \\ & | & \\ H- & N & -H \\ & | & \\ & H & \end{array} \qquad \textbf{Ammonium ion,}\quad NH_4^+$$

consisting of 4 hydrogen atoms covalently bonded to a single nitrogen (N) atom and having a net electrical charge of +1 for the whole cation.

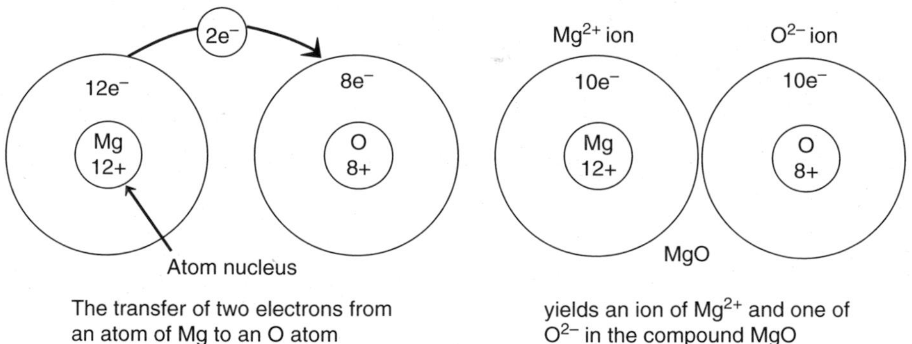

Figure 1.7. Ionic bonds are formed by the transfer of electrons and the mutual attraction of [illegible]ely charged ions in a crystalline lattice.

Summary of Chemical Compounds and the Chemical Bond

The preceding several pages have just covered some material on chemical compounds and bonds that are essential to understand chemistry. To summarize, these are as follows:

- Atoms of two or more different elements can form *chemical bonds* with each other to yield a product that is entirely different from the elements.
- Such a substance is called a *chemical compound.*
- The *formula* of a chemical compound gives the symbols of the elements and uses subscripts to show the relative numbers of atoms of each element in the compound.
- *Molecules* of some compounds are held together by *covalent bonds* consisting of shared electrons.
- Another kind of compound consists of *ions* composed of electrically charged atoms or groups of atoms held together by *ionic bonds* that exist because of the mutual attraction of oppositely charged ions.

Molecular Mass

The average mass of all molecules of a compound is its **molecular mass** (formerly called molecular weight). The molecular mass of a compound is calculated by multiplying the atomic mass of each element by the relative number of atoms of the element, then adding all the values obtained for each element in the compound. For example, the molecular mass of NH_3 is $14.0 + 3 \times 1.0 = 17.0$. As another example consider the following calculation of the molecular mass of ethylene (C_2H_4):

1. The chemical formula of the compound is C_2H_4.
2. Each molecule of C_2H_4 consists of 2 C atoms and 4 H atoms.
3. From the periodic table or Table 1.1, the atomic mass of C is 12.0 and that of H is 1.0.
4. Therefore, the molecular mass of C_2H_4 is

$$\underbrace{12.0 + 12.0}_{\text{From 2 C atoms}} + \underbrace{1.0 + 1.0 + 1.0 + 1.0}_{\text{From 4 H atoms}} = 28.0$$

1.7. CHEMICAL REACTIONS AND EQUATIONS

Chemical reactions occur when substances are changed to other substances through the breaking and formation of chemical bonds. For example, water is produced by the chemical reaction of hydrogen and oxygen:

Hydrogen plus oxygen yields water

Chemical reactions are written as **chemical** equations. The chemical reaction between hydrogen and water is written as the **balanced chemical equation**

$$2H_2 + O_2 \rightarrow 2H_2O \tag{1.7.1}$$

in which the arrow is read as "yields" and separates the hydrogen and oxygen **reactants** from the water **product**. Note that because elemental hydrogen and elemental oxygen occur as *diatomic molecules* of H_2 and O_2, respectively, it is necessary to write the equation in a way that reflects these correct chemical formulas of the elemental form. All correctly written chemical equations are **balanced**, in that *they must show the same number of each kind of atom on both sides of the equation.* The equation above is balanced because of the following:

On the left

- There are 2 H_2 *molecules,* each containing 2 H *atoms* for a total of 4 H atoms on the left.
- There is 1 O_2 *molecule,* containing 2 O *atoms* for a total of 2 O atoms on the left.

On the right

- There are 2 H_2O *molecules* each containing 2 H *atoms* and 1 O atom for a total of 4 H atoms and 2 O atoms on the right.

The process of balancing chemical equations is relatively straightforward for simple equations. It is discussed in Chapter 4.

1.8. NUMBERS IN CHEMISTRY: EXPONENTIAL NOTATION

An essential skill in chemistry is the ability to handle numbers, including very large and very small numbers. An example of the former is Avogadro's number, which is discussed in detail in Chapters 2 and 3. Avogadro's number is a way of expressing quantities of entities such as atoms or molecules and is equal to approximately 602 000 000 000 000 000 000 000. A number so large written in this decimal form is very cumbersome to express and very difficult to handle in calculations. It can be expressed much more conveniently in exponential notation. Avogadro's

Table 1.2. Numbers in Exponential and Decimal Form

Exponential Form of Number	Places Decimal Moved for Decimal Form	Decimal Form
$1.37 \times 10^5 = 1.37 \times 10 \times 10 \times 10 \times 10 \times 10$	→ 5 places	137 000
$7.19 \times 10^7 = 7.19 \times 10 \times 10 \times 10 \times 10 \times 10 \times 10 \times 10$	→ 7 places	71 900 000
$3.25 \times 10^{-2} = 3.25/(10 \times 10)$	← 2 places	0.0325
$2.6 \times 10^{-6} = 2.6/(10 \times 10 \times 10 \times 10 \times 10 \times 10)$	← 6 places	0.000 0026
$5.39 \times 10^{-5} = 5.39/(10 \times 10 \times 10 \times 10 \times 10)$	← 5 places	0.000 0539

number in exponential notation is 6.02×10^{23}. It is put into decimal form by moving the decimal in 6.02 to the right by 23 places. Exponential notation works equally well to express very small numbers, such as 0.000 000 000 000 000 087. In exponential notation, this number is 8.7×10^{-17}. To convert this number back to decimal form, the decimal point in 8.7 is simply moved 17 places to the left.

A number in exponential notation consists of a *digital number* equal to or greater than exactly 1 and less than exactly 10 (e.g., 1.00000, 4.3, 6.913, 8.005, and 9.99999) multiplied by a *power of 10* (e.g., 10^{-17}, 10^{13}, 10^{-5}, 10^3, and 10^{23}). Some examples of numbers expressed in exponential notation are given in Table 1.2. As can be seen in the second column of the table, a positive power of 10 shows the number of times that the digital number is multiplied by 10 and a negative power of 10 shows the number of times that the digital number is divided by 10.

Addition and Subtraction of Exponential Numbers

An electronic calculator keeps track of exponents automatically and with total accuracy. For example, getting the sum $7.13 \times 10^3 + 3.26 \times 10^4$ on a calculator simply involves the following sequence:

[7.13 EE3] [+] [3.26 EE4] [=] [3.97 EE4]

where 3.97 EE4 stands for 3.97×10^4. To do such a sum manually, the largest number in the sum should be set up in the standard exponential notation form and each of the other numbers should be taken to the same power of 10 as that of the largest number as shown below for the calculation of $3.07 \times 10^{-2} - 6.22 \times 10^{-3} + 4.14 \times 10^{-4}$:

3.07×10^{-2} (largest number, digital portion between 1 and 10)
-0.622×10^{-2} (same as 6.22×10^{-3})
$+0.041 \times 10^{-3}$ (same as 4.1×10^{-4})

Answer: 2.49×10^{-2}.

Multiplication and Division of Exponential Numbers

As with addition and subtraction, multiplication and division of exponential numbers on a calculator or computer are simply a matter of (correctly) pushing buttons. For example, to solve

$$\frac{1.39 \times 10^{-2} \times 9.05 \times 10^{8}}{3.11 \times 10^{4}}$$

on a calculator, the sequence below is followed:

[1.39 EE-2] [×] [9.05 EE8] [÷] [3.11 EE4] [=] [4.04 EE2] (same as 4.04×10^2)

In multiplication and division of exponential numbers, the digital portions of the numbers are handled conventionally. For the powers of 10 in multiplication exponents are added algebraically, whereas in division the exponents are subtracted algebraically. Therefore, in the preceding example,

$$\frac{1.39 \times 10^{-2} \times 9.05 \times 10^{8}}{3.11 \times 10^{4}}$$

the digital portion is

$$\frac{1.39 \times 9.05}{3.11} = 4.04$$

and the exponential portion is

$$\frac{10^{-2} \times 10^{8}}{10^{4}} = 10^{2}$$

So the answer is 4.04×10^2.

Example: Perform the calculation below without using the exponential notation feature of the calculator:

$$\frac{7.39 \times 10^{-2} \times 4.09 \times 10^{5}}{2.22 \times 10^{4} \times 1.03 \times 10^{-3}}$$

Answer:

$$\text{Exponent of answer} = \underbrace{-2 + 5}_{\text{Algebraic addition of exponents in the numerator}} - \underbrace{(4 - 3)}_{\text{Algebraic subtraction of exponents in the denominator}} = 2$$

Answer $= 13.2 \times 10^2 = 1.32 \times 10^3$

Example: Solve

$$\frac{3.49 \times 10^{3}}{3.26 \times 10^{18} \times 7.47 \times 10^{-5} \times 6.18 \times 10^{-8}}$$

Answer: 2.32×10^{-4}.

1.9. SIGNIFICANT FIGURES AND UNCERTAINTIES IN NUMBERS

The preceding section illustrated how to handle very large and very small numbers with *exponential notation.* This section considers **uncertainties** in numbers, taking into account the fact that numbers are known only to a certain degree of **accuracy**. The accuracy of a number is shown by how many **significant figures** or **significant digits** it contains. This can be illustrated by considering the atomic masses of elemental boron and sodium. The atomic mass of boron is given as 10.81. Written in this way, the number expressing the atomic mass of boron contains four significant digits—the 1, the 0, the 8, and the 1. It is understood to have an uncertainty of + or −1 in the last digit, meaning that it is really 10.81 ± 0.01. The atomic mass of sodium is given as 22.98977, a number with seven significant digits understood to mean 22.98977 ± 0.00001. Therefore, the atomic mass of sodium is known with more *certainty* than that of boron. The atomic masses in Table 1.1 reflect the fact that there is less variability in atomic masses of some elements as they occur in nature (due to variable distributions of elemental isotopes), so that their atomic masses are known with much more certainty (e.g., fluorine, 18.998403) than is the case for others (e.g., calcium listed with an atomic mass of 40.08).

The rules for expressing significant digits are summarized in Table 1.3. It is important to express numbers to the correct number of significant digits in chemical calculations and in the laboratory. The use of too many digits implies an accuracy in the number that does not exist and is misleading. The use of too few significant digits does not express the number to the degree of accuracy to which it is known.

Exercise: Referring to Table 1.3, give the number of significant digits and the rule(s) upon which they are based for each of the following numbers:

(a) 17.000 (b) 9.5378 (c) 7.001 (d) $50
(e) 0.00300 (f) 7400 (g) 6.207×10^{-7} (h) 13.5269184
(i) 0.05029

Answers: (a) 5, Rule 4; (b) 5, Rule 1; (c) 4, Rule 2; (d) exact number; (e) 3, Rules 3 and 4; (f) uncertain, Rule 5; (g) 4, Rule 6; (h) 9, Rule 1; (i) 4 Rules 2 and 3.

Significant Figures in Calculations

After numbers are obtained by a laboratory measurement, they are normally subjected to mathematical operations to get the desired final result. It is important that the answer have the correct number of significant figures. It should not have so few that accuracy is sacrificed or so many that an unjustified degree of accuracy

Table 1.3. Rules for Use of Significant Digits

Example Number	Number of Significant Digits	Rule
11.397	5	1. Nonzero digits in a number are always significant. The 1, 1, 3, 9, and 7 in this number are each significant
140.039	6	2. Zeros between nonzero digits are significant. The 1, 4, 0, 0, 3, and 9 in this number are each significant
0.00329	3	3. Zeros to the left of the first nonzero digit are not significant, because they are used only to locate the decimal point. Only 3, 2, and 9 in this number are significant
70.00	4	4. Zeros to the right of a decimal point that are preceded by a significant figure are significant. All three 0s, as well as the 7, are significant
32 000	Uncertain	5. The number of significant digits in a number with zeros to the left but not to the right of a decimal point (1700, 110 000) may be uncertain. Such numbers should be written in exponential notation
3.20×10^3	3	6. The number of significant digits in a number written in exponential notation is equal to the number of significant digits in the decimal portion
Exactly 50	Unlimited	7. Some numbers, such as the amount of money that one expects to receive when cashing a check or the number of children claimed for income tax exemptions, are defined as exact numbers without any uncertainty.

is implied. The two major rules that apply, one for addition/subtraction, the other for multiplication/division, are the following:

1. In addition and subtraction, the number of digits retained to the right of the decimal point should be the same as those in the number in the calculation with the fewest such digits.

   For example, $273.591 + 1.00327 + 229.13 = 503.72427$ is rounded to 503.72 because 229.13 has only two significant digits beyond the decimal; and $313.4 + 11.0785 + 229.13 = 553.6085$ is rounded to 553.6 because 313.4 has only one significant digit beyond the decimal.

2. The number of significant figures in the result of multiplication/division should be the same as that in the number in the calculation having the fewest significant figures. For example,

$$\frac{3.7218 \times 4.019 \times 10^{-3}}{1.48} = 1.0106699 \times 10^{-2}$$

is rounded to 1.01 (3 significant figures because 1.48 has only 3 significant figures); and

$$\frac{5.28721 \times 10^7 \times 7.245 \times 10^{-5}}{1.00732} = 3.802747533 \times 10^3$$

is rounded to 3.803×10^3 (4 significant figures because 7.245 has only 4 significant figures).

It should be noted that an exact number is treated in calculations as though it has an unlimited number of significant figures.

Exercise: Express each of the following to the correct number of significant figures:

(a) $13.1 + 394.0000 + 8.1937$ (b) $1.57 \times 10^{-4} \times 7.198 \times 10^{-2}$

(c) $189.2003 - 13.47 - 2.563$ (d) $221.9 \times 54.2 \times 123.008$

(e) $\dfrac{603.9 \times 21.7 \times 0.039217}{87}$ (f) $\dfrac{3.1789 \times 10^{-3} \times 7.000032 \times 10^4}{27.130921}$

(g) $100 \times 0.7428 \times 6.82197$ (where 100 is an exact number)

Answers: (a) 415.3; (b) 1.13×10^{-5}; (c) 173.17; (d) 1.48×10^6; (e) 5.9; (f) 8.2019; (g) 506.7.

Rounding Numbers

With an electronic calculator, it is easy to obtain a long string of digits that must be rounded to the correct number of significant figures. The rules for doing this are the following:

1. If the digit to be dropped is 0, 1, 2, 3, or 4, leave the last retained digit unchanged:

Example: Round 4.17821 to 4 significant digits
Answer : 4.178
Last retained digit Digit to be dropped

2. If the digit to be dropped is 5, 6, 7, 8, or 9, increase the last retained digit by 1:

Example: Round 4.17821 to 3 significant digits
Answer: 4.18
Last retained digit Digit to be dropped

Use of Three Significant Digits

It is possible to become thoroughly confused about how many significant figures to retain in an answer. In such a case, it is often permissible to use 3 significant figures.

Generally, this gives sufficient accuracy without doing grievous harm to the concept of significant figures.

1.10. MEASUREMENTS AND SYSTEMS OF MEASUREMENT

The development of chemistry has depended strongly upon careful measurements. Historically, measurements of the quantities of substances reacting and produced in chemical reactions have allowed the explanation of the fundamental nature of chemistry. Exact measurements continue to be of the utmost importance in chemistry, and are facilitated by increasingly more sophisticated instrumentation. For example, atmospheric chemists can determine a small degree of stratospheric ozone depletion by measuring minute amounts of ultraviolet radiation absorbed by ozone with satellite-mounted instruments. Determinations of a part per trillion or less of a toxic substance in water may serve to trace the source of a hazardous pollutant. This section discusses the basic measurements commonly made in chemistry and environmental chemistry.

SI Units of Measurement

Several systems of measurement are used in chemistry and environmental chemistry. The most systematic of these is the **International System of Units (Système Internationale d'Unités)**, abbreviated **SI**, a self-consistent set of units based upon the metric system recommended in 1960 by the General Conference on Weights and Measures to simplify and make more logical the many units used in the scientific and engineering community. Table 1.4 gives the seven base SI units in terms of which all other units are derived.

Multiples of Units

Quantities expressed in science often range over many orders of magnitude (many factors of 10). For example, a mole of molecular diatomic nitrogen contains 6.02×10^{23} N_2 molecules and very small particles in the atmosphere may be only about 1×10^{-6} m in diameter. Therefore, **prefixes** are used that give the number of times that the basic unit is multiplied. Each prefix has a name and an abbreviation. Those that are used in this book, or that are most commonly encountered, are given in Table 1.5.

Metric and English Systems of Measurement

The **metric** system has long been the standard system for scientific measurement and is the one most commonly used in this book. It was the first to use multiples of 10 to designate units that differ by orders of magnitude from a basic unit. The **English** system is still employed for many measurements encountered in normal

Table 1.4. Units of the International System of Units, SI

Physical Quantity Measured	Unit Name	Unit Symbol	Definition
Base Units			
Length	meter[a]	m	Distance traveled by light in a vacuum in 1/299792458 second
Mass	kilogram	kg	Mass of a platinum–iridium block located at the International Bureau of Weights and Measures at Sèvres, France
Time	second	s	9 192 631 770 periods of a specified line in the microwave spectrum of the cesium-133 isotope
Temperature	kelvin	K	1/273.16 the temperature interval between absolute zero and the triple point of water at 273.16 K (0.01°C)
Amount of substance	mole	mol	Amount of substance containing as many entities (atoms, molecules) as there are atoms in exactly 0.012 kilogram of the carbon-12 isotope
Electric current	ampere	A	Based upon current required to generate a specified force between parallel conductors
Luminous intensity	candela	cd	Based upon power generated by a specified source of monochromatic light
Examples of Derived Units			
Force	newton	N	Force required to impart an acceleration of 1 m/s^2 to a mass of 1 kg
Energy (heat)	joule	J	Work performed by 1 newton acting over a distance of 1 meter
Pressure	pascal	Pa	Force of 1 newton acting on an area of 1 square meter

[a] In the publications of the SI, the British/European spelling "metre" is used.

everyday activities in the U.S., including some environmental engineering measurements. Bathroom scales are still calibrated in pounds, well depths may be given in feet, and quantities of liquid wastes are frequently expressed as gallons or barrels. Furthermore, English units of pounds, tons, and gallons are still commonly used in commerce, even in the chemical industry. Therefore, it is still necessary to have some familiarity with this system; conversion factors between it and metric units are given in this book.

Table 1.5. Prefixes Commonly Used to Designate Multiples of Units

Prefix	Basic Unit is Multiplied by	Abbreviation
Tera	1 000 000 000 000 (10^{12})	T
Giga	1 000 000 000 (10^{9})	G
Mega	1 000 000 (10^{6})	M
Kilo	1 000 (10^{3})	k
Hecto	100 (10^{2})	h
Deka	10 (10)	da
Deci	0.1 (10^{-1})	d
Centi	0.01 (10^{-2})	c
Milli	0.001 (10^{-3})	m
Micro	0.000 001 (10^{-6})	μ
Nano	0.000 000 001 (10^{9})	n
Pico	0.000 000 000 001 (10^{-12})	p

1.11. UNITS OF MASS

Mass expresses the degree to which an object resists a change in its state of rest or motion and is proportional to the amount of matter in the object. **Weight** is the gravitational force acting upon an object and is proportional to mass. An object weighs much less in the gravitational field on the Moon's surface than on Earth, but the object's mass is the same in both places (Figure 1.8). Although mass and weight are not usually distinguished from each other in everyday activities, it is important for the science student to be aware of the differences between them.

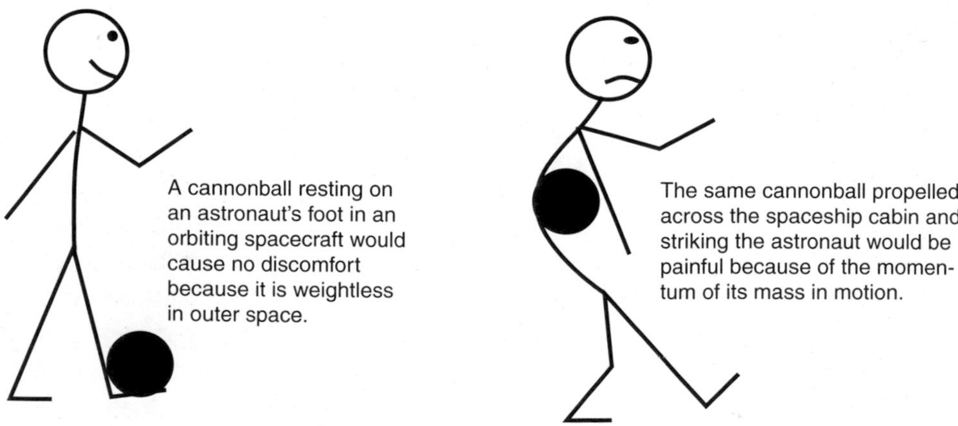

Figure 1.8. An object retains its mass even in the weightless surroundings of outer space.

Table 1.6. Metric Units of Mass

Unit of Mass	Abbreviation	Number of Grams	Definition
Megagram or metric ton	Mg	10^6	Quantities of industrial chemicals (1 Mg = 1.102 short tons)
Kilogram	kg	10^3	Body weight and other quantities for which the pound has been commonly used (1 kg = 2.2046 lb)
Gram	g	1	Mass of laboratory chemicals (1 ounce = 28.35 g and 1 lb = 453.6 g)
Milligram	mg	10^{-3}	Small quantities of chemicals
Microgram	μg	10^{-6}	Quantities of toxic pollutants

The **gram** (g), with a mass equal to 1/1000 that of the SI kilogram (see Table 1.4), is the fundamental unit of mass in the metric system. Although the gram is a convenient unit for many laboratory-scale operations, other units that are multiples of the gram are often more useful for expressing mass. The names of these are obtained by affixing the appropriate prefixes from Table 1.5 to "gram." Global burdens of atmospheric pollutants may be given in units of teragrams, each equal to 1×10^{12} grams. Significant quantities of toxic water pollutants may be measured in micrograms (1×10^{-6} grams). Large-scale industrial chemicals are marketed in units of megagrams (Mg). This quantity is also known as a metric ton, or tonne, and is somewhat larger (2205 lb) than the 2000 lb short ton still used in commerce in the United States. Table 1.6 summarizes some of the more commonly used metric units of mass and their relationship to some English units.

1.12. UNITS OF LENGTH

Length in the metric system is expressed in units based upon the **meter** (m) (see Table 1.4). A meter is 39.37 inches long, slightly longer than a yard (Figure 1.9). A kilometer (km) is equal to 1000 m and, like the mile, is used to measure relatively great distances. A centimeter (cm), equal to 0.01 m, is often convenient to designate lengths such as the dimensions of laboratory instruments. There are 2.540 cm per inch, and the cm is employed to express lengths that would be given in inches in the English system. The micrometer (μm) is about as long as a typical bacterial cell. The micrometer is also used to express wavelengths of infrared radiation by which Earth re-radiates solar energy back to outer space. (In the past, the micrometer was sometimes called the "micron," and abbreviated as simply "μ," but this term and abbreviation should no longer be used.) The nanometer (nm), equal to 10^{-9} m, is a convenient unit for the wavelength of visible light, which ranges from 400 to 800 nm. Atoms are even smaller than 1 nm; their dimensions are commonly given

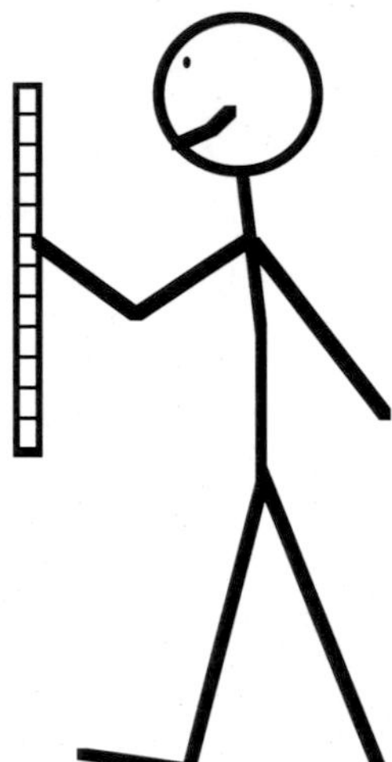

Figure 1.9. The meter stick is a common tool for measuring length.

in picometers (pm, 10^{-12} m). Table 1.7 lists common metric units of length, some examples of their use, and some related English units.

1.13. UNITS OF VOLUME

The basic metric unit of **volume** is the **liter** (L), which is defined in terms of metric units of length. As shown in Figure 1.10, a liter is the volume of a decimeter cubed, that is, 1 L = 1 dm^3 (a dm is 0.1 m, about 4 inches). A milliliter (mL) is the same volume as a centimeter cubed (cm^3 or cc—although this latter abbreviation should no longer be used), and a liter is 1000 cm^3. A kiloliter, or cubic meter (m^3), is a common unit of measurement for the volume of air. For example, standards for

Table 1.7. Metric Units of Length

Unit of Length	Abbreviation	Number of Meters	Definition
Kilometer	km	10^3	Distance (1 mile = 1.609 km)
Meter	m	1	Standard metric unit of length (1 m = 1.094 yards)
Centimeter	cm	10^{-2}	Used in place of inches (1 inch = 2.54 cm)
Millimeter	mm	10^{-3}	Same order of magnitude as sizes of letters on this page
Micrometer	μm	10^{-6}	Size of typical bacteria
Nanometer	nm	10^{-9}	Measurement of light wavelength
Picometer	pm	10^{-12}	Atomic dimensions

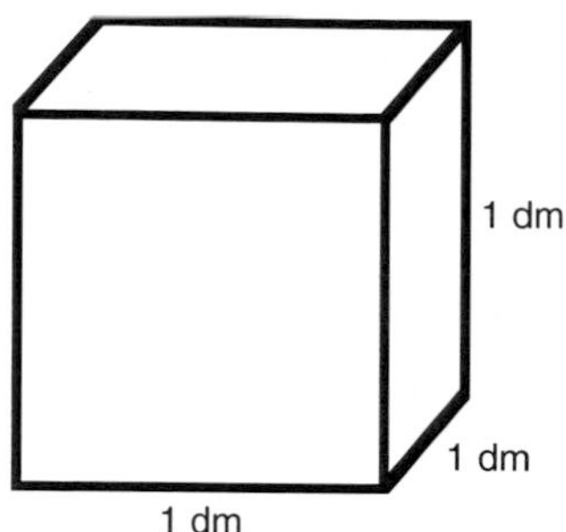

Figure 1.10. A cube that is 1 decimeter to a side has a volume of 1 liter.

human exposure to toxic substances in the workplace are frequently given in units of μg/m³. Table 1.8 gives some common metric units of volume. The measurement of volume is one of the more frequently performed routine laboratory measurements; Figure 1.11 shows some of the more common tools for laboratory volume measurement of liquids.

1.14. TEMPERATURE, HEAT, AND ENERGY

Temperature Scales

In chemistry, temperatures are usually expressed in metric units of **degrees Celsius** (°C), in which water freezes at 0°C and boils at 100°C. The **Fahrenheit** scale, still used for some non-scientific temperature measurements in the United States, defines the freezing temperature of water at 32 degrees Fahrenheit (°F) and boiling at 212°F, a range of 180°F. Therefore, each span of 100°C is equivalent to one of 180°F and each °C is equivalent to 1.8°F.

Table 1.8. Metric Units of Volume

Unit of Volume[a]	Abbreviation[b]	Number of Liters	Example of Use for Measurement
Kiloliter or cubic meter	kL	10^3	Volumes of air in air pollution studies
Liter	L	1	Basic metric unit of volume (1 liter = 1 dm³ = 1.057 quarts; 1 cubic foot = 28.32 L)
Milliliter	mL	10^{-3}	Equal to 1 cm³. Convenient unit for laboratory volume measurements
Microliter	μL	10^{-6}	Used to measure very small volumes for chemical analysis

[a] In the publications of the SI, the British/European spelling "litre" is used.

[b] The abbreviation "l" is also frequently used, although it can cause confusion with the number "one."

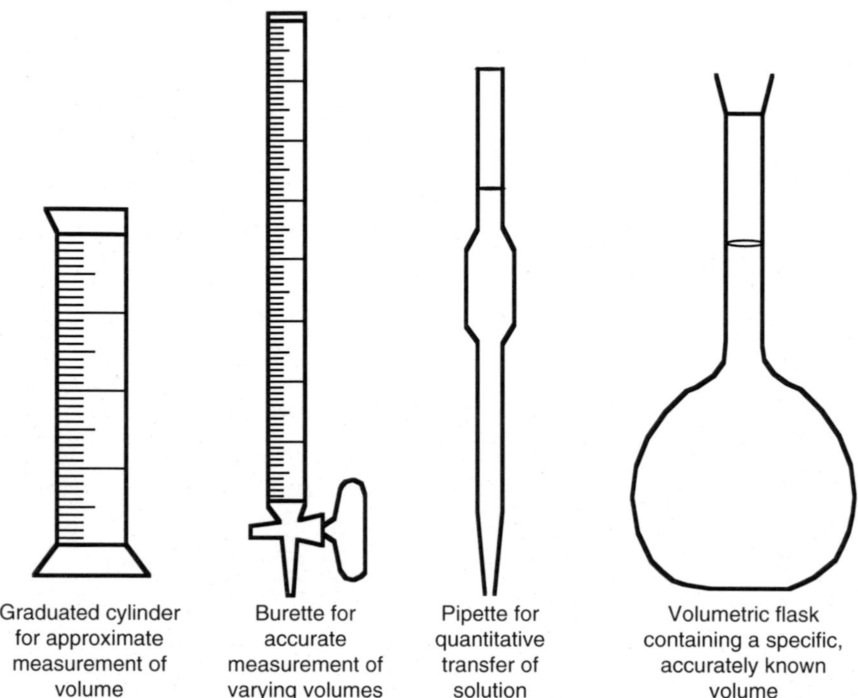

Figure 1.11. Glassware for volume measurement in the laboratory.

The most fundamental temperature scale is the **Kelvin** or **absolute** scale, for which zero is the lowest attainable temperature. A unit of temperature on this scale is equal to a degree Celsius, but it is called a **kelvin**, abbreviated K: Kelvin temperatures are designated as K, *not* °K. The value of absolute zero on the Kelvin scale is −273.15°C, so that the Kelvin temperature is always a number 273.15 (usually rounded to 273) higher than the Celsius temperature. Thus, water boils at 373 K and freezes at 273 K. The relationships among Kelvin, Celsius, and Fahrenheit temperatures are illustrated in Figure 1.12.

Converting from Fahrenheit to Celsius

With Figure 1.12 in mind, it is easy to convert from one temperature scale to another. Examples of how this is done are given below.

Example: What is the Celsius temperature equivalent to room temperature of 70°F?

Answer:

Step 1. Subtract 32°F from 70°F to get the number of degrees Fahrenheit above freezing. This is done because 0 on the Celsius scale is at the freezing point of water.

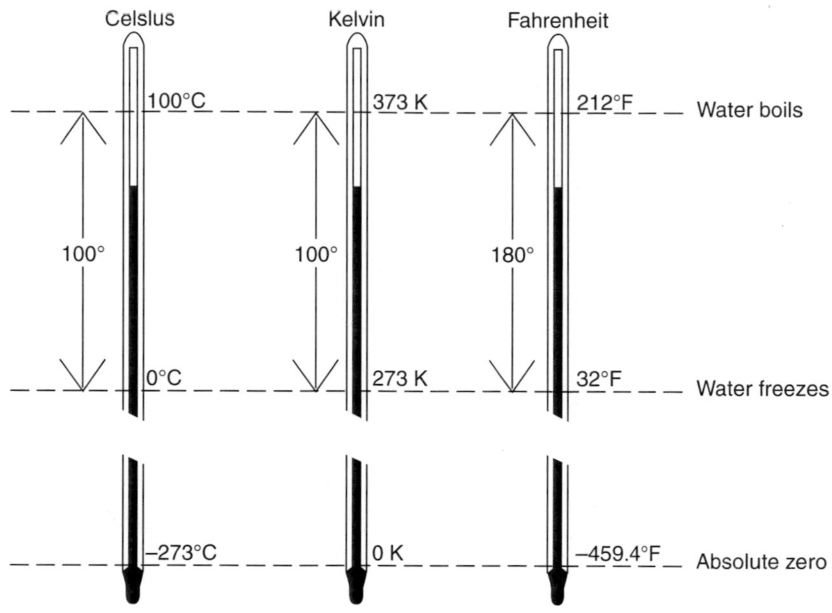

Figure 1.12. Comparison of temperature scales.

Step 2. Multiply the number of degrees Fahrenheit above the freezing point of water obtained above by the number of degrees Celsius per degree Fahrenheit:

$$°C = \underbrace{\frac{1.00°C}{1.80°F}}_{\text{Factor for conversion from °F to °C}} \times \underbrace{(70°F - 32°F)}_{\text{Number of °F above freezing}} = \frac{1.00°C}{1.80°F} \times 38°F = 21.1°C \quad (1.14.1)$$

In working the above example, it is first noted (as is obvious from Figure 1.11) that the freezing temperature of water, zero on the Celsius scale, corresponds to 32°F on the Fahrenheit scale. So 32°F is subtracted from 70°F to give the number of degrees Fahrenheit by which the temperature is above the freezing point of water. The number of degrees Fahrenheit above freezing is converted to degrees Celsius above the freezing point of water by multiplying by the factor 1.00°C/1.80°F. The origin of this factor is readily seen by referring to Figure 1.12 and observing that there are 100°C between the freezing and boiling temperatures of water and 180°F over the same range. Mathematically, the equation for converting from °F to °C is simply the following:

$$°C = \frac{1.00°C}{1.80°F} \times (°F - 32) \quad (1.14.2)$$

Example: What is the Celsius temperature corresponding to the normal body temperature of 98.6°F?

Answer: From Equation 1.14.2,

$$°C = \frac{1.00°C}{1.80°F} \times (98.6°F - 32°F) = 37.0°C \tag{1.14.3}$$

Example: What is the Celsius temperature corresponding to −5°F?

Answer: From Equation 1.14.2,

$$°C = \frac{1.00°C}{1.80°F} \times (-5°F - 32°F) = -20.6°C \tag{1.14.4}$$

Converting from Celsius to Fahrenheit

To convert from Celsius to Fahrenheit first requires multiplying the Celsius temperature by 1.80°F/1.00°C to get the number of Fahrenheit degrees above the freezing temperature of 32°F, then adding 32°F.

Example: What is the Fahrenheit temperature equivalent to 10°C?

Answer:

Step 1. Multiply 10°C by 1.80°F/1.00°C to get the number of degrees Fahrenheit above the freezing point of water.

Step 2. Since the freezing point of water is 32°F, add 32°F to the result of Step 1:

$$°F = \frac{1.80°F}{1.00°C} \times °C + 32°F = \frac{1.80°F}{1.00°C} \times 10°C + 32°F = 50°F$$

The formula for converting °C to °F is

$$°F = \frac{1.80°F}{1.00°C} \times °C + 32$$

To convert from °C to K, add 273 to the Celsius temperature. To convert from K to °C, subtract 273 from K. All of the conversions discussed here can be deduced without memorizing any equations by remembering that the freezing point of water is 0°C, 273 K, and 32°F, and the boiling point is 100°C, 373 K, and 212°F.

Melting Point and Boiling Point

In the preceding discussion, the melting and boiling points of water were both used in defining temperature scales. These are important thermal properties of any substance. For the present, **melting temperature** may be defined as the temperature at which a substance changes from a solid to a liquid. **Boiling** temperature is defined as the temperature at which a substance changes from a liquid to a gas.

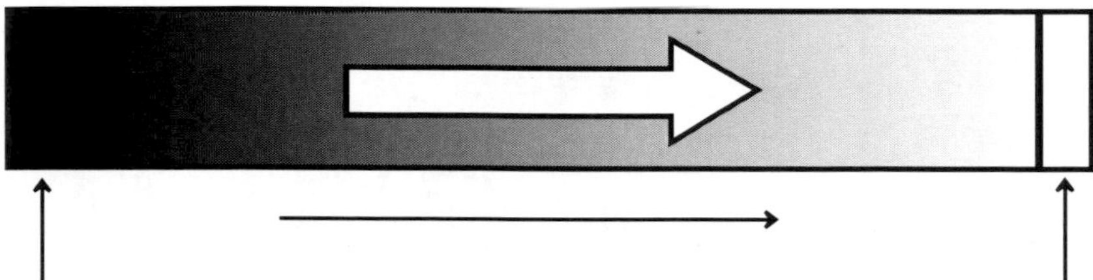

Figure 1.13. Heat energy flow from a hot to a colder object.

More-exacting definitions of these terms, particularly boiling temperature, are given later in the book.

Heat and Energy

As illustrated in Figure 1.13, when two objects at different temperatures are placed in contact with each other, the warmer object becomes cooler and the cooler one warmer until they reach the same temperature. This occurs because of a flow of energy between the objects. Such a flow is called **heat**.

The SI unit of heat is the joule (J) (Table 1.4). The kilojoule (1 kJ = 1000 J) is a convenient unit to use to express energy values in laboratory studies. The metric unit of energy is the calorie (cal), equal to 4.184 J. Throughout the liquid range of water, essentially 1 calorie of heat energy is required to raise the temperature of 1 g of water by 1°C. The "calories" that most people hear about are those used to express energy values of foods and are actually kilocalories (1 kcal = 4.184 kJ).

1.15. PRESSURE

Pressure is force per unit area. The SI unit of pressure is the pascal (Pa) (Table 1.4). The kilopascal (1 kPa = 1000 Pa) is often a more convenient unit of pressure to use than is the pascal.

Like many other quantities, pressure has been plagued with a large number of different kinds of units. One of the more meaningful and intuitive of these is the **atmosphere** (atm), and the average pressure exerted by air at sea level is 1 atmosphere. One atmosphere is equal to 101.3kPa or 14.7 lb/in^2. The latter means that an evacuated cube, 1 inch to a side, has a force of 14.70 lb exerted on each side due to atmospheric pressure. It is also the pressure that will hold up a column of liquid mercury metal 760mm long, as shown in Figure 1.14. Such a device used to measure atmospheric pressure is called a **barometer**, and the mercury barometer was the first instrument used to measure pressures with a high degree of accuracy. Consequently, the practice developed of expressing pressure in units of **millimeters of mercury** (mmHg, also called the **torr**).

Pressure is an especially important variable with gases, because the volume of a quantity of gas at a fixed temperature is inversely proportional to pressure. The temperature/pressure/volume relationships of gases (Boyle's law, Charles' law, and the general gas law) are discussed in Chapter 2.

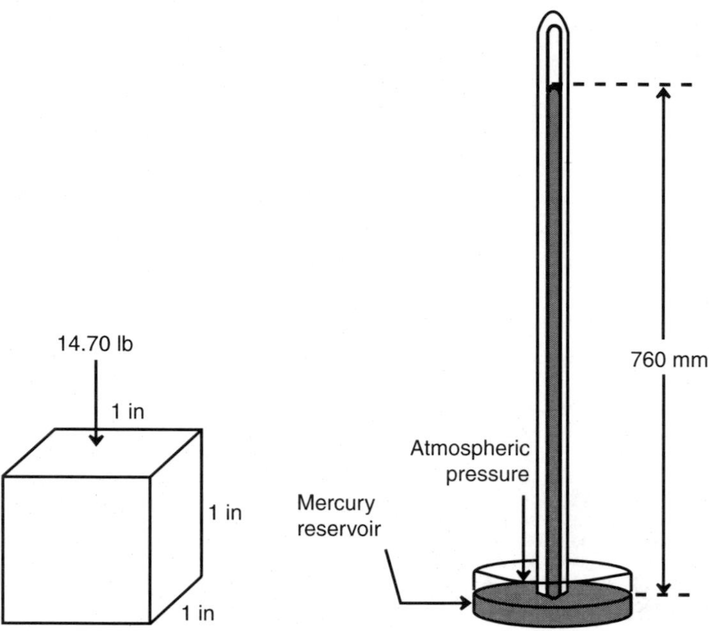

Figure 1.14. Average atmospheric pressure at sea level exerts a force of 14.7 pounds on an inch-square surface. This corresponds to a pressure sufficient to hold up a 760 mm column of mercury.

1.16. UNITS AND THEIR USE IN CALCULATIONS

Most numbers used in chemistry are accompanied by a **unit** that tells the type of quantity that the number expresses and the smallest whole portion of that quantity. For example, "36 liters" denotes that a volume is expressed and the smallest whole unit of the volume is 1 liter. The same quantity could be expressed as 360 deciliters, where the number is multiplied by 10 because the unit is only 1/10 as large.

Except in cases where the numbers express relative quantities, such as atomic masses relative to the mass of carbon-12 or specific gravity, it is essential to include units with numbers. In addition to correctly identifying the type and magnitude of the quantity expressed, the units are carried through mathematical operations. The wrong unit in the answer shows that something has been done wrong in the calculation and it must be checked.

Unit Conversion Factors

Most chemical calculations involve calculating one type of quantity, given another, or converting from one unit of measurement to another. For example, in the chemical reaction

$$2H_2 + O_2 \longrightarrow H_2O$$

someone might want to calculate the number of grams of H_2O produced when 3 g of H_2 react, or they might want to convert the number of grams of H_2 to ounces. These kinds of calculations are carried out with **unit conversion factors**. Suppose, for example, that the mass of a 160 lb person is to be expressed in kilograms; the person doing the calculation does not know the factor to convert from lb to kg, but does know that a 551 lb motorcycle has a mass of 250 kg. From this information, the needed unit conversion factor can be derived and the calculation completed as follows:

$$\text{Mass of person in kg} = 160 \text{ lb} \times \text{unit conversion factor} \quad \text{(problem to be solved)} \qquad (1.16.1)$$

$$250 \text{ kg} = 551 \text{ lb} \quad \text{(known relationship between lb and kg)} \qquad (1.16.2)$$

$$\frac{250 \text{ kg}}{551 \text{ 1b}} = \frac{551 \text{ kg}}{551 \text{ kg}} = 1 \quad \text{(the unit of kg is left on top because it is the unit needed; division is by 551 lb)} \qquad (1.16.3)$$

Although the concept may be a little difficult to understand, in calculating a unit conversion factor, the number involved is divided by itself, giving a numerical value of 1. In making conversions, any quantity multiplied by itself (the unit conversion factor) is still the same quantity, but just in different units. Dividing 250 kg by 551 lb gives the following conversion factor that can be used to calculate the mass of the person:

$$\frac{250 \text{ kg}}{551 \text{ 1b}} = \frac{0.454 \text{ kg}}{1.00 \text{ 1b}} \qquad (1.16.4)$$

$$\text{Mass of person} = 160 \text{ 1b} \times \frac{0.454 \text{ kg}}{1.00 \text{ 1b}} = 72.6 \text{ kg} \qquad (1.16.5)$$

As another example of the use of a unit conversion factor, calculate the number of liters of gasoline required to fill a 12-gallon fuel tank, given that there are 4 gallons in a quart and that a volume of 1 liter is equal to that of 1.057 quarts. This problem can be worked by first converting gallons to quarts, then quarts to liters. The two unit conversion factors required are the following

$$1 \text{ gal} = 4 \text{ qt} \qquad (1.16.6)$$

$$\frac{1 \text{ gal}}{1 \text{ gal}} = \frac{4 \text{ qt}}{1 \text{ gal}} = 1 \text{ (conversion from gallons to quarts)} \qquad (1.16.7)$$

$$1.057 \text{ qt} = 1 \text{ L} \qquad (1.16.8)$$

$$\frac{1 \text{ L}}{1.057 \text{ qt}} = \frac{1 \text{ L}}{1 \text{ L}} = 1 \text{ (conversion from quarts to liters)} \qquad (1.16.9)$$

Both unit conversion factors are used to calculate the capacity of the tank in liters:

$$\text{Tank capacity} = 12\ \text{gal} \times \frac{4\ \text{qt}}{1\ \text{gal}} \times \frac{1\ \text{L}}{1.057\ \text{qt}} = 45.4\ \text{L} \qquad (1.16.10)$$

Cancellation of Units

The preceding examples show that units are canceled in mathematical operations, just as numbers may be. When the same unit appears both above and below the line in a mathematical operation, the units cancel. An example of such an operation is shown for lb in the following, in which the unit of lb simply cancels, leaving kg as the unit remaining:

$$160\ \cancel{\text{lb}} \times \frac{0.454\ \text{kg}}{1.00\ \cancel{\text{lb}}}$$

Calculation of Some Unit Conversion Factors

Several values of units are given that enable conversion between metric and English units in Table 1.6 (mass), Table 1.7 (length), and Table 1.8 (volume). For example, Table 1.6 states that a megagram (Mg, metric ton) is equal to 1.102 short tons (T). By using this equality to give the correct unit conversion factors, it is easy to calculate the number of metric tons in a given number of short tons of material or vice versa. To do this, first write the known equality given that a megagram is equal to 1.102 short tons:

$$1\ \text{Mg} = 1.102\ \text{T} \qquad (1.16.11)$$

If the number of Mg is to be calculated given a mass in T, the unit conversion factor needed is

$$\frac{1\ \text{Mg}}{1.102\ \text{T}} = \frac{1.102\ \text{T}}{1.102\ \text{T}} = 1 \qquad (1.16.12)$$

leaving Mg on top. Suppose, for example, that the problem is to calculate the mass in Mg of a 3521 T shipment of industrial soda ash. The calculation involves simply multiplying the known mass in T times the unit conversion factor required to convert to Mg:

$$3521\ \text{T} \times \frac{1\ \text{Mg}}{1.102\ \text{T}} = 3195\ \text{Mg} \qquad (1.16.13)$$

If the problem had been to calculate the number of T in 789 Mg of copper ore, the following steps would be followed:

$$1.102\ \text{T} = 1\ \text{Mg} \quad \frac{1.102\ \text{T}}{1\ \text{Mg}} = \frac{1\ \text{Mg}}{1\ \text{Mg}} = 1 \qquad (1.16.14)$$

Table 1.9. Examples of Some Unit Conversion Factors

Equality	Conversion Factors	
1 kg = 2.2046 lb	$\frac{1\ \text{kg}}{2.2046\ \text{lb}} = 1$	$\frac{2.2046\ \text{lb}}{1\ \text{kg}} = 1$
1 oz = 28.35 g	$\frac{1\ \text{oz}}{28.35\ \text{g}} = 1$	$\frac{28.35\ \text{g}}{1\ \text{oz}} = 1$
1 mi = 1.609 km	$\frac{1\ \text{mi}}{1.609\ \text{km}} = 1$	$\frac{1.609\ \text{km}}{1\ \text{mi}} = 1$
1 in = 2.54 cm	$\frac{1\ \text{in}}{2.54\ \text{cm}} = 1$	$\frac{2.54\ \text{cm}}{1\ \text{in}} = 1$
1 L = 1.057 qt	$\frac{1\ \text{L}}{1.057\ \text{qt}} = 1$	$\frac{1.057\ \text{qt}}{1\ \text{L}} = 1$
1 cal = 4.184 J	$\frac{1\ \text{cal}}{4.184\ \text{J}} = 1$	$\frac{4.184\ \text{J}}{1\ \text{cal}} = 1$
1 atm = 101.4 kPa	$\frac{1\ \text{atm}}{101.4\ \text{kPa}} = 1$	$\frac{101.4\ \text{kPa}}{1\ \text{atm}} = 1$

$$789\ \text{Mg} \times \frac{1.102\ \text{T}}{1\ \text{Mg}} = 869\ \text{T copper one} \qquad (1.16.15)$$

Table 1.9 gives some unit conversion factors calculated from the information given in Tables 1.6–1.8 and in preceding parts of this chapter. Note that, in each case, two unit conversion factors are calculated; the one that is used depends upon the units that are required for the answer.

CHAPTER SUMMARY

The chapter summary below is presented in a programmed format to review the main points covered in this chapter. It is used most effectively by filling in the blanks, referring back to the chapter as necessary. The answers are given at the end of the summary.

During the early years of life on Earth, [1]____________________ greatly altered the atmosphere by releasing [2]____________________. Now human activities are altering climate by releasing [3]____________________ as the result of [4]____________________. The maintenance and enhancement of conditions that will enable humans and other organisms to exist on Planet Earth is the basis of [5]____________________. One of the earliest efforts to "save the environment" in the U.S. dating to around 1900 was [6]____________________ and the modern environmental movement is often dated from 1962 as the result of publication of [7]____________________. Chemistry is defined as [8]____________________. Environmental chemistry is defined as [9]____________________

____

____.

The practice of chemical science and engineering in ways that minimize consumption of materials and energy, produce the least possible amount of waste, and are inherently safe is called [10]________________________. All substances are composed of only about a hundred fundamental kinds of matter called [11]__________, each composed of very small entities called [12]________________. The three major constituents of atoms and their charges are [13]______________ ______________. Of these, the two that have relatively high masses are contained in the [14]__________ of the atom. The subatomic particles with a relatively low mass are contained in [15]______________________________ in the atom. The number of protons in the nucleus of each atom of an element is the [16]____________________ of the element. Each element is represented by an abbreviation called a [17]__________. In addition to atomic number, name, and chemical symbol, each element has a characteristic [18]______________________. Atoms of most elements consist of two or more isotopes that have different [19]______________________________. An arrangement of the elements in a manner that reflects their recurring behavior with increasing atomic number is the [20]__________________ in which elements with similar chemical properties are called [21]________________ and are contained in [22]______________ in the periodic table. Instead of existing as atoms, elemental hydrogen consists of [23]________________, each composed of [24]______________ linked by a [25]________________. Water is not an element, but is a [26]________ ______________, for which the [27]______________________is H_2O. Species consisting of electrically charged atoms or groups of atoms are called [28]__________. Those with positive charges are called [29]______________ and those with negative charges are [30]______________. Compounds made of these kinds of entities are held together by [31]__________________ bonds. The average mass of all molecules of a compound is its [32]______________, which is calculated by [33]______________

____

____________________________. [34]____________________ occur when substances are changed to other substances through the breaking and formation of chemical bonds and are written as [35]____________________. To be correct, these must be [36]__________________. In them, the arrow is read as [37]______________ and separates the [38]__________________ from the [39]________________. Very large or small numbers are conveniently expressed in [40]________________, which is the product of a [41]______________ with a value equal to or greater than [42]__________ and less than [43]__________ multiplied times a [44]______________. In such a notation, 3 790 000 is expressed as [45]________________ and 0.000 000 057 is expressed as [46]______________. The accuracy of a number is shown by how many [47]______________________________ it contains. Nonzero digits in a number are always [48]______________________________. Zeros between nonzero digits are [49]__________. Zeros to the left of the first nonzero digit are [50]__________. Zeros to the right of a decimal point that are preceded by a significant figure are [51]__________. The number of significant digits in a number written in exponential notation is equal to [52]__________. Some numbers such as the amount of money that one expects to receive when cashing a check are defined as [53]__________. In addition and subtraction, the number of digits retained to the right of the decimal point should be [54]__________. The number of significant figures in the result of multiplication/division should be [55]__________. In rounding numbers, if the digit to be dropped is 0, 1, 2, 3, or 4, [56]__________, whereas if the digit to be dropped is 5, 6, 7, 8, or 9, [57]__________. A self-consistent set of units based upon the metric system is the [58]__________. [59]__________ is proportional to the amount of matter in an object, the metric unit for which is the [60]__________. Length in the metric system is expressed in units based upon the [61]__________. The basic metric unit of volume is the [62]__________. In °C, °F, and K, respectively, water freezes at [63]__________ and boils at [64]__________. Boiling temperature is defined as [65]__________. Energy that flows from a warmer to a colder object is called [66]__________, commonly expressed in units of [67]__________ or [68]__________. [69]__________ is force per unit area, some of the common units for which are [70]__________. A unit after a number tells the [71]__________ that the number expresses and the smallest [72]__________. The quantity 0.454 kg/1 lb is an example of [73]__________.

Answers to Chapter Summary

1. cyanobacteria
2. elemental oxygen
3. carbon dioxide
4. combustion of fossil fuels
5. sustainability
6. the soil conservation movement
7. Rachel Carson's *Silent Spring*
8. the science of matter

9. that branch of chemistry that deals with the origins, transport, reactions, effects, and fates of chemical species in the water, air, earth, and living environments and the influence of human activities thereon
10. green chemistry
11. elements
12. atoms
13. positively charged protons, negatively charged electrons, and uncharged (neutral) neutrons
14. nucleus
15. a cloud of negative charge
16. atomic number
17. chemical symbol
18. atomic mass
19. numbers of neutrons in their nuclei
20. periodic table
21. groups of elements
22. vertical columns
23. molecules
24. 2 H atoms
25. chemical bond
26. chemical compound
27. chemical formula
28. ions
29. cations
30. anions
31. ionic
32. molecular mass
33. multiplying the atomic mass of each element by the relative number of atoms of the element, then adding all the values obtained for each element in the compound

34. Chemical reactions
35. chemical equations
36. balanced
37. yields
38. reactants
39. products
40. exponential notation
41. digital number
42. exactly 1
43. exactly 10
44. power of 10
45. 3.79×10^6
46. 5.7×10^{-8}
47. significant figures or significant digits
48. significant
49. significant
50. not significant
51. significant
52. the number of significant digits in the decimal portion
53. exact numbers
54. the same as that in the number in the calculation with the fewest such digits
55. the same as that in the number in the calculation having the fewest significant figures
56. leave the last digit retained unchanged
57. increase the last retained digit by 1
58. International System of Units, SI
59. Mass

60. gram
61. meter
62. liter
63. 0°C, 32°F, and 273 K
64. 100°C, 212°F, and 373 K
65. the temperature at which a substance changes from a liquid to a gas
66. heat
67. joules
68. calories
69. Pressure
70. atmosphere, torr, mm Hg, lb/in^2
71. type of quantity
72. whole portion of that quantity
73. a unit conversion factor

LITERATURE CITED

1. Manahan, Stanley E. *Environmental Chemistry*, 8th ed. Taylor & Francis, Boca Raton, FL, 2004.
2. Manahan, Stanley E. *Toxicological Chemistry*, 3rd ed. Taylor & Francis, Boca Raton, FL, 2003.
3. Manahan, Stanley E. *Green Chemistry and the Ten Commandments of Sustainability*, 3rd ed. ChemChar Research, Columbia, MO, 2008.
4. Manahan, Stanley E. Industrial Ecology for Sustainable Resource Utilization. Chapter 17 in *Environmental Science and Technology: A Sustainable Approach to Green Science and Technology*, 2nd ed. Taylor & Francis, Boca Raton, FL, 2007, pp. 525–64.
5. Allen, David T., and David A. Shonnard. *Green Engineering: Environmentally Conscious Design of Chemical Processes*. Prentice Hall, Upper Saddle River, NJ, 2002.

QUESTIONS AND PROBLEMS

1. Using Internet resources, look up cyanobacteria and their influence on Earth's early atmosphere. What were these microorganisms formerly called? Describe how their activities made possible the emergence of millions of new species. Describe a modern-day environmental pollution problem caused by cyanobacteria.

2. Based on the definition of environmental chemistry at the beginning of this chapter, suggest five environmental spheres. Which of these is the anthrosphere, and of what does it consist?

3. From its name, suggest the meaning of a "command-and-control" approach to maintaining environmental quality. What are some of the limitations of such an approach? In which respects is the practice of green chemistry a more "natural" way of maintaining environmental quality?

4. Based upon what you know or can learn about making, using, and disposing of paper, attempt to fit the whole process into the picture of green chemistry illustrated in Figure 1.1 and suggest how the process can be carried out with maximum sustainability.

5. Consider the following atom:

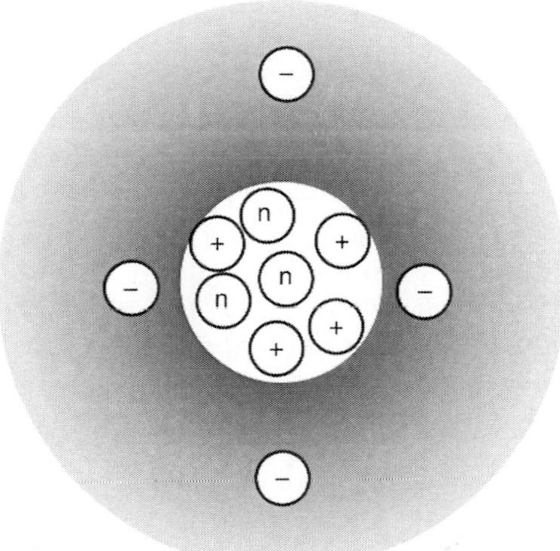

   How many electrons, protons, and neutrons does it have? What is its atomic number? Give the name and chemical symbol of the element of which it is composed.

6. Give the numbers of electrons, protons, and neutrons in each of the atoms represented below. Give the atomic numbers, names, and chemical symbols of each of the elements represented.

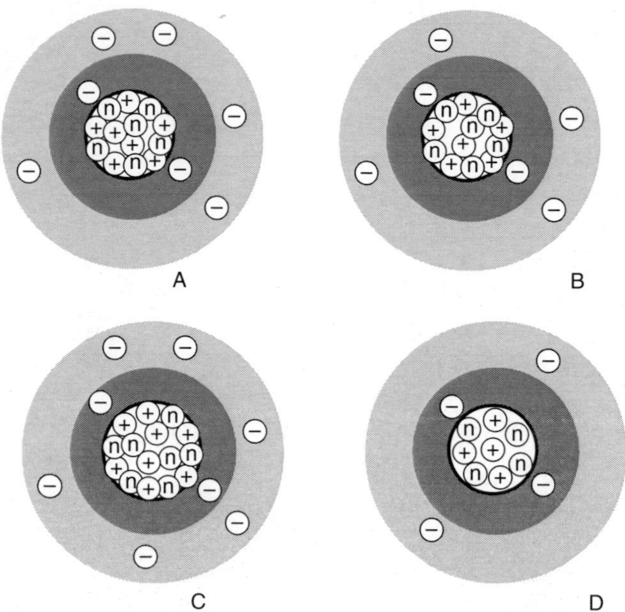

7. What distinguishes a radioactive isotope from a "normal" stable isotope?
8. Why is the periodic table so named?
9. Match the following:

A. O_2	1. Element consisting of individual atoms
B. NH_3	2. Element consisting of chemically bonded atoms
C. Ar	3. Ionic compound
D. NaCl	4. Covalently bound compound

10. After examining Figure 1.7, consider what might happen when an atom of sodium (Na), atomic number 11, loses an electron to an atom of fluorine (F), atomic number 9. What kinds of particles are formed by this transfer of a negatively charged electron? Is a chemical compound formed? What is it called?
11. Give the chemical formula and molecular mass of the molecule represented below:

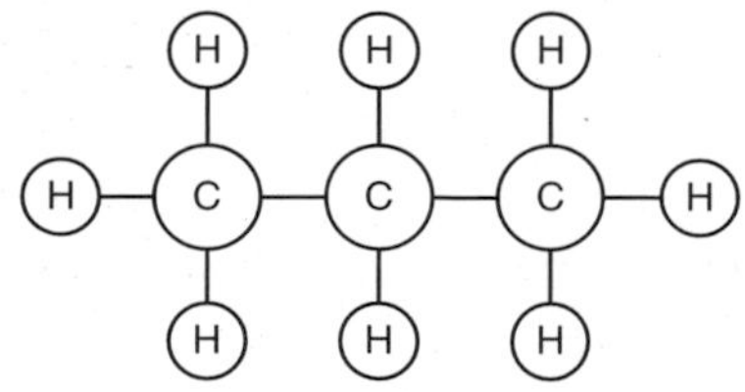

12. Calculate the molecular masses of (a) C_2H_2; (b) N_2H_4; (c) Na_2O; (d) O_3 (ozone); (e) PH_3; (f) CO_2; (g) C_3H_9O.

13. Match the following:

A. Periodic table	1. Has groups
B. Elemental hydrogen	2. Consists of ions
C. Water H_2O	3. Molecule containing two different elements
D. Magnesium oxide, MgO	4. Consists of diatomic, covalently bonded molecules

14. Is the equation, $H_2 + O_2 \rightarrow H_2O$, a balanced chemical equation? Explain. Point out the reactants and products in the equation.

15. Write each of the following in exponential form to three significant figures:

    (a) 321 000, (b) 0.000 005 29, (c) 5170, (d) 000 000 000 000 784,

    (e) 86 300 000 000 000.

16. Write each of the following in decimal form: (a) 7.49×10^3; (b) 9.6×10^{-5}; (c) 1.16×10^{21}; (d) 4.47×10^{-17}; (e) 2.93×10^{13}.

17. Without using a calculator, calculate the following sums expressed to three significant figures in the correct exponential notation:

    (a) $4.13 \times 10^3 + 8.76 \times 10^2 + 1.22 \times 10^4$

    (b) $4.13 \times 10^{-4} + 8.76 \times 10^{-3} + 1.22 \times 10^{-2}$

18. Without using a calculator for the exponential portions, calculate each of the following:

    (a) $1.39 \times 10^{-2} \times 9.05 \times 10^8 \times 3.11 \times 10^4$

    (b) $9.05 \times 10^{-6} \times 3.19 \times 10^3$

    (c) $4.02 \times 10^5 \times 1.93 \times 10^{-7}$

19. Match the following numbers, with the significant figures for each given in parentheses, with the rule for assigning significant figures that applies to each:

A. 0.00027 (2)	1. Nonzero digits in a number are always significant.
B. 7.28139 (6)	2. Zeros between nonzero digits are significant.
C. 7.4×10^3 (2)	3. Zeros on the left of the first nonzero digit are not significant.

D. \$50 (infinite) — 4. Zeros to the right of a decimal point that are preceded by a significant figure are significant.

E. 81.000 (5) — 5. The number of significant digits in a number written in exponential notation is equal to the number of significant digits in the decimal portion.

F. 40.007 (5) — 6. Some numbers are defined as exact numbers without any uncertainty.

20. Using the appropriate rules, round each of the following to the correct number of significant digits: (a) 923.527 + 3.02891 + 729.29; (b) 273.591 + 12.72489 + 0.1324; (c) 473 + 9.3827 + 349.17; (d) 693.59102 + 9.00327 + 229.461853.

21. Using the appropriate rules, round each of the following to the correct number of significant digits: (a) $3.52 \times 8.02891 \times 729$; (b) $4.52 \times 10^3 \times 8.021 \times 0.5779$;

(c) $\dfrac{7.7218 \times 10^7 \times 4.019 \times 10^{-3}}{4.019 \times 10^{-5}}$; (d) $\dfrac{7.8 \times 6.028 \times 10^{-3}}{4.183 \times 10^{-5} \times 2.19 \times 10^5}$

22. Round each of the following properly: (a) 7.32987 to 3 places; (b) 1.193528 to 4 places; (c) 7.1382×10^3 to 2 places; (d) 9.04557×10^{-17} to 4 places; (e) 7.1235.801 to 3 places; (f) 5.8092355 to 3 places.

23. How many significant digits does each of the following numbers have? (a) 7.8231; (b) 7.63×10^5; (c) 0.004; (d) 20.071.

24. Match the abbreviations below with the numbers that they represent:

A. Centi	1. $\times 10^{-3}$
B. Kilo	2. $\times 10^{-6}$
C. Deka	3. $\times 1000$
D. Micro	4. $\times 10$
E. Milli	5. $\times 0.01$

25. Match the following units of measurement with their definitions:

A. Mole	1. Metric unit of volume
B. Metre	2. Metric unit of mass
C. Gram	3. SI unit for amount of substance
D. Kelvin	4. Distance traveled by light in a vacuum in $3.335640952 \times 10^{-9}$ s

E. Liter 5. 1/273.16 the temperature interval between absolute zero and the triple point of water at 273.16 K (0.01°C)

26. Denote each of the following as characteristic of mass (*m*) or characteristic of weight (*w*):

(a) Varies with gravity

(b) Degree to which an object resists a change in its state of rest or motion

(c) Direct measure of the amount of matter in the object

(d) Different on the Moon's surface than on Earth

27. Match the following units with the quantity that they are most likely to be used to express:

A. Mg	1. Quantities of toxic pollutants
B. μg	2. Quantities of large-scale industrial chemicals
C. kg	3. Quantities of laboratory chemicals
D. g	4. Global burdens of atmospheric pollutants
E. Teragrams	5. Body mass

28. Calculate (a) the number of grams in 1.56 pounds; (b) the number of kilograms in a 2000 pound ton; (c) the number of g in 2.14 kg; (d) the number of atmospheric dust particles, each weighing an average of 2.56 μg, to make up an ounce of dust particles.

29. Distinguish between a meter and a metre.

30. How tall is a 6 foot person in cm?

31. Estimate approximately how many bacterial cells would have to be laid end-to-end to reach an inch.

32. Match the following units with the quantity that they are most likely to be used to measure:

A. km	1. Distance between this line and the line directly below
B. m	2. Distance run by an athlete in 5 seconds
C. Nanometer	3. Distance traveled by an automobile in 1 hour
D. Centimeter	4. Dimensions of this book
E. mm	5. Wavelength of visible light

33. Explain how metric units of volume can be defined in terms of length.

34. Recalling the appropriate formula from elementary geometry, what is the volume in liters of a round tank with a radius of 39.0 cm and a depth of 15.0 cm?

35. Consider gasoline at a price of $4.29 per gallon. What is its equivalent price in dollars per liter ($/L)?

36. Match the following units with the quantity that they are most likely to be used to measure:

A. Milliliter	1. Volume of milk purchased in a supermarket
B. Kiloliter	2. Volume of a laboratory chemical
C. Microliter	3. Volume of air in air pollution studies
D. Liter	4. Volume of chemical reagent in a syringe for chemical analysis

37. Give the volume in liters of cubes that are (a) 12 cm, (b) 14 cm, and (c) 17.2 cm to a side.

38. Convert each of the following Fahrenheit temperatures to Celsius and Kelvin: (a) 237°F; (b) 105°F; (c) 17°F; (d) 2°F; (e) −32°F; (f) −5°F; (g) 31.2°F.

39. Convert each of the following Celsius temperatures to Fahrenheit: (a) 237°C; (b) 75°C; (c) 17°C; (d) 100°C; (e) −32°C; (f) −40°C; (g) −11°C.

40. Convert each of the following Celsius temperatures to Kelvin: (a) 237°C; (b) 75°C; (c) 48°C; (d) 100°C; (e) −0°C; (f) −40°C; (g) −200°C.

41. Calculate the value of temperature in °F that is numerically equal to the temperature in °C.

42. The number 1.8 can be used in making conversions between Fahrenheit and Celsius temperatures. Explain how it is used and where it comes from in this application.

43. Calculate how many calories there are in 1 joule.

44. Match each of the following pertaining to units of pressure:

A. pascal	1. Based on a column of liquid
B. atm	2. Takes 14.7 to equal 1 atm

C. mmHg 3. SI unit

D. lb/in^2 4. Essentially 1 for air at sea level

45. Try to explain why pressure is a more important variable for gases than for liquids.

46. The pressure in a typical automobile tire is supposed to be 35 lb/in^2 (above normal atmospheric pressure). Calculate the equivalent pressure in (a) pascal, (b) atm, and (c) torr.

47. Knowing that there are 12 inches per foot, calculate the normal pressure of the atmosphere in lb/ft^2.

48. Atmospheric pressure readings on weather reports in the U.S. used to be given as 29–30 "inches." Speculate on what such a reading might mean.

49. Using unit conversion factors, calculate the following:

    (a) A pressure in inches of mercury equivalent to 1 atm pressure

    (b) The mass in metric tons of 760 000 tons of contaminated soil

    (c) The number of cubic meters in a cubic mile of atmospheric air

    (d) The cost of 100 liters of gasoline priced at $3.55/gallon

    (e) The number of kilograms of cheese in 200 ounces of this food

    (f) The pressure in kPa equivalent to 5.00 atm pressure

50. An analyst reported some titration data as "34.52 mL." What two things are stated by this expression?

51. Why is it not wrong to give the atomic mass of aluminum as 26.98, even though a unit is not specified for the mass?

52. Explain what is meant by a unit conversion factor. How is such a factor used? Why may quantities be multiplied by a proper unit conversion factor without concern about changing the magnitude of the quantity?

53. Consider only the following information: *An object with a mass of 1 kg also has a mass of 2.2046 lb. A pound contains 16 ounces. A piece of chalk 2 inches long is also 5.08 cm long. There are 36 inches in 1 yard. A 1-liter volumetric flask contains 1.057 quarts. A cubic centimeter is the same volume as a milliliter.* From this information, show how to calculate unit conversion factors to convert the following: (a) from yards to meters; (b) from ounces to grams; (c) from quarts to deciliters; (d) from cubic inches to cubic centimeters.

54. A quantity of colored water was poured into a 50 mL graduated cylinder, raising it to the level shown on the left in the figure. Next, an object with a mass of 25.2 g was placed in the water, raising the level to that shown on the right. What was the density of the object in g/mL?

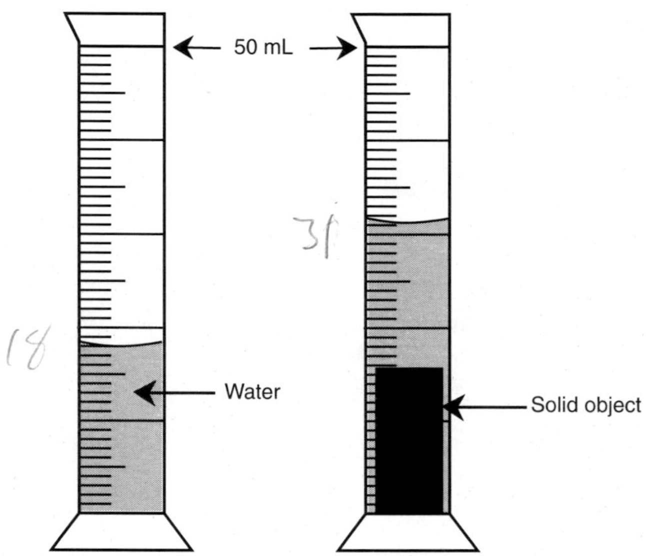

2. MATTER AND MATERIALS

2.1. WHAT IS MATTER AND WHY DOES IT MATTER FOR SUSTAINABILITY AND GREEN CHEMISTRY?

In Chapter 1, chemistry was defined as the science of **matter**—*anything that has mass and occupies space.* This chapter deals specifically with matter: what it is, how it acts, what its properties are (Figure 2.1). Environmental chemistry and green chemistry are all about matter. The wrong kind of matter in the wrong place can be a serious pollutant. Human efforts to obtain matter in a desired form have profound influences upon Earth and its support systems. Minerals and fuels are dug from or pumped from Earth, often with profound environmental consequences. Wars are fought over the availability of matter, such as essential metals and petroleum. Melting of polar icecaps induced by global warming—water as solid matter—is raising sea levels, with profound effects for some human populations. Transforming matter from atmospheric carbon dioxide and water in soil to biomass occupies vast areas of Earth's surface for agricultural enterprises, with pronounced environmental effects and profound implications for sustainability. A huge part of sustainability is the ability to transform matter to desired forms in a sustainable fashion. Recycling as much matter as possible is essential for the maintenance of sustainability. As supplies of various kinds of matter become limited, sustainable substitutes must be found, in a challenge to green chemistry. The list could go on and on; suffice it to say that matter matters—a lot.

Most of this book is concerned with chemical processes—those in which chemical bonds are broken and formed to produce different substances—and the *chemical properties* that they define. However, before going into chemical phenomena in any detail, it is helpful to consider matter in its bulk form, aside from its chemical behavior. Nonchemical aspects include physical state—whether a substance is a solid, liquid, or gas—color, hardness, extent to which matter dissolves in water, melting temperature, and boiling temperature. These kinds of *physical properties* are essential in describing the nature of matter and the chemical changes that it undergoes. They are also essential in understanding and describing its environmental chemical and biological behavior. For example, the temperature–density

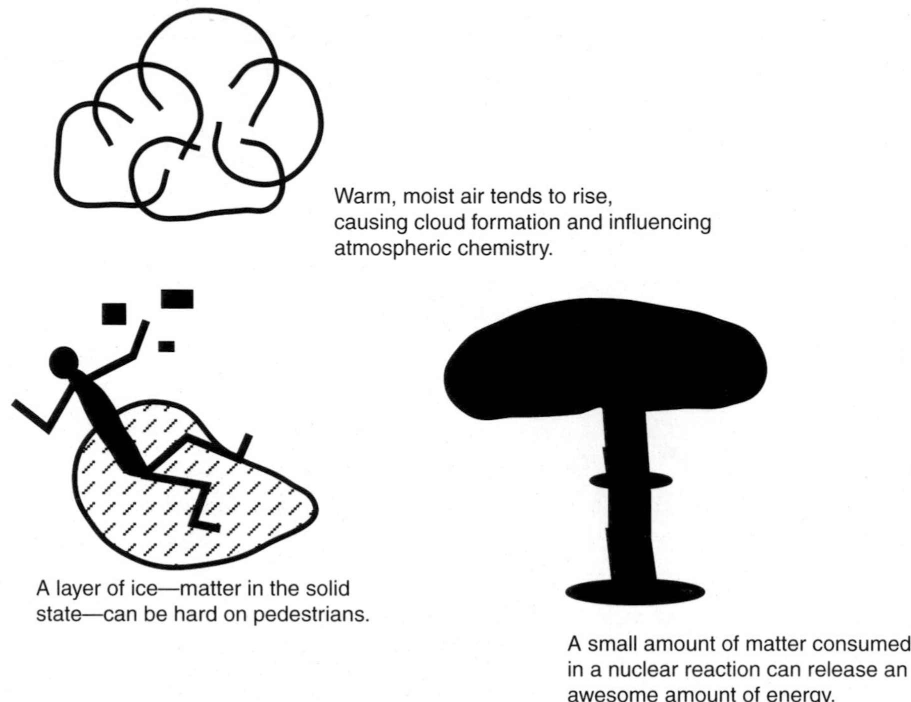

Figure 2.1. Different kinds of matter have a vast variety of properties that determine what matter does and how it is used.

relationships of water result in stratification of bodies of water such that very different chemical and biochemical processes occur at different depths in a pond, lake, or reservoir. An analogous stratification of the atmosphere profoundly influences atmospheric chemistry.

2.2. CLASSIFICATION OF MATTER

Matter exists in the form of either *elements* or *compounds* (see Sections 1.5 and 1.6, respectively). Recall that compounds consist of atoms of two or more elements bonded together. Chemical changes in which chemical bonds are broken or formed are involved when a compound is produced from two or more elements or from other compounds, or when one or more elements are isolated from a compound. These kinds of changes, which are illustrated by the examples in Table 2.1, profoundly affect the properties of matter. For example, the first reaction in the table shows that sulfur (S), a yellow, crumbly solid, reacts with oxygen, a colorless, odorless gas that is essential for life, to produce sulfur dioxide, a toxic gas with a choking odor. In the second reaction in Table 2.1, the hydrogen atoms in methane (CH_4), a highly flammable, light gas, are replaced by chlorine atoms to produce carbon tetrachloride (CCl_4), a dense liquid so nonflammable that it has been used as a fire extinguisher.

Table 2.1. Chemical Processes Involving Elements and Compounds

Type of Process	Chemical Reaction
Two elements combining to form a chemical compound: sulfur plus oxygen yields sulfur dioxide	$S + O_2 \rightarrow SO_2$
A compound reacting with an element to form two different compounds: methane plus chlorine yields carbon tetrachloride plus hydrogen chloride	$CH_4 + 4Cl_2 \rightarrow CCl_4 + 4HCl$
Two compounds reacting to form a different compound: calcium oxide plus water yields calcium hydroxide	$CaO + H_2O \rightarrow Ca(OH)_2$
A compound plus an element reacting to produce another compound and another pure element, iron oxide plus carbon yields carbon monoxide plus elemental iron	$Fe_2O_3 + 3C \rightarrow 3CO + 2Fe$
A compound breaking down to its constituent elements: passing an electrical current through water yields hydrogen and oxygen	$2H_2O \rightarrow 2H_2 + O_2$

Some General Types of Matter

In discussing matter, some general terms are employed that are encountered frequently enough that they should be mentioned here. More-exact definitions are given later in the text.

Elements are divided between metals and nonmetals; several elements with properties of both metals and nonmetals are called **metalloids**. **Metals** are elements that are generally solid, shiny in appearance, electrically conducting, and malleable—that is, they can be pounded into flat sheets without disintegrating. Examples of metals are iron, copper, and silver. Most metallic objects that are commonly encountered are not composed of just one kind of elemental metal, but are alloys consisting of homogeneous mixtures of two or more metals. **Nonmetals** often have a dull appearance and are not at all malleable. In contrast to metals, all of which except mercury exist as solids at room temperatures, nonmetals frequently occur as gases or liquids. Colorless oxygen gas, green chlorine gas (transported and stored as a liquid under pressure), and brown bromine liquid are common nonmetals. In a very general sense, we tend to regard as nonmetals substances that actually contain metals in a chemically combined form. One would classify table salt, sodium chloride, as a nonmetal even though it contains chemically bound sodium metal.

Another general classification of matter is in the categories of inorganic and organic substances. **Organic substances** consist of virtually all compounds that contain carbon, including substances made by life processes (e.g., wood, flesh, cotton, and wool), petroleum, natural gas (methane), solvents (dry cleaning fluids), synthetic fibers, and plastics. The rest of the chemical kingdom is composed of **inorganic substances** made up of virtually all substances that do not contain carbon. These include metals, rocks, table salt, water, sand, and concrete.

Mixtures and Pure Substances

Matter consisting of only one compound or of only one form of an element is a **pure substance**; all other matter is a **mixture** of two or more substances. The compound water with nothing dissolved in it is a pure substance. Highly purified helium gas is an elemental pure substance. The composition of a pure substance is defined and constant and its properties are always the same under specified conditions. Therefore, pure water at −1°C is always a solid (ice), whereas a mixture of water and salt might be either a liquid or a solid at that temperature, depending upon the amount of salt dissolved in the water. Air is a mixture of elemental gases and compounds, predominantly nitrogen, oxygen, argon, carbon dioxide, and water vapor. Drinking water is a mixture containing calcium ion (Ca^{2+}, see ions in Section 1.6), hydrogen carbonate ion (bicarbonate, HCO_3^-), nitrogen gas, carbon dioxide gas, and other substances dissolved in the water. Mixtures can be separated into their constituent pure substances by **physical processes**. Liquified air is distilled to isolate pure oxygen (used in welding, industrial processes, for some specialized wastewater treatment processes, and for breathing by people with emphysema), liquid nitrogen (used for quick-freezing frozen foods), and argon (used in specialized types of welding because of its chemically nonreactive nature).

Homogeneous and Heterogeneous Mixtures

A **heterogeneous mixture** is one that is not uniform throughout and possesses readily distinguishable constituents (matter in different phases, see Section 2.5) that can be isolated by means as simple as mechanical separation. Concrete (Figure 2.2) is such a mixture; individual grains of sand and pieces of gravel can be separated from concrete with a pick or other tool. **Homogeneous mixtures** are uniform throughout; to observe different constituents would require going down to molecular

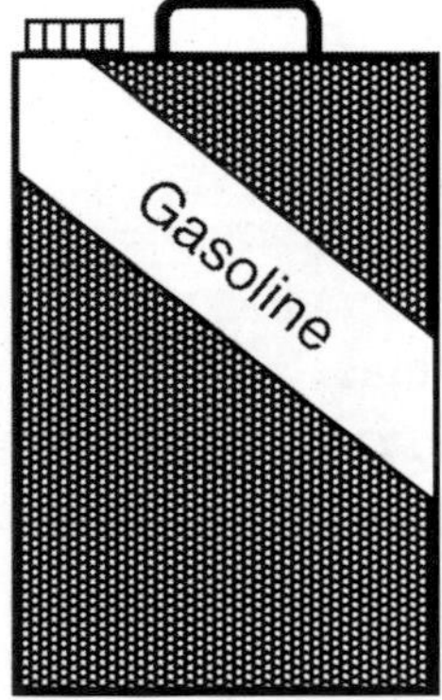

Figure 2.2. Concrete (left) is a heterogeneous mixture with readily visible grains of sand, pieces of gravel, and cement dust. Gasoline is a homogeneous mixture consisting of a solution of petroleum hydrocarbon liquids, dye, and additives such as ethanol to improve its fuel qualities.

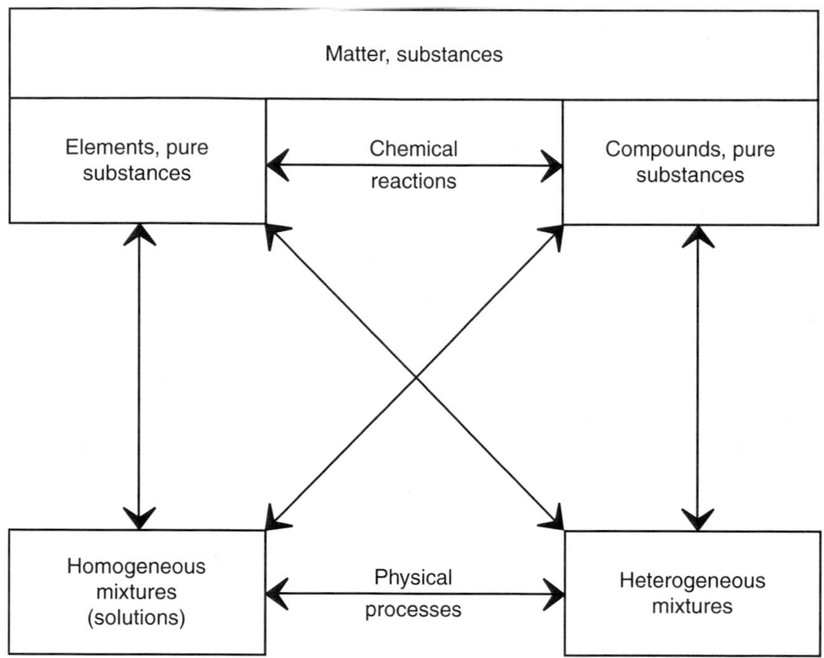

Figure 2.3. Classification of matter.

levels. Homogeneous mixtures may have varying compositions and consist of two or more chemically distinct constituents, so they are not pure substances. Air, well filtered to remove particles of dust, pollen, or smoke, is a homogeneous mixture. Homogeneous mixtures are also called **solutions**, a term that is usually applied to mixtures composed of gases, solids, and other liquids dissolved in a liquid. A hazardous waste leachate is a solution containing—in addition to water—contaminants such as dissolved acids, iron, heavy metals, and toxic organic compounds. Whereas mechanical means, including centrifugation and filtration, can be used to separate the solids from the liquids in a heterogeneous mixture of solids and liquids, the isolation of components of homogeneous mixtures requires processes such as distillation and freezing.

Summary of the Classification of Matter

As discussed above, all matter consists of compounds or elements. These may exist as pure substances or mixtures; the latter may be either homogeneous or heterogeneous. These relationships are summarized in Figure 2.3.

2.3. QUANTITY OF MATTER: THE MOLE

One of the most fundamental characteristics of a specific body of matter is the quantity of it. In discussing the quantitative chemical characteristics of matter, it is essential to have a way of expressing quantity in a way that is proportional to the number of individual entities of the substance—that is, atoms, molecules, or ions—in numbers that are readily related to the properties of the atoms or molecules of the substance. The simplest way to do this would be as individual atoms, molecules, or ions, but for laboratory quantities, these would number the order of 10^{23}, which is far too large to be used routinely in expressing quantities of matter. Instead, such quantities are readily expressed as moles of substance. A mole is defined in terms of specific entities, such as atoms of Ar, molecules of H_2O, or Na^+ and Cl^- ions, each pair of which composes a "molecule" of NaCl. A **mole** is defined as *the quantity of substance that contains the same number of specified entities as there are atoms of C in exactly 0.012 kg (12 g) of carbon-12.* It is easier to specify the quantity of a substance equivalent to its number of moles than it is to define the mole. To do so, simply state the atomic mass (of an element) or the molecular mass (of a compound) and affix "grams" to it the **molar mass** is then given by this value, with units of g/mol. This is illustrated by the following examples:

- *A mole of argon, which always exists as individual Ar atoms*: The atomic mass of Ar is 40.0. Therefore, exactly one mole of Ar is 40.0 g of argon, and the molar mass of argon is 40.0 g/mol.
- *A mole of molecular elemental hydrogen, H_2*: The atomic mass of H is 1.0, the molecular mass of H_2 is, therefore, 2.0, and a mole of H_2 is 2.0 g of H_2; the molar mass of H_2 is 2.0 g/mol.
- *A mole of methane, CH_4*: The atomic mass of H is 1.0 and that of C is 12.0, so the molecular mass of CH_4 is 16.0. Therefore, a mole of methane has a mass of 16.0 g; the molar mass of methane is 16.0 g/mol.

The Mole and Avogadro's Number

In Section 1.8, Avogadro's number was mentioned as an example of a huge number, 6.02×10^{23}. **Avogadro's number** *is the number of specified entities in a mole of substance.* The "specified entities" may consist of atoms or molecules or they may be groups of ions making up the smallest possible unit of an ionic compound, such as 2 Na^+ ions and one S^{2-} in Na_2S. (It is not really correct to refer to Na_2S as a molecule, because the compound consists of ions arranged in a crystalline structure such that there are 2 Na^+ ions for each S^{2-} ion.) A general term that covers all these possibilities is the **formula unit**. The average mass of a formula unit is called the **formula mass**. Examples of the terms defined in this section are given in Table 2.2.

2.4. PHYSICAL PROPERTIES OF MATTER

Physical properties of matter are those that can be measured without altering the chemical composition of the matter. A typical physical property is color, which can be observed without changing matter at all. Malleability of metals, the degree to

Table 2.2. Relationships Involving Moles of Substance

Substance	Formula Unit	Formula Mass	Mass of 1 mole (g)	Numbers and Kinds of Individual Entities in 1 mole
Helium	He atom	4.003[a]	4.003	6.02×10^{23} (Avogadro's number) of He atoms
Fluorine	F_2 molecule	38.00[b]	38.00	6.02×10^{23} F_2 molecules: $2 \times 6.02 \times 10^{23}$ F atoms
Methane	CH_4 molecule	16.04[b]	16.04	6.02×10^{23} CH_4 molecules: 6.02×10^{23} C atoms $4 \times 6.02 \times 10^{23}$ H atoms
Sodium oxide	Na_2O	62.00[c]	62.00	6.02×10^{23} Na_2O formula units: $2 \times 6.02 \times 10^{23}$ Na^+ ions 6.02×10^{23} O^{2-} ions

[a] Specifically, atomic mass.
[b] Specifically, molecular mass.
[c] Reference should be made to formula mass for this compound because it consists of Na^+ and S^{2-} ions in a ratio of 2/1.

which they can be pounded into thin sheets, certainly alters the shape of an object but does not change it chemically. On the other hand, observation of a chemical property, such as whether sugar burns when ignited in air, potentially involves a complete change in the chemical composition of the substance tested.

Physical properties are important in describing and identifying particular kinds matter. For example, if a substance is liquid at room temperature, has a lustrous metallic color, conducts electricity well, and has a very high density (mass per unit volume) of 13.6 g/cm^3, it is doubtless elemental mercury metal. Physical properties are very useful in assessing the hazards and predicting the fates of environmental pollutants. An organic substance that readily forms a vapor (a volatile organic compound) will tend to enter the atmosphere or to pose an inhalation hazard. A brightly colored water-soluble pollutant may cause deterioration of water quality by adding water "color." Much of the health hazard of asbestos is its tendency to form extremely small-diameter fibers that are readily carried far into the lungs and that can puncture individual cells.

Several important physical properties of matter are discussed in this section. The most commonly considered of these are density, color, and solubility. Thermal properties are addressed separately in Section 2.9.

Density

Density (d) is defined as mass per unit volume and is expressed by the formula

$$d = \frac{\text{mass}}{\text{volume}} \tag{2.4.1}$$

Density is useful for identifying and characterizing pure substances and mixtures. For example, the density of a mixture of automobile system antifreeze and water, used to prevent the engine coolant from freezing in winter or boiling in summer, varies with the composition of the mixture. Its composition and, therefore, the degree of protection that it offers against freezing can be estimated by measuring its density and relating that through a table to the freezing temperature of the mixture.

Density may be expressed in any units of mass or volume. The densities of liquids and solids are normally given in units of grams per cubic centimeter (g/cm^3, the same as grams per milliliter, g/mL). These values are convenient because they are of the order of 1 g/cm^3 for common liquids and solids.

The volume of a given mass of substance varies with temperature, so the density is a function of temperature. This variation is relatively small for solids, greater for liquids, and very high for gases. Densities of gases vary a great deal with pressure, as well. The temperature-dependent variation of density is an important property of water, and results in stratification of bodies of water, which greatly affects the environmental chemistry that occurs in lakes and reservoirs. The density of liquid water has a maximum value of 1.0000 g/mL at 4°C, is 0.9998 g/mL at 0°C, and is 0.9970 g/mL at 25°C. The combined temperature/pressure relationship for the density of air causes air to become stratified into layers, particularly the troposphere near the surface and the stratosphere from about 13 to 50 km altitude.

Whereas liquids and solids have densities of the order of 1 to several g/cm^3, gases at atmospheric temperature and pressure have densities of only about one one-thousandth as much (see Table 2.3 and Figure 2.4).

Specific Gravity

Often densities are expressed by means of **specific gravity**, defined as the ratio of the density of a substance to that of a standard substance. For solids and liquids, the standard substance is usually water; for gases, it is usually air. For example, the density of ethanol (ethyl alcohol) at 20°C is 0.7895 g/mL. The specific gravity of ethanol at 20°C referred to water at 4°C is given by

$$\text{Specific gravity} = \frac{\text{density of ethanol}}{\text{density of water}} = \frac{0.7895\text{ g/mL}}{1.0000\text{ g/mL}} = 0.7895 \qquad (2.4.2)$$

Table 2.3. Densities of Some Solids, Liquids, and Gases

Solids	*d* (g/cm^3)	Liquids	*d* (g/cm^3)	Gases	*d* (g/L)[a]
Sugar	1.59	Benzene	0.879	Helium	0.178
Sodium chloride	2.16	Water (4°C)	1.0000	Methane	0.714
Iron	7.86	Carbon tetrachloride	1.595	Air	1.293
Lead	11.34	Sulfuric acid	1.84	Chlorine	3.17

[a] At 0°C and 1 atm pressure.

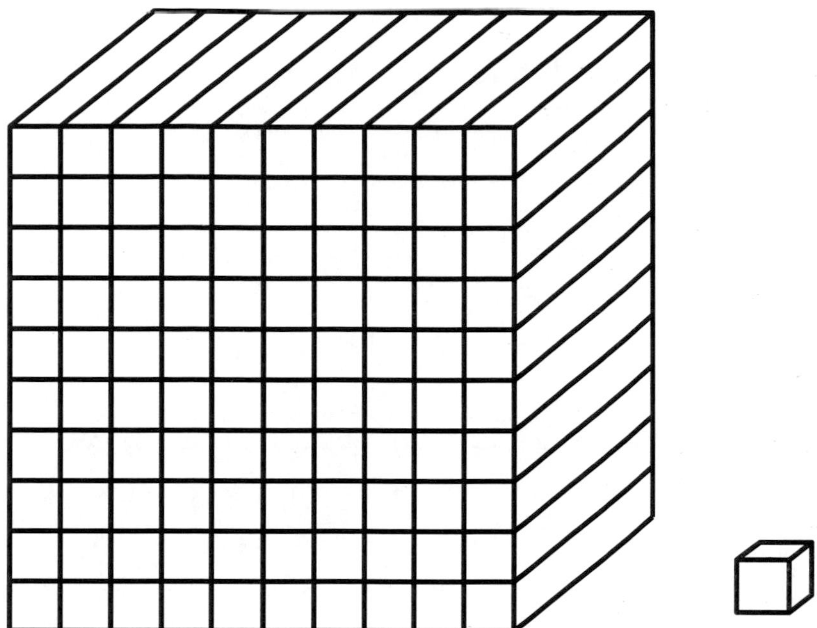

Figure 2.4. A quantity of air that occupies a volume of 1000 cm^3 at room temperature and atmospheric pressure (left) has about the same mass as 1 cm^3 of liquid water (right).

For an exact value of specific gravity, the temperatures of the substances should be specified. In this case, the notation "specific gravity of ethanol at 20°C/4°C" shows that the specific gravity is the ratio of the density of ethanol at 20°C to that of water at 4°C.

Color

Color is one of the more useful properties for identifying substances without doing any chemical or physical tests. A violet vapor, for example, is characteristic of iodine. A red/brown gas could well be bromine or nitrogen dioxide (NO_2); a practiced eye can distinguish between the two. Just a small amount of potassium permanganate ($KMnO_4$) in solution provides an intense purple color. A characteristic yellow/brown color in water may be indicative of organically bound iron.

The human eye responds to colors of **electromagnetic radiation** ranging in wavelength from about 400 nanometers (nm) to somewhat over 700 nm. Within this wavelength range, humans see **light**; immediately below 400 nm is **ultraviolet radiation**, and somewhat above 700 nm is **infrared radiation** (Figure 2.5). Light with a mixture of wavelengths throughout the visible region, such as sunlight, appears white to the eye. Light over narrower wavelength regions has the following colors: 400–450 nm, blue; 490–550 nm, green; 550–580 nm, yellow; 580–650 nm, orange, 650 nm–upper limit of visible region, red. Solutions are colored because of the light they absorb. Red, orange, and yellow solutions absorb violet and blue

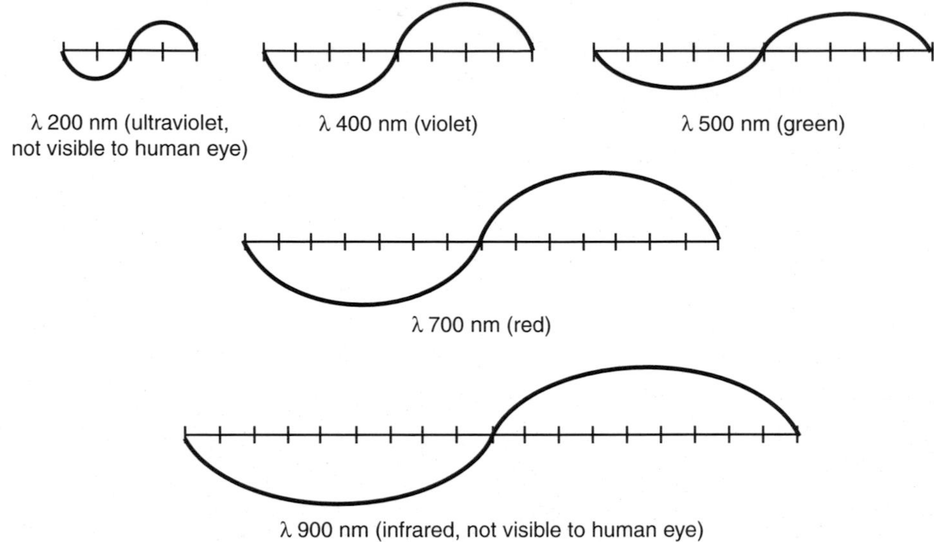

Figure 2.5. Electromagnetic radiation of different wavelengths. Each division represents 50 nm.

light; purple solutions absorb green and yellow light; and blue and green solutions absorb orange and red light. Solutions that do not absorb light are colorless (clear); solids that do not absorb light are white.

Electromagnetic Radiation and Green Chemistry

As shown by the descriptions above and discussed at greater length in Section 3.11, electromagnetic radiation can be visualized as waves, but it also represents a means of energy transfer. This is because electromagnetic radiation consists of **quanta**, which can be visualized as packets of energy. The shorter the wavelength of electromagnetic radiation, the more energetic these quanta are. Visualize, for example, the tail wagging on a dog. If the tail is wagging slowly from side to side, it is analogous to long-wavelength electromagnetic radiation with relatively few wags per unit time, or low **frequency**. There is relatively little energy involved in a slowly wagging tail. But an excited dog may wag its tail vigorously, a high frequency and "shorter wavelength." A rapidly wagging tail entails relatively more energy.

The interaction of electromagnetic radiation with matter is very important, particularly in the transfer of energy. Such radiation in the form of visible light from the Sun is absorbed by matter in the form of molecules in the atmosphere, soil on the ground, and plants growing on the soil, so it is the main means by which the enormous energy of the sun is transferred to Earth. Energy from the warmed Earth surface is radiated back into space as relatively longer-wavelength infrared radiation. This radiation impinges on molecules of carbon dioxide and water vapor in the atmosphere and its energy is transferred to these molecules, causing them to move more vigorously and raising the temperature of the atmosphere. This is the phenomenon that keeps Earth's atmosphere at a temperature hospitable to life but, in excess, is responsible for undesirable global warming.

Many chemical processes, such as chemical syntheses by which new compounds are formed, require input of energy—a major cost in the chemical industry and one with significant implications for sustainability and the environmental effects of chemical processing. Normally, the required energy is supplied by heating. As an alternative to heating, one of the most effective ways to transfer energy directly to matter in a chemical reaction is through the use of ultraviolet radiation, electromagnetic radiation with wavelengths shorter and energy per photon higher than visible light. Therefore, in some cases, ultraviolet radiation is used in the practice of green chemistry to transfer energy directly and specifically to molecules in green chemical synthetic processes. The ultraviolet radiation is transferred directly to molecules involved in chemical reactions, energizing them, but not heating the whole reaction mixture. In favorable cases, less energy is required and processes are safer because of the lower overall reaction temperatures.

2.5. STATES OF MATTER

Figure 2.6 illustrates the three **states** in which matter can exist. **Solids** have a definite shape and volume. **Liquids** have an indefinite shape and take on the shape of the container in which they are contained. Solids and liquids are not significantly compressible, which means that a specific quantity of a substance has a definite volume and cannot be squeezed into a significantly smaller volume. **Gases** take on both the shape and volume of their containers. A quantity of gas can be compressed to a very small volume and will expand to occupy the volume of any container into which it is introduced.

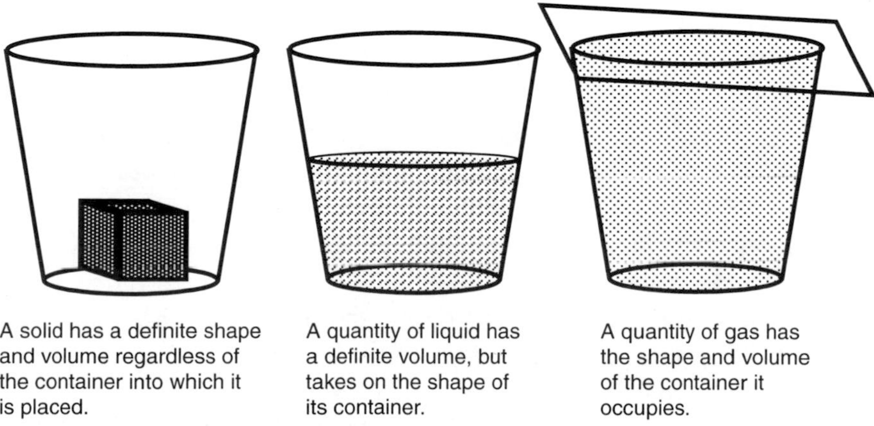

Figure 2.6. Representations of the three states of matter.

Each of these separate phases is discussed in separate sections in this chapter. Everyone is familiar with the three states of matter for water:

- Gas: water vapor in a humid atmosphere, steam
- Liquid: water in a lake, groundwater
- Solid: ice in polar ice caps, snow in snowpack

Changes in matter from one phase to another are very important in the environment. For example, water vapor changing from the gas phase to liquid results in cloud formation or precipitation. Water is desalinated by producing water vapor from sea water, leaving the solid salt behind, and re-condensing the pure water vapor as a salt-free liquid. Some organic pollutants are extracted from water for chemical analysis by transferring them from the water to another organic phase that is immiscible with water. The condensation of organic pollutants from gaseous materials to solid products in the atmosphere is responsible for the visibility-reducing particulate matter in photochemical smog. Additional examples of phase changes are discussed in Section 2.10 and some of the energy relationships involved are covered in Section 2.9.

2.6. GASES

We live at the bottom of a "sea" of gas—Earth's atmosphere. It is composed of a mixture of gases, the most abundant of which are nitrogen, oxygen, argon, carbon dioxide, and water vapor. Although these gases have neither color nor odor, so that we are not aware of their presence, they are crucial to the well-being of life on Earth. Deprived of oxygen, an animal loses consciousness and dies within a short time. Nitrogen extracted from the air is converted to chemically bound forms that are crucial to plant growth. Plants require carbon dioxide and water vapor condenses to produce rain.

To understand chemistry and matter, it is important to have a basic understanding of the nature and behavior of gases. Physically, gases are the "loosest" form of matter. A quantity of gas has neither a definite shape nor a definite volume, so that it takes on the shape and volume of the container in which it is held. The reason for this behavior is that gas molecules move independently and at random, bouncing off each other as they do so. They move very rapidly: at 0°C, the average molecule of hydrogen gas moves at 3600 miles per hour, 6–7 times as fast as a passenger jet airplane. Gas molecules colliding with container walls exert **pressure**. An underinflated automobile tire has a low pressure because there are relatively few gas molecules in the tire to collide with its walls, which feel relatively soft when pushed. As more air is added, the pressure is increased, the tire feels harder, and its internal pressure may become so high that it can no longer be contained by the tire structure, so that the tire wall bursts.

A quantity of a gas is mostly empty space. That this is so is rather dramatically illustrated by comparison of the high volume of gas compared with the same amount of material in a liquid or solid. For example, at 100°C, a mole (18 g) of liquid water occupies a little more than 18 mL of volume, equivalent to just a few teaspoonfuls. When enough heat energy is added to the water to convert it all into gaseous steam at 100°C, the volume becomes 30 600 mL, which is 1700 times the original volume. This huge increase in volume occurs because of the large distances separating gas molecules compared with their own diameters. It is this great distance that allows gas to be compressed (pushed together). Therefore, air with a volume of more than twice that of a tire can be forced into the tire when it becomes pressurized because the air molecules can be forced closer together under the pressure in the tire when it is inflated.

The rapid, constant motion of gas molecules explains the phenomenon of gas **diffusion** in which gases move large distances from their sources. Diffusion is responsible for much of the hazard of volatile, flammable liquids, such as gasoline. Molecules of gasoline evaporated from an open container of this liquid can spread from their source. If the gaseous gasoline reaches an ignition source, such as an open flame, it may ignite and cause a fire or explosion.

The Gas Laws

To describe the physical and chemical behavior of gases, it is essential to have a rudimentary understanding of the relationships among the following: quantity of gas in numbers of moles (n, see Section 2.3), volume (V), temperature (T), and pressure (P). These relationships are expressed by the **gas laws**. The gas laws as discussed here apply to **ideal gases**, for which it is assumed that the molecules of gas have negligible volume, that there are no forces of attraction between them, and that they collide with each other in a perfectly elastic manner. Real gases such as H_2, O_2, He, and Ar behave in a nearly ideal fashion at moderate pressures and all but very low temperatures. The gas laws were derived from observations of the effects on gas volume of changing pressure, temperature, and quantity of gas, as stated by Boyle's law, Charles' law, and Avogadro's law, respectively. These three laws are described below:

Boyle's Law

As illustrated in Figure 2.7, doubling the pressure on a quantity of a gas halves its volume. This illustrates **Boyle's law**, according to which, *at constant temperature, the volume of a fixed quantity of gas is inversely proportional to the pressure of the gas.* Boyle's law may be stated mathematically as

$$V = (\text{a constant}) \times \frac{1}{P} \qquad (2.6.1)$$

This inverse relationship between the volume of a gas and its pressure is a consequence of the compressibility of matter in the gas phase (recall that solids and liquids are not significantly compressible).

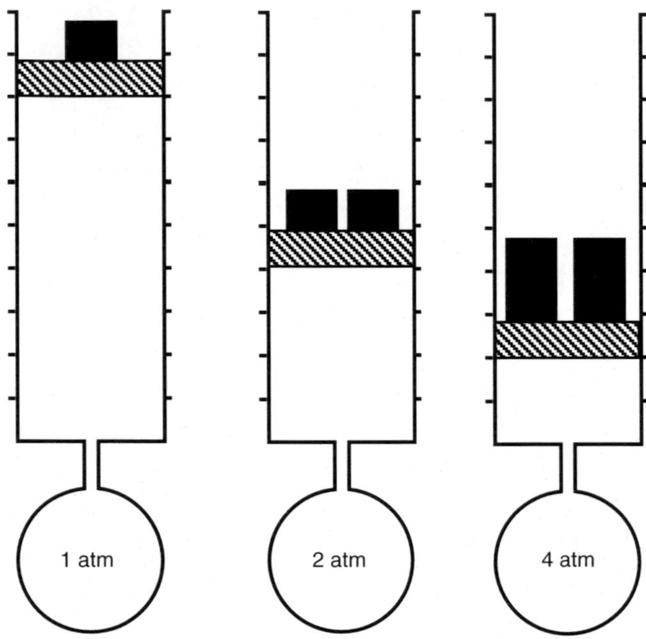

Figure 2.7. In accordance with Boyle's law, doubling the pressure of a specific quantity of gas at a constant temperature halves its volume.

Charles' Law

Charles' law gives the relationship between gas volume and temperature, and states that *the volume of a fixed quantity of gas is directly proportional to the absolute temperature (°C + 273) at constant pressure*. This law may be stated mathematically as

$$V = (\text{a constant}) \times T \tag{2.6.2}$$

where T is the temperature in K (see Section 1.12). According to this relationship, doubling the absolute temperature of a fixed quantity of gas at constant pressure doubles the volume.

Avogadro's Law

The third of the three fundamentally important gas laws is **Avogadro's law**, relating the volume of gas to its quantity in moles. This law states that *at constant temperature and pressure, the volume of a gas is directly proportional to the number of molecules of gas, commonly expressed as moles*. Mathematically, this relationship is

$$V = (\text{a constant}) \times n \tag{2.6.3}$$

where n is the number of moles of gas.

The General Gas Law

The three gas laws just defined may be combined into a **general gas law** stating that *the volume of a quantity of ideal gas is proportional to the number of moles of gas and its absolute temperature and inversely proportional to its pressure.* Mathematically, this law is

$$V = (\text{a constant}) \times \frac{nT}{P} \tag{2.6.4}$$

Designating the proportionality constant as the **ideal gas constant**, *R*, yields the **ideal gas equation**:

$$V = \frac{RnT}{P} \text{ or } PV = nRT \tag{2.6.5}$$

The units of *R* depend upon the way in which the ideal gas equation is used. For calculations involving volume in liters and pressures in atmospheres, the value of *R* is 0.0821 L atm/(K mol).

The ideal gas equation shows that at a chosen temperature and pressure a mole of any gas should occupy the same volume. A temperature of 0°C (273.15 K) and 1 atm pressure have been chosen as **standard temperature and pressure (STP)**. *At STP, the volume of 1 mole of ideal gas is 22.4 L.* This volume is called the **molar volume of a gas**. It should be remembered that at a temperature of *273 K* and a pressure of *1 atm*, the volume of *1 mole* of ideal gas is *22.4 L*. Knowing these values, it is always possible to calculate the value of *R* as follows:

$$R = \frac{PV}{nT} = \frac{1 \text{ atm} \times 22.4 \text{ L}}{1 \text{ mole} \times 273 \text{ K}} = 0.0821 \frac{\text{L} \times \text{atm}}{\text{K} \times \text{mol}} \tag{2.6.6}$$

Gas Law Calculations

A common calculation is that of the volume of a gas, starting with a particular volume and changing temperature and/or pressure. This kind of calculation follows logically from the ideal gas equation, but can also be reasoned out, knowing that an *increase* in temperature causes an *increase* in volume, whereas an *increase* in pressure causes a *decrease* in volume. Therefore, a second volume, V_2, is calculated from an initial volume, V_1, by the relationship

$$V_2 = V_1 \times \frac{\square}{\square} \times \frac{\square}{\square}$$

↑ **Ratio of temperatures** ↑ **Ratio of pressures** (2.6.7)

The use of this relationship in doing gas law calculations requires only that one remember the following:

- If the temperature *increases* (if $T_2 > T_1$), the volume *increases*. Therefore, T_2 is always placed over T_1 in the ratio of temperatures.
- If the pressure *increases* (if $P_2 > P_1$), the volume *decreases*. Therefore, P_1 is always placed over P_2 in the ratio of pressures.
- If there is no change in temperature or pressure, the corresponding ratio remains 1.

These relationships can be illustrated by several examples.

Charles' Law Calculation

When the temperature of a quantity of gas changes while the pressure stays the same, the resulting calculation is a *Charles' law calculation*. As an example of such a calculation, calculate the volume of a gas with an initial volume of 10.0 L when the temperature changes from −11.0°C to 95.0°C at constant pressure. The first step in solving any gas law problem involving a temperature change is to convert Celsius temperatures to kelvin:

$$T_1 = 273 + (-11) = 262\text{ K}, \qquad T_2 = 273 + 95 = 368\text{ K}$$

Substitution into Equation 2.6.7 gives

$$V_2 = V_1 \times \frac{368\text{ K}}{262\text{ K}} \times 1 \text{ (factor for pressure = 1, because } P_2 = P_1) \tag{2.6.8}$$

$$V_2 = 10.0\text{ L} \times \frac{368\text{ K}}{262\text{ K}} = 14.0\text{ L}$$

Note that in this calculation, it is seen that the temperature increases; this increases the volume, so the higher temperature is placed over the lower. Furthermore, since $P_2 = P_1$, the factor for the pressure ratio simply drops out.

Calculate next the volume of a gas with an initial volume of 10.0 L after the temperature decreases from 111.0°C to 2.0°C at constant pressure:

$$T_1 = 273 + (111) = 384\text{ K}, \qquad T_2 = 273 + 2 = 275\text{ K}$$

Substitution into Equation 2.6.6 gives

$$V_2 = V_1 \times \frac{275\text{ K}}{384\text{ K}} = 10.0\text{ L} \times \frac{275\text{ K}}{384\text{ K}} = 7.16\text{ L} \tag{2.6.9}$$

In this calculation, the temperature decreases; this decreases the volume, so the lower temperature is placed over the higher (mathematically, it is still T_2 over T_1 regardless of which way the temperature changes).

Boyle's Law Calculation

When the pressure of a quantity of gas changes while the temperature stays the same, the resulting calculation is a *Boyle's law calculation*. As an example, calculate the new volume that results when the pressure of a quantity of gas initially occupying 12.0 L is changed from 0.856 atm to 1.27 atm at constant temperature. In this case, the pressure has increased; this decreases the volume of the gas so that V_2 is given by the following:

$$V_2 = V_1 \times \frac{0.856 \text{ atm}}{1.27 \text{ atm}} = \times 1 \text{ (because } T_2 = T_1\text{)} \tag{2.6.10}$$

$$V_2 = 12.0 \text{ L} \times \frac{0.856 \text{ atm}}{1.27 \text{ atm}} = 8.09 \text{ L}$$

Calculate next the volume of a gas with an initial volume of 15.0 L when the pressure decreases from 1.71 atm to 1.07 atm at constant temperature. In this case, the pressure decreases; this increases the volume, so the appropriate factor to multiply by is 1.71/1.07:

$$V_2 = V_1 \times \frac{1.71 \text{ atm}}{1.07 \text{ atm}} = 15.0 \text{ L} \times \frac{1.71 \text{ atm}}{1.07 \text{ atm}} = 24.0 \text{ L} \tag{2.6.11}$$

Calculations Using the Ideal Gas Law:

The ideal gas law Equation (2.6.5)

$$PV = nRT$$

may be used to calculate any of the variables in it if the others are known. As an example, calculate the volume of 0.333mol of gas at 300 K under a pressure of 0.950 atm:

$$V = \frac{nRT}{P} = \frac{0.333 \text{ mol} \times 0.0821 \text{ L atm/(K mol)} \times 300 \text{ K}}{0.950 \text{ atm}} = 8.63 \text{ L} \tag{2.6.12}$$

As another example, calculate the temperature of 2.50 mol of gas that occupies a volume of 52.6 L under a pressure of 1.15 atm:

$$T = \frac{PV}{nR} = \frac{1.15 \text{ atm} \times 52.6 \text{ L}}{2.50 \text{ mol} \times 0.0821 \text{ L atm/(K mol)}} = 295 \text{ K} \tag{2.6.13}$$

Other examples of gas law calculations are given in the problem section at the end of this chapter.

Where subscripts 1 and 2 denote parameters before and after any change, respectively, the following relationship can be derived from the fact that *R* is a constant in the ideal gas law equation:

$$\frac{P_2V_2}{n_2T_2} = \frac{P_1V_1}{n_1T_1} \tag{2.6.14}$$

By holding the appropriate parameters constant, this equation can be used to derive the mathematical relationships used in calculations involving Boyle's law, Charles' law, or Avogadro's law, as well as any combination of these laws.

Exercise: Derive an equation that will give the volume of a gas after all three of the parameters P, T, and n have been changed.

Answer: Rearrangement of Equation 2.6.14 gives the relationship needed:

$$V_2 = V_1 \frac{P_1 n_2 T_2}{P_2 n_1 T_1} \tag{2.6.15}$$

2.7. LIQUIDS AND SOLUTIONS

Whereas the molecules in gases are relatively far apart and have minimal attractive forces among them (none in the case of ideal gases, discussed in the preceding section), molecules of liquids are close enough together that they can be regarded as "touching" and are strongly attracted to each other. Like those of gases, however, the molecules of liquids move freely relative to each other. These characteristics give rise to several significant properties of liquids. Whereas gases are mostly empty space, most of the volume of a liquid is occupied by molecules; therefore, the densities of liquids are much higher than those of gases. Because there is little unoccupied space between molecules of liquid, liquids are only slightly *compressible*; doubling the pressure on a liquid causes a barely perceptible change in volume, whereas it halves the volume of an ideal gas. Liquids do not expand to fill their containers as do gases. Essentially, therefore, a given quantity of liquid occupies a fixed volume. Because of the free movement of molecules relative to each other in a liquid, it takes on the shape of that portion of the container that it occupies.

Evaporation and Condensation of Liquids

Consider the surface of a body of water as illustrated in Figure 2.8. The molecules of liquid water are in constant motion relative to each other. They have a distribution of energies such that there are more higher-energy molecules at higher temperatures. Some molecules are energetic enough that they escape the attractive forces of the other molecules in the mass of liquid and enter the gas phase. This phenomenon is called **evaporation**.

Molecules of liquid in the gas phase, such as water vapor molecules above a body of water, may come together or strike the surface of the liquid and reenter the liquid phase. This process is called **condensation**. It is what happens, for example, when water vapor molecules in the atmosphere strike the surface of very small water droplets in clouds and enter the liquid water phase, causing the droplets to grow to sufficient size to fall as precipitation.

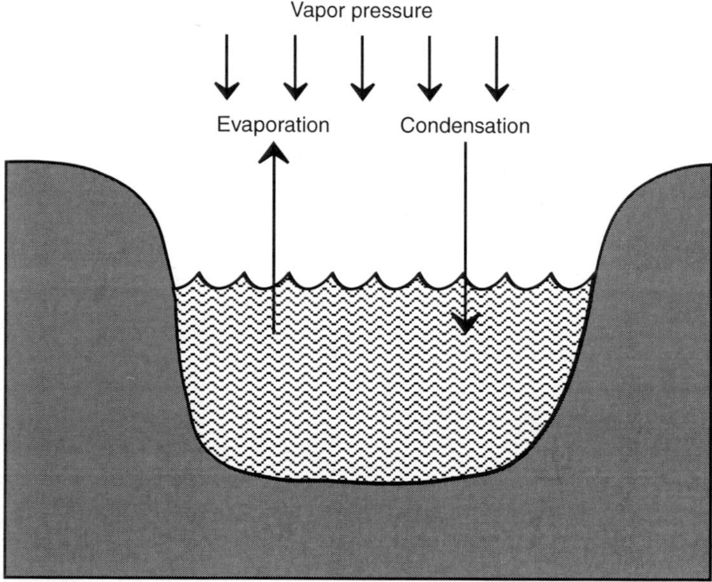

Figure 2.8. Evaporation and condensation of liquid.

Vapor Pressure

In a confined area above a liquid, equilibrium is established between the evaporation and condensation of molecules from the liquid. For a given temperature, this results in a steady-state level of the vapor, which can be described as a pressure. Such a pressure is called the **vapor pressure** of the liquid (Figure 2.8).

Vapor pressure is very important in determining the fates and effects of substances, including hazardous substances, in the environment. Loss of a liquid by evaporation to the atmosphere increases with increasing vapor pressure and is an important mechanism by which volatile pollutants enter the atmosphere. High vapor pressure of a flammable liquid can result in the formation of explosive mixtures of vapor above a liquid, such as in a storage tank.

Solutions

Solutions were mentioned in Section 2.2 as homogeneous mixtures. Here, they are regarded as liquids in which quantities of gases, solids, or other liquids are dispersed as individual ions or molecules. Rainwater, for example, is a solution containing molecules of N_2, O_2, and CO_2 from air dispersed among the water molecules. Rainwater from polluted air may contain harmful substances such as sulfuric acid (H_2SO_4) in small quantities among the water molecules. The predominant liquid constituent of a solution is called the **solvent**. A substance dispersed in it is said to be **dissolved** in the solvent and is called the **solute**.

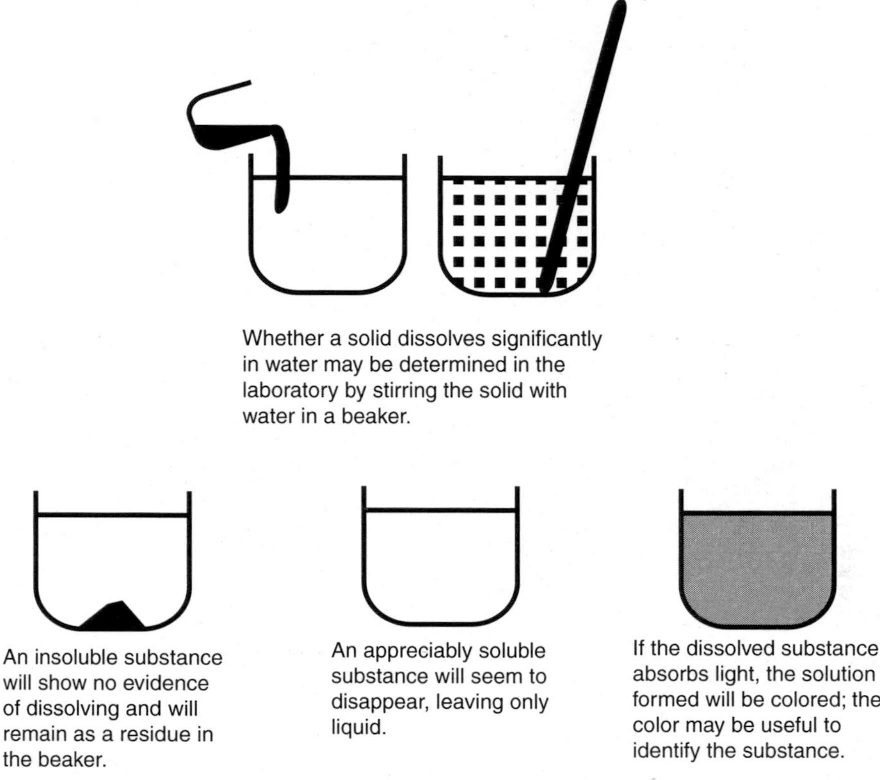

Figure 2.9. Illustration of whether a substance dissolves in a liquid.

Solubility

Mixing and stirring some solids, such as sugar, with water results in a rapid and obvious change in which all or part of the substance seems to disappear as it dissolves (Figure 2.9). When this happens, the substance is said to *dissolve*, and one that does so to a significant extent is said to be **soluble** in the solvent. Substances that do not dissolve to a perceptible degree are called **insoluble**. *Solubility* is a significant physical property. It is a relative term. For example, polytetrafluoroethylene (Teflon) is insoluble in water, limestone is slightly soluble, and ethanol (ethyl alcohol) is infinitely so in that it can be mixed with water in any proportion ranging from pure water to pure ethanol with only a single liquid phase consisting of an ethanol/water solution being observed.

Solution Concentration

Solutions are addressed specifically and in more detail in Chapter 7. At this point, however, it is useful to have some understanding of the quantitative expression of the amount of solute dissolved in a specified amount of solution. In a relative sense, a solution that has comparatively little solute dissolved per unit volume of solution is called a **dilute solution**, whereas one with an amount of solute of the same order of magnitude as that of the solvent is a **concentrated solution.** Water saturated with air at 25°C contains only about 8 mg of oxygen dissolved in 1 L of water; this composes a *dilute solution* of oxygen. A typical engine coolant solution contains about as much ethylene glycol (antifreeze) as it does water, so it is a *concentrated solution.* A solution that is at equilibrium with excess solute so that it contains the maximum amount of solute that it can dissolve is called a **saturated solution.** One that can still dissolve more solute is called an **unsaturated solution.**

For the chemist, the most useful way to express the concentration of a solution is in terms of the number of moles of solute dissolved per liter of solution. The **molar concentration,** *M*, of a solution is *the number of moles of solute dissolved per liter of solution.* Mathematically,

$$M = \frac{\text{moles of solute}}{\text{number of liters of solution}} \tag{2.7.1}$$

Example: Exactly 34.0 g of ammonia, NH_3, were dissolved in water and the solution was made up to a volume of exactly 0.500 L. What was the molar concentration, *M*, of ammonia in the resulting solution?

Answer: The molar mass of NH_3 is 17.0 g/mol. Therefore,

$$\text{Number of moles of } NH_3 = \frac{34.0 \text{ g}}{17.0 \text{ g/mol}} = 2.00 \text{ mol} \tag{2.7.2}$$

The molar concentration of the solution is calculated from Equation 2.7.1:

$$M = \frac{2.00 \text{ mol}}{0.500 \text{ mol}} = 4.00 \text{ mol/L} \tag{2.7.3}$$

Equations 2.7.1 and 2.7.2 are useful in doing calculations that relate mass of solute to solution volume and solution concentration in chemistry (see "stoichiometry" in Chapter 5).

2.8. SOLIDS

Matter in the form of *solids* is said to be in the **solid state.** The solid state is the most organized form of matter in that the atoms, molecules, and ions in it are in essentially fixed relative positions and are strongly attracted to each other. Therefore, solids have a definite shape, maintain a constant volume, are virtually incompressible under pressure, and expand and contract only slightly with changes in temperature. Like liquids, solids have generally very high densities relative to gases. (Some solids, such as those made from Styrofoam, appear to have very low densities, but that is because such materials are composed mostly of bubbles of air in a solid matrix.) Because of the strong attraction of the atoms, molecules, and ions of solids for each other, solids do not enter the vapor phase very readily at all; the phenomenon by which this happens to a limited extent is called **sublimation.**

Some solids (e.g., quartz and sodium chloride) have very well-defined geometric shapes and form characteristic crystals. These are called **crystalline solids** and occur because the molecules or ions of which they are composed assume well-defined, specific positions relative to each other. Other solids have indefinite shapes because their constituents are arranged at random; glass is such a solid. These are **amorphous solids**.

2.9. THERMAL PROPERTIES

The behavior of a substance when heated or cooled defines several important physical properties of it, including the temperature at which it melts or vaporizes. Also included are the amounts of heat required by a given mass of the substance to raise its temperature by a unit of temperature, to melt it, or to vaporize it.

Melting Point

The **melting point** of a pure substance is the temperature at which it changes from a solid to a liquid. At the melting temperature, pure solid and pure liquid composed of the substance may be present together in a state of equilibrium. An impure substance does not have a single melting temperature. Instead, as heat is applied to the substance, it begins to melt at a temperature below the melting temperature of the pure substance; melting proceeds as the temperature increases until no solid remains. Therefore, melting temperature measurements serve two purposes in characterizing a substance. The melting point of a pure substance is indicative of the identity of the substance. The melting behavior—whether melting occurs at a single temperature or over a temperature range—is a measure of substance purity.

Boiling Point

Boiling occurs when a liquid is heated to a temperature such that bubbles of vapor of the substance are evolved. The boiling temperature depends upon pressure, and reduced pressures can be used to cause liquids to boil at lower temperatures. When the surface of a pure liquid substance is in contact with the pure vapor of the substance at 1 atm pressure, boiling occurs at a temperature called the **normal boiling point** (Figure 2.10). Whereas a pure liquid remains at a constant temperature during boiling until it has all turned to vapor, the temperature of an impure liquid increases during boiling. As with melting point, the boiling point is useful for identifying liquids. The extent to which the boiling temperature is constant as the liquid is converted to vapor is a measure of the purity of the liquid.

Specific Heat

As the temperature of a substance is raised, energy must be put into it to enable the molecules of the substance to move more rapidly relative to each other and to

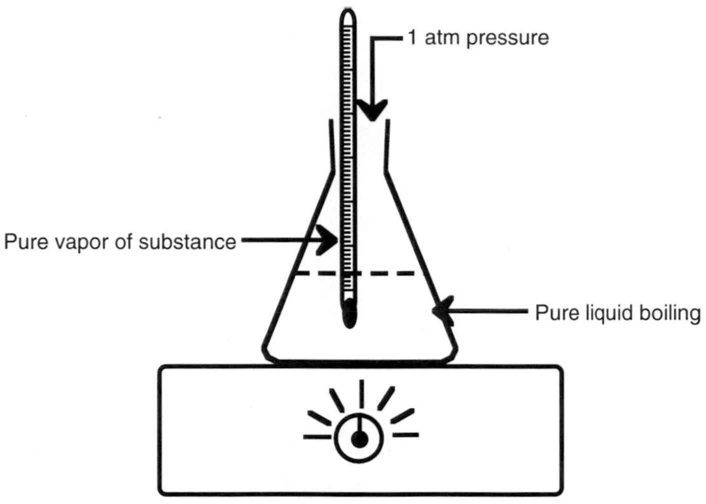

Figure 2.10. Illustration of normal boiling liquid.

overcome the attractive forces between them. The amount of heat energy required to raise the temperature of a unit mass of a solid or liquid substance by a degree of temperature varies with the substance. Of the common liquids, the most heat is required to raise the temperature of water. This is because of the molecular structure of water, the molecules of which are strongly attracted to each other by electrical charges between them (each water molecule has a relatively positive and a relatively negative end). Furthermore, water molecules tend to be held in molecular networks by special kinds of chemical bonds called hydrogen bonds, in which hydrogen atoms form bridges between O atoms in different water molecules. The very high amount of energy required to increase the temperature of water has some important environmental implications, because it stabilizes the temperatures of bodies of water and of geographic areas close to bodies of water. In contrast to water, hydrocarbon liquids such as those in gasoline require relatively little heat for warming, because the molecules interact much less with each other than do those of water.

The heat required to raise the temperature of a substance is called the **specific heat**, defined as the amount of heat energy required to raise the temperature of 1 g of substance by 1°C. It can be expressed by the equation

$$\text{Specific heat} = \frac{\text{heat energy absorbed, J}}{(\text{mass, g})(\text{increase in temperature, °C})} \tag{2.9.1}$$

where the heat energy is in units of joules (J). The most important value of specific heat is that of water, 4.18 J/(g°C). From Equation 2.9.1, the amount of heat, q, required to raise the mass, m, of a particular substance over a temperature range of ΔT is

$$q = (\text{specific heat}) \times m \times \Delta T \tag{2.9.2}$$

This equation can be used to calculate quantities of heat used in increasing water temperature or released when water temperature decreases. As an example, consider the amount of heat required to raise the temperature of 11.6 g of liquid water from 14.3°C to 21.8°C:

$$q = 4.18\,\frac{\text{J}}{\text{g°C}} \times 11.6\text{ g} \times (21.8\text{°C} - 14.3\text{°C}) = 364\text{ J} \tag{2.9.3}$$

Heat of Vaporization

Large amounts of heat energy may be required to convert liquid to vapor compared with the specific heat of the liquid. This is because changing a liquid to a gas requires breaking the molecules of liquid away from each other, not just increasing the rates at which they move relative to one another. The vaporization of a pure liquid occurs at a constant temperature (normal boiling temperature, see above). For example, the temperature of a quantity of pure water is increased by heating to 100°C, where it remains until all the water has evaporated. The **heat of vaporization** is the quantity of heat taken up in converting a unit mass of liquid entirely to vapor at a constant temperature. The heat of vaporization of water is 2260 J/g (2.26 kJ/g) for water boiling at 100°C at 1 atm pressure. This amount of heat energy is about 540 times that required to raise the temperature of 1 g of liquid water by 1°C. The fact that water's heat of vaporization is the largest of any common liquid has significant environmental effects. It means that enormous amounts of heat are required to produce water vapor from liquid water in bodies of water. This means that the temperature of a body of water is relatively stable when it is exposed to high temperatures, because so much heat is dissipated when a fraction of the water evaporates. When water vapor condenses, similar enormous amounts of heat energy, called **heat of condensation**, are released. This occurs when water vapor forms precipitation in storm clouds, and is the driving force behind the tremendous releases of heat that occur in thunderstorms and hurricanes.

The heat of vaporization of water can be used to calculate the heat required to evaporate a quantity of water. For example, the heat, q, required to evaporate 2.50 g of liquid water is

$$\begin{aligned} q &= \text{(heat of vaporization)(mass of water)} \\ &= 2.26\text{ kJ/g} \times 2.50\text{ g} = 5.65\text{ kJ} \end{aligned} \tag{2.9.4}$$

and that released when 5.00 g of water vapor condenses is

$$\begin{aligned} q &= -\text{(heat of vaporization)(mass of water)} \\ &= -2.26\text{ kJ/g} \times 5.00\text{g} = -11.3\text{ kJ} \end{aligned} \tag{2.9.5}$$

The latter value is negative to express the fact that heat is *released.*

Heat of Fusion

Heat of fusion is the quantity of heat taken up in converting a unit mass of solid entirely to liquid at a constant temperature. The heat of fusion of water is 330 J/g for ice melting at 0°C. This amount of heat energy is 80 times that required to raise the temperature of 1 g of liquid water by 1°C. When liquid water freezes, an exactly equal amount of energy per unit mass is released, but is denoted as a negative value.

The heat of fusion of water can be used to calculate the heat required to melt a quantity of ice. For example, the heat, q, required to melt 2.50 g of ice is

$$q = (\text{heat of fusion})(\text{mass of water}) = 330\ \text{J/g} \times 2.50\ \text{g} = 825\ \text{J} \qquad (2.9.6)$$

and that released when 5.00 g of liquid water freezes is

$$q = -(\text{heat of fusion})(\text{mass of water}) = -330\ \text{J/g} \times 5.00\ \text{g} = -1650\ \text{J} \qquad (2.9.7)$$

Phase Change Materials in Green Technology

When water changes from vapor to liquid at 100°C or vice versa, or when ice changes to liquid water at 0°C or vice versa, it is acting as a **phase change material**, one that can absorb large amounts of heat energy at a constant temperature. The way that this works can be visualized is with an ice chest. When solid ice is put into such a chest, the interior cools to 0°C, the temperature at which ice melts. As the ice melts, heat energy is absorbed from the interior of the chest and through its walls, maintaining the temperature at 0°C until all the ice is melted. (Actually, a thermometer suspended inside the chest would read a little over 0°C, because heat being absorbed through the chest walls tends to keep the interior a little warmer than the 0°C of the ice.) This kind of phenomenon can be used to utilize energy with relatively higher efficiency. Some buildings have been constructed in which the air-conditioning refrigeration system is used to convert liquid water to ice at night when lower outside temperatures make the process of cooling relatively more efficient and electricity rates are off-peak and lower. During the heat of daytime, the ice melts as air blowing over it is cooled and directed into the building for cooling.

An interesting possibility for the use of phase change materials in buildings is the incorporation of these materials into wallboard in dwelling walls. Such materials that change phase at around normal room temperature are available in the form of some hydrocarbons and renewable vegetable and animal lipid (fat/oil) materials, which may be chemically modified to give them desired thermal properties. Up to almost one-quarter the mass of wallboard may be impregnated with phase change materials that absorb or release heat at a comfortable room temperature. During daytime, when heating is aided by solar heat, these materials change from solid to liquid at a relatively constant temperature. At night, the reverse occurs as the liquefied phase change materials convert back to solids at their melting temperature, absorbing heat and keeping the structure at a comfortable temperature.

A potentially useful application of phase change materials is in heating systems for hybrid vehicles. Automobiles are heated by hot coolant circulated through an internal combustion engine as it runs. Particularly with future plug-in hybrids that use house current to charge a battery when the vehicle is not in use, the vehicle might be driven for many kilometers before the internal combustion engine even comes on. A mass of phase change material held in a very well insulated container and heated by the automobile engine when it is running or by resistance heating from the house current during recharging could provide a heat source for the automotive heating system.

2.10. SEPARATION AND CHARACTERIZATION OF MATTER

A very important aspect of the understanding of matter is the separation of mixtures of matter into their constituent pure substances. Consider the treatment of an uncharacterized waste sludge found improperly disposed of in barrels. Typically, the sludge will appear as a black goo with an obnoxious odor from several constituents. In order to do a proper analysis of this material, it often needs to be separated into its constituent components, such as by extracting water-soluble substances into water, extracting organic materials into an organic solvent, and distilling off volatile constituents. The materials thus isolated can be analyzed to determine what is present. With this knowledge and with information about the separation itself, a strategy can be devised to treat the waste in an effective, economical manner.

Many kinds of processes are used for separations, and it is beyond the scope of this chapter to go into detail about them here. However, several of the more important separation operations are discussed briefly below.

Distillation

Distillation consists of evaporating a liquid by heating and cooling the vapor so that it condenses back to the liquid in a different container, as shown for water in Figure 2.11. Less-volatile constituents such as solids are left behind in the distillation flask; more-volatile impurities such as volatile organic compound pollutants in water will distill off first and can be placed in different containers before the major liquid constituent is collected.

The apparatus shown in Figure 2.11 could be used, for example, to prepare fresh water from seawater, which can be considered a solution of sodium chloride in water. The seawater to be purified is placed in a round-bottomed distillation flask heated with an electrically powered heating mantle. The seawater is boiled and pure water vapor in the gaseous state flows into the condenser, where it is cooled and condensed back to a purified salt-free product in the receiving flask. When most of the seawater has been evaporated, the distillation flask and its contents are cooled and part of the sodium chloride separates out as crystals that can be removed.

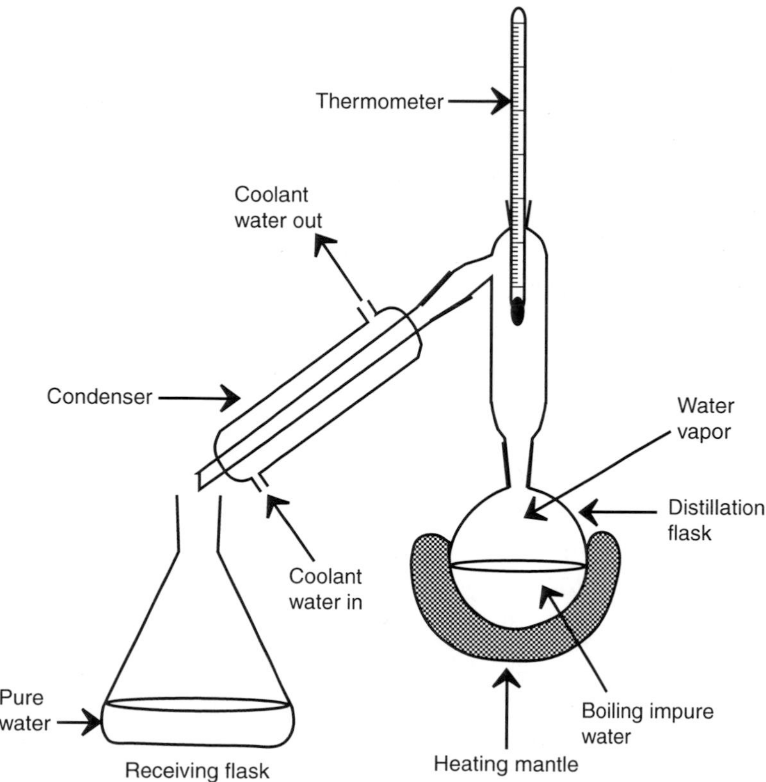

Figure 2.11. Separation by distillation.

Sophisticated versions of this distillation process are used in some arid regions of the world to produce potable (drinking) water from seawater.

Distillation is widely used in the petroleum and chemical industries. A sophisticated distillation apparatus is used to separate the numerous organic components of crude oil, including petroleum ether, gasoline, kerosene, and jet fuel fractions. This requires **fractional distillation**, in which part of the vapor recondenses in a **fractionating column** that would be mounted vertically on top of the distillation flask in the apparatus shown in Figure 2.11. The most-volatile components enter into the condenser and are condensed back to liquid products; less-volatile liquid constituents return to the distillation flask. The net effect is that of numerous distillations, which gives a very efficient separation.

In some cases, the residues left from distillation, **distillation bottoms** ("still bottoms"), are waste materials. Several important categories of hazardous wastes consist of distillation bottoms, which pose disposal problems.

Separation in Waste Treatment

In addition to distillation, several other kinds of separation procedures are used for waste treatment. These are mentioned here as examples of separation processes.

Phase Transitions

Distillation is an example of a separation in which one of the constituents being separated undergoes a phase **transition** from one phase to another, and then back to the same phase. In distillation, for example, a liquid is converted to the vapor state, and then recondensed as a liquid. Several other kinds of separations by phase transition as applied to waste treatment are the following:

Evaporation is usually employed to remove water from an aqueous waste to concentrate it. A special case of this technique is **thin-film evaporation**, in which volatile constituents are removed by heating a thin layer of liquid or sludge waste spread on a heated surface.

Drying—removal of solvent or water from a solid or semisolid (sludge) or the removal of solvent from a liquid or suspension—is a very important operation, because water is often the major constituent of waste products, such as sludges. In **freeze drying**, the solvent, usually water, is sublimed from a frozen material. Hazardous-waste solids and sludges are dried to reduce the quantity of waste, to remove solvent or water that might interfere with subsequent treatment processes, and to remove hazardous volatile constituents.

Stripping is a means of separating volatile components from less-volatile ones in a liquid mixture by the partitioning of the more-volatile materials to a gas phase of air or steam (steam stripping). The gas phase is introduced into the aqueous solution or suspension containing the waste in a stripping tower that is equipped with trays or packed to provide maximum turbulence and contact between the liquid and gas phases. The two major products are condensed vapor and a stripped bottoms residue. Examples of two volatile components that can be removed from water by air stripping are the organic solvents benzene and dichloromethane.

Physical precipitation is used here as a term to describe processes in which a dissolved solute comes out of solution as a solid as a result of a physical change in the solution. The major changes that can cause physical precipitation are cooling the solution, evaporation of solvent, or alteration of solvent composition. The most common type of physical precipitation by alteration of solvent composition occurs when a water-miscible organic solvent is added to an aqueous (water) solution of a salt, so that the solubility of the salt is lowered below its concentration in the solution.

Phase Transfer

Phase transfer consists of the transfer of a solute in a mixture from one phase to another. An important type of phase transfer process is **solvent extraction** (Figure 2.12), a process in which a substance is transferred from solution in one solvent (usually water) to another (usually an organic solvent) without any chemical change taking place. When solvents are used to leach substances from solids or

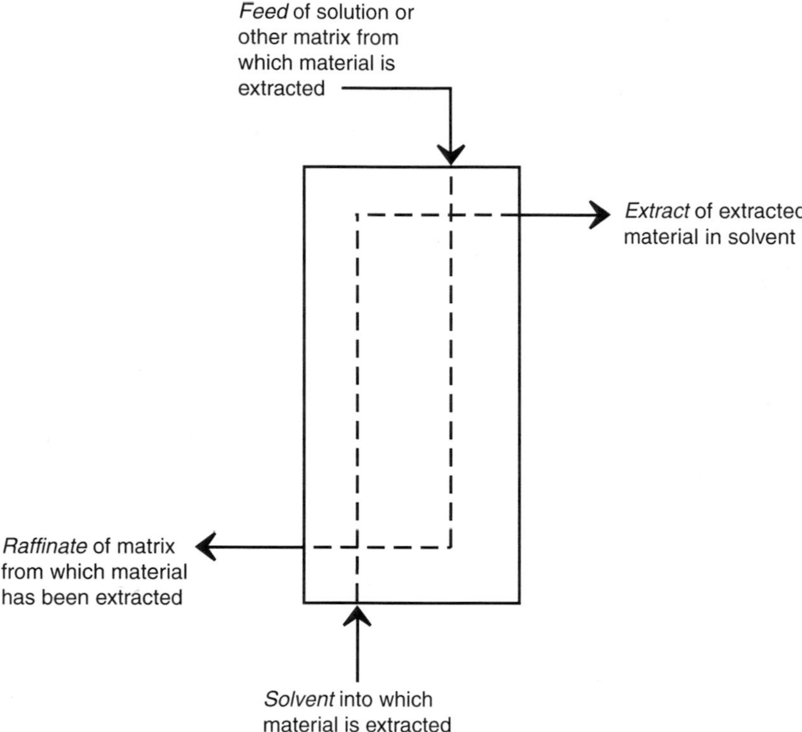

Figure 2.12. Outline of solvent extraction/leaching process, with important terms in italics.

sludges, the process is called **leaching**. One of the more promising approaches to solvent extraction and leaching of hazardous wastes is the use of supercritical fluids, most commonly CO_2, as extraction solvents. A supercritical fluid is one that has characteristics of both liquid and gas and consists of a substance above its supercritical temperature and pressure (31.1°C and 73.8 atm, respectively, for CO_2). After a substance has been extracted from a waste into a supercritical fluid at high pressure, the pressure can be released, resulting in separation of the substance extracted. The fluid can then be compressed again and recirculated through the extraction system. Some possibilities for treatment of hazardous wastes by extraction with supercritical CO_2 include removal of organic contaminants from wastewater, extraction of organohalide pesticides from soil, extraction of oil from emulsions used in aluminum and steel processing, and regeneration of spent activated carbon.

Transfer of a substance from a solution to a solid phase is called **sorption**. The most important sorbent used in waste treatment is **activated carbon**. Activated carbon can also be applied to pretreatment of waste streams going into other processes in order to improve treatment efficiency and reduce fouling. Activated carbon sorption is most effective for removing from water organic materials that are poorly water-soluble and that have high molecular masses.

Molecular Separation

A third major class of physical separation is **molecular separation**, often based upon **membrane processes** in which dissolved contaminants or solvent pass through a size-selective membrane under pressure. **Reverse osmosis** is the most widely used of the membrane techniques. It operates by virtue of a membrane that is selectively permeable to water and excludes ionic solutes. Reverse osmosis uses high pressures to force permeate through the membrane, producing a concentrate containing high levels of dissolved salts.

Reverse osmosis is very important in environmental preservation and sustainability. One of its applications is to the acquisition of fresh water from seawater, and several arid countries now obtain significant water supplies by reverse osmosis. Irrigation water picks up dissolved salts from fertilizer and other sources, so that runoff from irrigation water may be too saline. Reverse osmosis has been used to desalinate such water. Wastewater that has been through municipal water systems picks up significant amounts of sodium chloride (table salt) and other dissolved salts. Therefore, reverse osmosis is an important step in the renovation of such water to drinking water standards.

2.11. THE GREEN CHEMISTRY OF MATTER

The way in which matter is handled by humans is of utmost importance for environmental chemistry, green chemistry, and sustainability, as indicated by some of the reasons listed below. Much of the rest of this book deals with these aspects of matter and materials.

- Mining of minerals and extraction of fossil fuels can involve severe disruption of Earth's geosphere; the dumping of residues from the processing of mined minerals can cause major environmental problems.
- Shortages of particular kinds of matter, such as essential metals, have important implications for sustainability.
- Even small amounts of the wrong kind of matter in the wrong places constitute air and water pollution.
- Dangerous byproducts of chemical processing discarded to the geosphere are responsible for troublesome hazardous waste sites.
- Production of biomass matter in farm crops and forests involves much of the surface of the geosphere.
- The movement of large quantities of matter in the form of goods, materials, and fuels has important environmental implications.

Given factors such as those listed above, it is little wonder that the acquisition, processing, recycling, and disposal of matter are very important considerations in environmental chemistry and the practice of green chemistry. This is illustrated by Figure 1.1 in Chapter 1 showing the definition of green chemistry. To the extent possible, green chemistry emphasizes the use of matter from renewable resources. Water and oxygen and nitrogen from the atmosphere are renewable resources. Other than these, the most widely used of potential renewable resources consists of organic materials (those containing the element carbon). In modern societies, large amounts of organic substances are made from nonrenewable petroleum hydrocarbons. The practice of green chemistry calls for the maximum use of biological materials. These are continually being renewed by photosynthesis carried out by green plants, which extract carbon from atmospheric carbon dioxide and use solar energy in the form of light to convert this carbon to organic materials.

Some matter is inherently nonrenewable, consisting of depleting resources of which fixed amounts are available from the geosphere. Some of these are elements so abundant that they can be considered as almost renewable; iron and aluminum fall into this category. Others, such as platinum and palladium metals used in automobile exhaust pollution control catalysts, are in very limited supply. For these kinds of materials, **recycling** is of the utmost importance. Recycling also has the benefit of reducing the amounts of wastes produced from, and saving energy involved in, the processing of relatively abundant organic matter and metals such as aluminum and iron. Paper is a prime example of commonly recycled renewable organic matter. Whereas processing of iron and aluminum from their ores requires large amounts of energy, much less overall is required to melt aluminum and iron scrap to reprocess these metals. Organic materials that cannot be economically reprocessed for the materials that they contain can be used to produce energy. One of the ways in which this is done is through burning these materials as fuel in power plants. Biodegradable organic matter buried in landfills produces methane gases by biological processes. Methane released to the atmosphere from municipal waste disposal sites has a very high atmospheric warming effect per molecule. Modern facilities now collect this methane for use as a very clean fuel in power plants.

CHAPTER SUMMARY

The chapter summary below is presented in a programmed format to review the main points covered in this chapter. It is used most effectively by filling in the blanks, referring back to the chapter as necessary. The correct answers are given at the end of the summary.

Matter is defined as [1]______________________________ Chemical properties of matter are those that are observed [2]____________________________, whereas physical properties are those that are observed [3]____________________________. Electromagnetic radiation in the form of ultraviolet radiation can be used in the practice of green chemistry to [4]________________________________. Matter consists of only a few [5]________________ and a large number of substances formed from them called [6]____________________. Some general properties of metals are [7]_______________________________________. Some common properties of nonmetals are [8]________________________________. [9]____________ substances consist of virtually all compounds that contain carbon, whereas [10]________________ substances make up virtually all matter that does not contain carbon. Matter consisting of [11]____________________________ is a pure substance; any other form of matter is a [12]______________ of two or more substances. A [13]________________ is not uniform throughout and possesses readily distinguishable constituents, whereas [14]___________________ are uniform throughout. A mole is defined as the quantity of substance that [15]____. A mole of H_2O contains [16]__________ g of this compound. Avogadro's number is the number of specified entities [17]____________________ and it has a value of [18]__________________. A general term that includes atoms and molecules as well as groups of ions making up the smallest possible quantity of an ionic compound is the [19]_________________. The average mass of a formula unit is called the [20]__________________. Density is defined as [21]___________________ and expressed by the formula [22]__________________. Relative densities are expressed by means of a number called [23]__________________, defined as [24]____. Some solutions are colored because they [25]________________________. Solids, liquids, and gases are the three [26]______________ of matter. A solid has a definite [27]____________ and [28]___________________ regardless of the container in which it is placed, a liquid has a definite [29]_____________, but takes on the [30]______________ of its container, and a gas has the same [31]___________________________ as its container. Gas molecules move [32]_______________________, colliding with container walls to exert [33]____________________. A quantity of a gas has a [34]______________________ compared with the same amount of material in a liquid or solid. [35]_________________________ enables gas molecules to move large distances from their sources. The gas laws relate the four parameters of [36]____ of gas. Boyle's law states that [37]____. Charles' law states that [38]____. Avogadro's law states that [39]____. The general gas law states that [40]____. The ideal gas law is given mathematically as [41]_____________________. At a temperature of 0°C and 1 atm pressure, known as [42]____, the volume of 1 mol of ideal gas is [43]____________. Molecules of liquids move [44]___________ relative to each other. Because there is little unoccupied space between molecules of liquid, liquids are only slightly [45]_____________________. The phenomenon by which molecules of liquid enter the gas phase is called [46]________________, whereas the opposite phenomenon is termed [47]______________________. A steady-state level of the vapor of a liquid above the liquid surface is called the [48]_____________________ of the liquid.

Solutions are regarded as liquids in which [49]____. The predominant liquid constituent of a solution is called the [50]______________; a substance dispersed in it is said to be [51]______________ in the solvent and is called the [52]______________. A substance that dissolves to a significant extent is said to be [53]______________ in the solvent. Substances that do not dissolve to a perceptible degree are classified as [54]______________. A solution that has comparatively little solute dissolved per unit volume of solution is called a [55]______________ solution, whereas one with an amount of solute of the same order of magnitude as that of the solvent is a [56]________________ solution. The molar concentration, M, of a solution is defined as [57]______________________________ and is given by the formula [58]____________________________. The solid state is the most [59]__________ form of matter, because the atoms, molecules, and ions in it are [60]__________________________. The phenomenon by which solids enter the gaseous state directly is called [61]________________. Solids that form pieces with well-defined geometric shapes are called [62]____________________, and those that have indefinite shapes because their constituents are arranged at random are called [63]__________________. The melting point of a pure substance is the temperature at which the substance changes [64]______________________________. At this temperature, pure solid and pure liquid composed of the substance [65]____________________________. An impure substance does not have [66]____________________________. The two purposes served by melting temperature measurements in characterizing a substance are [67]____ and [68]____.

When a pure liquid is heated such that the surface of the liquid is in contact with the pure vapor of the substance at 1 atm pressure, boiling occurs at a temperature called the [69]______________________. The extent to which the boiling temperature is constant as the liquid is converted to vapor is a measure [70]__________________. The amount of heat energy required to raise the temperature of a unit mass of a solid or liquid substance by a degree of temperature is called the [71]________________ of the substance, and can be expressed by the equation [72]__________________________. The heat of vaporization is defined as [73]____. Mathematically, the heat, q, required to evaporate a specified mass of liquid water is [74]______________________________________. Heat of fusion is defined as [75]____.

Distillation consists of [76]____. Fractional distillation is a separation process in which part of the vapor [77]____________________________. Residues left from distillation are called [78]____________________. A process in which a solvent is sublimed from a frozen material is called [79]________________________. Stripping is a means of separating volatile components from less-volatile ones in a liquid mixture by [80]____. Solvent extraction is a process in which [81]____. The most important sorbent used in waste treatment is [82]______________.

A water purification process called [83]______________ uses high pressures to force permeate through a membrane that is selectively permeable to water and excludes ionic solutes. Insofar as possible, green chemistry emphasizes the use of matter from [84]__________________. Recycling is particularly important for utilization of materials that are [85]______________. In addition to conserving materials, recycling conserves [86]_____________ and reduces [87]___________.

Answers to Chapter Summary

1. anything that has mass and occupies space
2. by chemical change or the potential for chemical change
3. without any chemical change
4. put energy into matter involved in chemical processes
5. elements
6. compounds
7. generally solid, shiny in appearance, electrically conducting, and malleable
8. a dull appearance, not at all malleable, and frequently present as gases or liquids
9. Organic
10. inorganic
11. only one compound or of only one form of an element
12. mixture
13. heterogeneous mixture
14. homogeneous mixtures
15. contains the same number of specified entities as there are atoms of C in exactly 0.012 kg (12 g) of carbon-12
16. 18
17. in a mole of substance
18. 6.02×10^{23}
19. formula unit
20. formula mass
21. mass per unit volume

22. d = mass/volume
23. specific gravity
24. the ratio of the density of a substance to that of a standard substance
25. absorb light
26. states
27. shape
28. volume
29. volume
30. shape
31. shape and volume
32. independently and at random
33. pressure
34. high volume
35. Diffusion
36. moles (n), volume (V), temperature (T), and pressure (P)
37. at constant temperature, the volume of a fixed quantity of gas is inversely proportional to the pressure of the gas
38. the volume of a fixed quantity of gas is directly proportional to the absolute temperature at constant pressure
39. at constant temperature and pressure, the volume of a gas is directly proportional to the number of molecules of gas, commonly expressed as moles
40. the volume of a quantity of ideal gas is proportional to the number of moles of gas and its absolute temperature and inversely proportional to its pressure
41. $PV = nRT$
42. standard temperature and pressure (STP)
43. 22.4 L
44. freely
45. compressible
46. evaporation

47. condensation
48. vapor pressure
49. quantities of gases, solids, or other liquids are dispersed as individual ions or molecules
50. solvent
51. dissolved
52. solute
53. soluble
54. insoluble
55. dilute
56. concentrated
57. the number of moles of solute dissolved per liter of solution
58. $M = \frac{\text{moles of solute}}{\text{number of liters of solution}}$
59. organized
60. in essentially fixed relative positions and are highly attracted to each other
61. sublimation
62. crystalline solids
63. amorphous solids
64. from a solid to a liquid
65. may be present together in a state of equilibrium
66. a single melting temperature
67. the melting point of a pure substance is indicative of the identity of the substance
68. the melting behavior—whether melting occurs at a single temperature or over a temperature range—is a measure of substance purity
69. normal boiling point
70. of the purity of the liquid
71. specific heat

72. specific heat $= \dfrac{\text{heat energy absorbed, J}}{\text{(mass, g)(increase in temperature, °C)}}$
73. the quantity of heat taken up in converting a unit mass of liquid entirely to vapor at a constant temperature
74. q = (heat of vaporization) × (mass of water)
75. the quantity of heat taken up in converting a unit mass of solid entirely to liquid at a constant temperature
76. evaporating a liquid by heating and cooling the vapor so that it condenses back to the liquid in a different container
77. recondenses in a fractionating column
78. distillation bottoms
79. freeze drying
80. the partitioning of the more-volatile materials to a gas phase of air or steam
81. a substance is transferred from solution in one solvent to another without any chemical change taking place
82. activated carbon
83. reverse osmosis
84. renewable sources
85. non-renewable
86. energy
87. pollution

QUESTIONS AND PROBLEMS

1. What is the definition of matter?
2. Classify each of the following as chemical processes (c) or physical processes (p):

   ( ) Ice melts

   ( ) $CaO + H_2O \rightarrow Ca(OH)_2$

   ( ) A small silver sphere pounded into a thin sheet

( ) A copper-covered roof turns green in the atmosphere

( ) Wood burns

( ) Oxygen isolated by the distillation of liquid air

3. List (a) three common characteristics of metals and (b) three common characteristics of nonmetals.

4. Classify each of the following as organic substances (o) or inorganic substances (i):

( ) Crude oil ( ) A dime coin ( ) A shirt ( ) Concrete

( ) Asphalt isolated as a residue of petroleum distillation

( ) A compound composed of silicon, oxygen, and aluminum

5. Sandy sediment from the bottom of a river was filtered to isolate a clear liquid, which was distilled to give a liquid compound that could not be broken down further by non-chemical means. Explain how these operations illustrate heterogeneous mixtures, homogeneous mixtures, and pure substances.

6. Explain why a solution is defined as a homogeneous mixture.

7. Justify the statement that "exactly 12 g of carbon-12 contain 6.02×10^{23} atoms of the isotope of carbon that contains 6 protons and 6 neutrons in its nucleus." What two terms relating to quantity of matter are illustrated by this statement?

8. Where *X* is a number equal to the formula mass of a substance, complete the following formula: Mass of a mole of a substance = ____________.

9. Fill in the blanks in the table below:

Substance	Formula Unit	Formula Mass	Mass of 1 mole	Numbers and Kinds of Individual Entities in 1 mole
Neon	Ne atom	______	_____ g	_________ Ne atoms
Nitrogen gas	N_2 molecule	______	_____ g	_________ N_2 molecules
				_________ N atoms
Ethene	C_2H_4 molecules	______	_____ g	_________ C_2H_4 molecules
				_________ C atoms
				_________ H atoms

10. Calculate the density of each of the following substances, for which the mass of a specified volume is given:

    (a) 93.6 g occupying 15 mL (b) 0.992 g occupying 1.005 mL

    (c) 13.7 g occupying 11.4 mL (d) 16.8 g occupying 3.19 mL

11. Calculate the volume in mL occupied by 156.3 g of water at 4°C.

12. From densities given in Table 2.3, calculate each of the following:

    (a) Volume of 105 g of iron (b) Mass of 157 cm^3 of benzene

    (c) Volume of 1932 g of air (d) Mass of 13.2 cm^3 of sugar

    (e) Volume of 10.0 g of helium at 0°C and 1 atm pressure (f) Mass of 0.525 cm^3 of carbon tetrachloride

13. From information given in Table 2.3, calculate the specific gravities of (a) benzene, (b) carbon tetrachloride, and (c) sulfuric acid relative to water at 4°C.

14. From information given in this chapter, calculate the density of pure nitrogen gas, N_2, at STP. To do so will require some knowledge of the mole and of the gas laws.

15. Different colors are due to different ______________________ of light.

16. Explain how the various forms of water observed in the environment illustrate the three states of matter.

17. Match each gas law designated with a letter, with the phenomenon that it explains:

    A. Charles' law B. General gas law C. Boyle's law D. Avogadro's law

    1. The volume of a mole of ideal gas at STP is 22.4 L.
    2. Tripling the number of moles of gas at constant temperature and pressure triples the volume.
    3. Raising the absolute temperature of a quantity of gas at constant pressure from 300 K to 450 K increases the gas volume by 50%.
    4. Doubling the pressure on a quantity of gas at constant temperature halves the volume.

18. How might an individual molecule of a gas describe its surroundings? What might it see, feel, and do?

19. Explain pressure and diffusion in terms of the behavior of gas molecules.

20. Use the gas law to calculate whether the volume of a mole of water vapor at 100°C and 1 atm pressure is 30 600 mL as stated in this chapter.

21. Using the appropriate gas laws, calculate the quantities denoted by the blanks in the following table:

Conditions Before				Conditions After			
n_1 (mol)	V_1 (L)	P_1 (atm)	T_1 (K)	n_2 (mol)	V_2 (L)	P_2 (atm)	T_2 (K)
1.25	13.2	1.27	298	1.25	(a) __	0.987	298
1.25	13.2	1.27	298	1.25	(b) __	1.27	407
1.25	13.2	1.27	298	4.00	(c) __	1.27	298
1.25	13.2	1.27	298	1.36	(d) __	1.06	372
1.25	13.2	1.27	298	1.25	30.5	(e) __	298

22. In terms of distances between molecules and movement of molecules relative to each other, distinguish between liquids and gases and liquids and solids. Compare and explain the shape and compressibility of gases, liquids, and solids.

23. Define (a) evaporation, (b) condensation, and (c) vapor pressure as related to liquids.

24. Define (a) solvent, (b) solute, (c) solubility, (d) saturated solution, (e) unsaturated solution, and (f) insoluble solute.

25. From information given in Section 2.7, calculate the molar concentration of O_2 in water saturated with air at 25°C.

26. Calculate the molar concentration of solute in each of the following solutions:

    (a) Exactly 7.59 g of NH_3 dissolved in 1.500 L of solution.

    (b) Exactly 0.291 g of H_2SO_4 dissolved in 8.65 L of solution.

    (c) Exactly 85.2 g of NaOH dissolved in 5.00 L of solution.

    (d) Exactly 31.25 g of NaCl dissolved in 2.00 L of solution.

    (e) Exactly 6.00 mg of atmospheric N_2 dissolved in water in a total solution volume of 2.00 L.

    (f) Exactly 1000 g of ethylene glycol, $C_2H_6O_2$, dissolved in water and made up to a volume of 2.00 L.

27. Why is the solid state called the most organized form of matter?

28. As related to solids, define (a) sublimation, (b) crystalline solids, and (c) amorphous solids.

29. Experimentally, how could a chemist be sure that (a) a melting point temperature was being observed, (c) the melting point observed was that of a pure substance, and (c) the melting point observed was that of a mixture.

30. Calculate the specific heat of substances for which:

    (a) 22.5 J of heat energy was required to raise the temperature of 2.60 g of the substance from 11.5 to 16.2°C

    (b) 45.6 J of heat energy was required to raise the temperature of 1.90 g of the substance from 8.6 to 19.3°C

    (c) 1.20 J of heat energy was required to raise the temperature of 3.00 g of the substance from 25.013 to 25.362°C

31. Calculate the amount of heat required to raise the temperature of 7.25 g of liquid water from 22.7 to 29.2°C.

32. Distinguish between specific heat and (a) heat of vaporization and (b) heat of fusion.

33. Calculate the heat, q, (a) required to evaporate 5.20 g of liquid water and (b) released when 6.50 g of water vapor condenses.

34. Calculate the heat, q, (a) required to melt 5.20 g of ice and (b) released when 6.50 g of liquid water freezes.

35. Suppose that salt were removed from seawater by distillation. In reference to the apparatus shown in Figure 2.11, explain where the pure water product would be isolated and where the salt would be found.

36. How does fractional distillation differ from simple distillation?

37. What are distillation bottoms and what do they have to do in some cases with wastes?

38. Describe solvent extraction as an example of a phase transfer process.

39. For what kind of treatment process is activated carbon used?

40. Name and describe a purification process that operates by virtue of "a membrane that is selectively permeable to water and excludes ionic solutes."

3. ATOMS AND ELEMENTS

3.1. ATOMS AND ELEMENTS

At the very beginning of this book, chemistry was defined as the science of matter. Chapter 2 examined the nature of matter, largely at a macroscopic level and primarily through its physical properties. Today, the learning of chemistry is simplified by taking advantage of what is known about matter at its most microscopic and fundamental level—atoms and molecules. Although atoms have been viewed so far in this book as simple, indivisible particles, they are in fact complicated bodies. Subtle differences in the arrangements and energies of electrons that make up most of the volume of atoms determine the chemical characteristics of the atoms. In particular, the electronic structures of atoms cause the periodic behavior of the elements, as described briefly in Section 1.5 and summarized in the periodic table in Figure 1.4. This chapter addresses atomic theory and atomic structure in more detail to provide a basis for the understanding of chemistry.

3.2. THE ATOMIC THEORY

The nature of atoms in relation to chemical behavior is summarized in the **atomic theory**. This theory in essentially its modern form was advanced in 1808 by John Dalton, an English schoolteacher, taking advantage of a substantial body of chemical knowledge and the contributions of others. It has done more than any other concept to place chemistry on a sound, systematic theoretical foundation. Those parts of Dalton's atomic theory that are consistent with current understanding of atoms are summarized in Figure 3.1.

Laws That Are Explained by Dalton's Atomic Theory

The atomic theory outlined in Figure 3.1 explains the following three laws that are of fundamental importance in chemistry:

1. **Law of Conservation of Mass:** *There is no detectable change in mass in an ordinary chemical reaction.* (This law was first stated in 1798 by "the

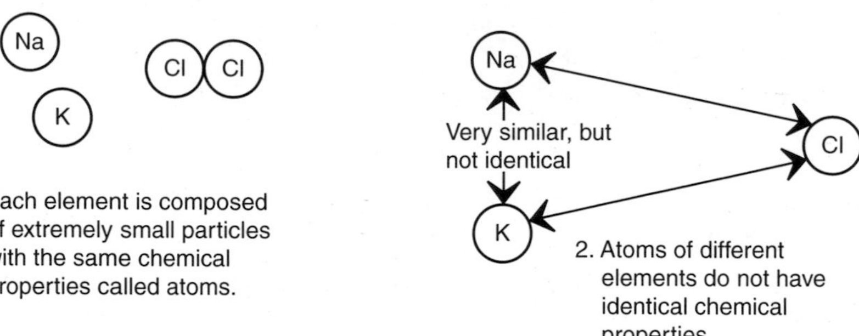

1. Each element is composed of extremely small particles with the same chemical properties called atoms.

2. Atoms of different elements do not have identical chemical properties.

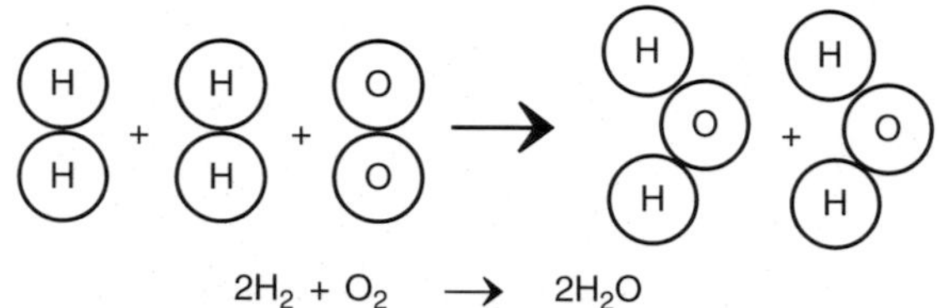

$$2H_2 + O_2 \longrightarrow 2H_2O$$

3. Chemical compounds are formed by the combination of atoms of different elements in definite, constant ratios that usually can be expressed as integers or simple fractions.

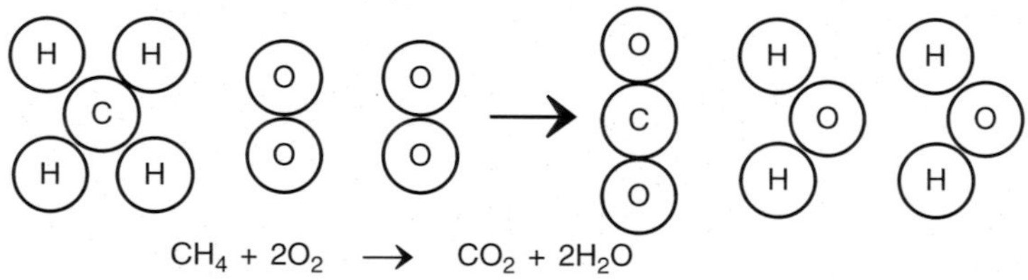

$$CH_4 + 2O_2 \longrightarrow CO_2 + 2H_2O$$

4. Chemical reactions involve the separation and combination of atoms, as in this example where bonds are broken between C and H in CH_4 and between O and O in O_2, and bonds are formed between C and O in CO_2 and between H and O in H_2O.

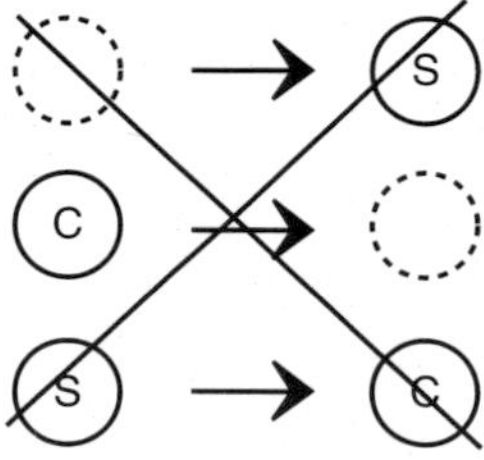

5. During the course of ordinary chemical reactions, the phenomena illustrated above do not occur: Atoms are not created, destroyed, or changed to atoms of other elements.

Figure 3.1. Illustration of Dalton's atomic theory.

father of chemistry," the Frenchman Antoine Lavoisier. Since, as shown in Item 5 of Figure 3.1, no atoms are lost, gained, or changed in chemical reactions, mass is conserved.)

2. **Law of Constant Composition:** *A specific chemical compound always contains the same elements in the same proportions by mass.* (If atoms always combine in definite, constant ratios to form a particular chemical compound, as implied in Item 3 of Figure 3.1, the elemental composition of the compound by mass always remains the same.)

3. **Law of Multiple Proportions:** *When two elements combine to form two or more compounds, the masses of one combining with a fixed mass of the other are in ratios of small whole numbers* (as illustrated for two compounds composed only of carbon and hydrogen below).

For CH_4: Relative mass of hydrogen = 4 × 1.0 = 4.0 (because there are 4 atoms of H, atomic mass 1.0)

Relative mass of carbon = 1 × 12.0 = 12.0 (because there is 1 atom of C, atomic mass 12.0)

$$\text{CH ratio for } CH_4 = \frac{\text{Mass of C}}{\text{Mass of H}} = \frac{12.0}{4.0} = 3.0$$

For C_2H_6: Relative mass of hydrogen = 6 × 1.0 = 6.0 (because there are 6 atoms of H, atomic mass 1.0)

Relative mass of carbon = 2 × 12.0 = 24.0 (because there are 2 atoms of C, atomic mass 12.0)

$$\text{CH ratio for } C_2H_6 = \frac{\text{Mass of C}}{\text{Mass of H}} = \frac{24.0}{6.0} = 4.0$$

$$\text{Comparing } C_2H_6 \text{ and } CH_4\text{:} \ \frac{\text{CH ratio for } C_2H_6}{\text{CH ratio for } CH_4} = \frac{4.0}{3.0} = 4/3$$

Note that this is a ratio of small whole numbers.

Small Size of Atoms

It is difficult to imagine just how small an individual atom is. An especially small unit of mass, the *atomic mass unit*, u, is used to express the masses of atoms. This unit is a mass equal to exactly 1/12 that of an atom of the carbon-12 isotope. An atomic mass unit is only 1.66×10^{-24} g. An average atom of hydrogen, the lightest element, has a mass of only 1.0079 u. The average mass of an atom of uranium, the heaviest naturally occurring element, is 238.03 u. To place these values in perspective, consider that a signature written by ballpoint pen on a piece of paper

Figure 3.2. The ballpoint pen ink in a typical signature might have a mass of 6×10^{19} u.

(Figure 3.2) typically has a mass of 0.1 mg (1×10^{-4} g). This mass is equal to 6×10^{19} u. It would take almost 60 000 000 000 000 000 000 hydrogen atoms to equal the mass of ink in such a signature!

Atoms visualized as spheres have diameters of around $1-3 \times 10^{-10}$ m (100–300 pm). By way of comparison, a small marble has a diameter of around 1 cm (1×10^{-2} m), which is about 100 000 000 times that of a typical atom.

Atomic Mass

The *atomic mass* of an element is the average mass of all atoms of the element relative to carbon-12 taken as exactly 12. Since atomic masses are *relative* quantities, they can be expressed without units. Or atomic masses may be given in atomic mass units. For example, an atomic mass of 14.0067 for nitrogen means that the average mass of all nitrogen atoms is 14.0067/12 as great as the mass of the carbon-12 isotope and is also 14.0067 u.

3.3. SUBATOMIC PARTICLES

Small as atoms are, they in turn consist of even smaller entities called **subatomic particles**. Although physicists have found several dozen of these, chemists need consider only three—*protons*, *neutrons*, and *electrons* (these were introduced briefly in Section 1.5). These subatomic particles differ in mass and charge. Like the atom, their masses are expressed in atomic mass units.

The proton, p, has a mass of 1.007277 u and a unit charge of +1. This charge is equal to 1.6022×10^{-19} coulombs, where a coulomb is the amount of electrical charge involved in a flow of electrical current of 1 ampere for 1 second.

The neutron, n, has no electrical charge and a mass of 1.009665 u. The proton and neutron each have a mass of essentially 1 u and are said to have a *mass number* of 1. (Mass number is a useful concept expressing the total number of protons and neutrons, as well as the approximate mass, of a nucleus or subatomic particle.)

Table 3.1. Properties of Protons, Neutrons, and Electrons

Subatomic Particle	Symbol[a]	Unit Charge	Mass Number	Mass (u)	Mass (g)
Proton	p	+1	1	1.007277	1.6726×10^{-24}
Neutron	n	0	1	1.008665	1.6749×10^{-24}
Electron	e	−1	0	0.000549	9.1096×10^{-28}

[a] The mass number and charge of each of these kinds of particles can be indicated by a superscript and subscript, respectively in the symbols ${}^{1}_{1}p$, ${}^{1}_{0}n$, ${}^{0}_{-1}e$.

The electron, e, has an electrical charge of −1. It is very light, however, with a mass of only 0.00054859 u, about 1/1840 that of the proton or neutron. Its mass number is 0. The properties of protons, neutrons, and electrons are summarized in Table 3.1.

Although it is convenient to think of the proton and neutron as having the same mass, and each is assigned a mass number of 1, it can be seen from Table 3.1 that their exact masses differ slightly from each other. Furthermore, the mass of an atom differs slightly from the sum of the masses of subatomic particles composing the atom: because of the energy relationships involved in holding the subatomic particles together in an atom, the masses of the atom's constituent subatomic particles do not add up to exactly the mass of the atom.

3.4. THE BASIC STRUCTURE OF THE ATOM

As mentioned in Section 1.5, protons and neutrons are located in the *nucleus* of an atom; the remainder of the atom consists of a cloud of rapidly moving electrons. Since protons and neutrons have much higher masses than electrons, essentially all the mass of an atom is in its nucleus. However, the electron cloud makes up virtually all of the volume of the atom, and the nucleus is very small.

Atomic Number, Isotopes, and Mass Number of Isotopes

All atoms of the same element have the same number of protons and electrons, equal to the *atomic number* of the element. (When reference is made to atoms here, it is understood that they are electrically neutral atoms and not ions consisting of atoms that have lost or gained one or more electrons.) Thus, all helium atoms have 2 protons in their nuclei, and all nitrogen atoms have 7 protons.

In Section 1.5, isotopes were defined as atoms of the same element that differ in the number of neutrons in their nuclei. It is noted above that both the proton and neutron have a *mass number* of exactly 1. Mass number is commonly used to denote isotopes. The **mass number of an isotope** *is the sum of the number of protons and neutrons in the nucleus of the isotope.* Since atoms with the same number of protons—that is, atoms of the same element—may have different numbers of neutrons, a particular element may have several isotopes. The naturally occurring forms of some elements consist of only one isotope, but other isotopes of these elements can be made artificially, usually by exposing the elements to neutrons produced by a nuclear reactor.

It is convenient to have a symbol that clearly designates an isotope of an element. Borrowing from nuclear science, a superscript in front of the symbol for the element is used to show the mass number and a subscript before the symbol designates the number of protons in the nucleus (atomic number). Using this notation, carbon-12 is denoted by ${}^{12}_{6}C$.

Exercise: Fill in the blanks designated with letters in the following table:

Element	Atomic Number	Mass Number of Isotope	Number of Neutrons in Nucleus	Isotope Symbol
Nitrogen	7	14	7	${}^{14}_{7}N$
Chlorine	17	35	(a) _____	${}^{35}_{17}N$
Chlorine	(b) _____	37	(c) _____	(d) _____
(e) _____	6	(f) _____	7	(g) _____
(h) _____	(i) _____	(j) _____	(k) _____	${}^{11}_{5}B$

Answers: (a) 18, (b) 17, (c) 20, (d) ${}^{37}_{17}Cl$, (e) carbon, (f) 13, (g) ${}^{13}_{6}C$, (h) boron, (i) 5, (j) 11, (k) 6

Electrons in Atoms

In Section 1.5, it was noted that electrons form a cloud of negative charge around the nucleus of an atom. The energy levels, orientations in space, and behavior of electrons vary with the number of them contained in an atom. In a general sense, the arrangements of electrons in atoms are described by *electron configuration*, a term discussed in some detail later in this chapter.

Attraction between Electrons and the Nucleus

The electrons in an atom are held around the nucleus by the attraction between their negative charges and the positive charges of the protons in the nucleus. Opposite electrical charges attract, and like charges repel. The forces of attraction and repulsion are expressed quantitatively by **Coulomb's law**:

$$F = \frac{Q_1 Q_2}{d} \qquad (3.4.1)$$

where F is the force of interaction, Q_1 and Q_2 are the electrical charges on the bodies involved, and d is the distance between the bodies.

Coulomb's law explains the attraction between negatively charged electrons and the positively charged nucleus in an atom. However, the law does not explain why electrons move around the nucleus, rather than coming to rest on it. The behavior of electrons in atoms—as well as the numbers, types, and strengths of chemical bonds formed between atoms—is explained by *quantum theory*, a concept that is discussed in some detail in Sections 3.11–3.16.

3.5. DEVELOPMENT OF THE PERIODIC TABLE

The *periodic table*, which was mentioned briefly in connection with the elements in Section 1.5, was first described by the Russian chemist Dmitri Mendeleev in 1867. It was based upon Mendeleev's observations of periodicity in chemical behavior without any knowledge of atomic structure. The periodic table is discussed in more detail in this chapter, along with the development of the concepts of atomic structure. Elements are listed in the periodic table in an ordered, systematic way that correlates with their electron structures. The elements are placed in **rows** or **periods** of the periodic table in order of increasing atomic number such that there is a periodic repetition of elemental properties across the periods. The rows are arranged such that elements in vertical columns called **groups** have similar chemical properties, reflecting similar arrangements of the outermost electrons in their atoms.

With some knowledge of the ways in which electrons behave in atoms, it is much easier to develop the concept of the periodic table. This is done for the first 20 elements in the following sections. After these elements are discussed, they can be placed in an abbreviated 20-element version of the periodic table. Such a table is shown in Figure 3.9.

3.6. HYDROGEN, THE SIMPLEST ATOM

The simplest atom is that of hydrogen, as it has only one positively charged proton in its nucleus, which is surrounded by a cloud of negative charge formed by only one electron. By far the most abundant kind of hydrogen atom has no neutrons in its nucleus. Having only one proton with a mass number (see Section 3.3) of 1, and 1 electron with a mass number of 0, the mass number of this form of hydrogen is $1 + 0 = 1$. There are, however, two other forms of hydrogen atoms having, in addition to the proton, 1 and 2 neutrons, respectively, in their nuclei. The three different forms of elemental hydrogen are *isotopes* of hydrogen that all have the same number of protons but different numbers of neutrons in their nuclei. Only about 1 of 7000 hydrogen atoms is $^{2}_{1}H$ deuterium, mass number 2 (from 1 proton + 1 neutron). The mass number of *tritium*, which has 2 neutrons, is 1 (from 1 proton + 2 neutrons) = 3.

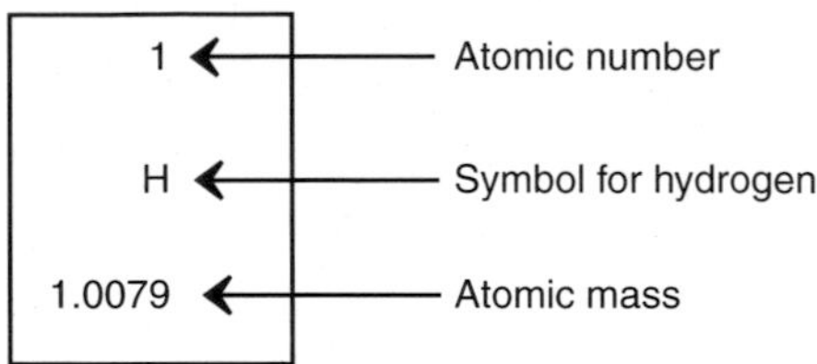

Figure 3.3. Designation of hydrogen in the periodic table.

These three forms of hydrogen atoms can be designated as $^{1}_{1}\mathbf{H}$, $^{2}_{1}\mathbf{H}$, and $^{3}_{1}\mathbf{H}$. The meaning of this notation is reviewed below:

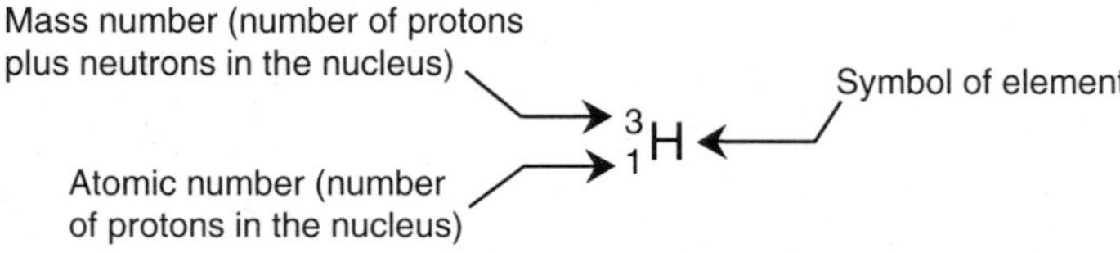

Designation of Hydrogen in the Periodic Table

The atomic mass of hydrogen is 1.0079. This means that the average atom of hydrogen has a mass of 1.0079 u; hydrogen's atomic mass is simply 1.0079 relative to the carbon-12 isotope taken as exactly 12. With this information, it is possible to place hydrogen in the periodic table with the designation shown in Figure 3.3.

Showing Electrons in Hydrogen Atoms and Molecules

It is useful to have some simple way of showing the hydrogen atom's single electron in chemical symbols and formulas. This is accomplished with **electron-dot symbols** or **Lewis symbols** (after G. N. Lewis), which use dots around the symbol of an element to show outer electrons (those that may become involved in chemical bonds). The Lewis symbol for hydrogen is

H•

As mentioned in Section 1.3, elemental hydrogen consists of molecules made up of 2 H atoms held together by a chemical bond consisting of two shared electrons. Just as it is useful to show electrons in atoms with a Lewis symbol, it is helpful to visualize molecules and the electrons in them with **electron-dot formulas** or **Lewis formulas** as shown for the H_2 molecule in Figure 3.4.

Properties of Elemental Hydrogen

Pure hydrogen is colorless and odorless. It is a gas at all but very low temperatures; liquid hydrogen boils at −253°C and solidifies at −259°C. Hydrogen gas has

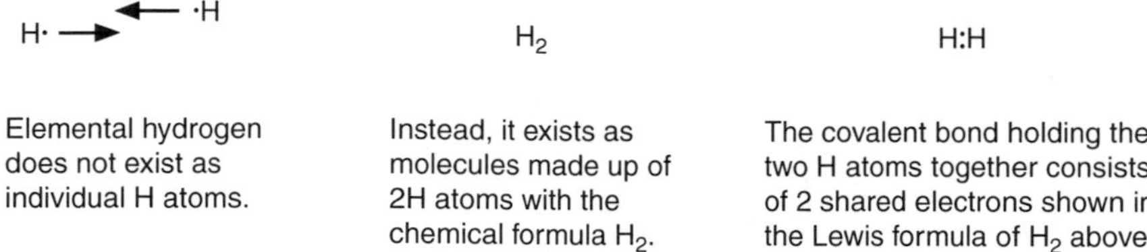

Figure 3.4. Lewis formula for the hydrogen molecule.

the lowest density of any pure substance. It reacts chemically with a large number of elements, and hydrogen-filled balloons that float readily in the atmosphere also explode convincingly when touched with a flame because of the following very rapid reaction with oxygen in the air:

$$2H_2 + O_2 \rightarrow 2H_2O \tag{3.6.1}$$

Production and Uses of Elemental Hydrogen

Elemental hydrogen is one of the more widely produced industrial chemicals. For use in chemical synthesis and other industrial applications, it is commonly made by **steam reforming** of methane (natural gas, CH_4) under high-temperature, high-pressure conditions:

$$CH_4 + H_2O \xrightarrow[30\,\text{atm}\,F]{800\,°CT} CO + 3H_2 \tag{3.6.2}$$

Hydrogen is used to manufacture a number of chemicals. One of the most important of these is ammonia, NH_3. Methanol (methyl alcohol), CH_3OH, is a widely used industrial chemical and solvent synthesized by the following reaction between carbon monoxide and hydrogen:

$$CO + 2H_2 \rightarrow CH_3OH \tag{3.6.3}$$

Methanol made by this process can be blended with gasoline to yield a fuel that produces relatively less pollutant carbon monoxide; such "oxygenated gasoline additives" are now required for use in some cities and urban areas. Gasoline is upgraded by the chemical addition of hydrogen to some petroleum fractions. Synthetic petroleum can be made by the addition of hydrogen to coal at high temperatures and pressures. A once widely used process in the food industry is the addition of hydrogen to unsaturated vegetable oils in the synthesis of margarine and other hydrogenated fats and oils, a practice that has lost favor because of health concerns over the "trans-fats" products.

3.7. HELIUM, THE FIRST ATOM WITH A FILLED ELECTRON SHELL

In Section 3.4, it was mentioned that the *electron configurations* of atoms determine their chemical behavior. Electrons in atoms occupy distinct *energy levels.*

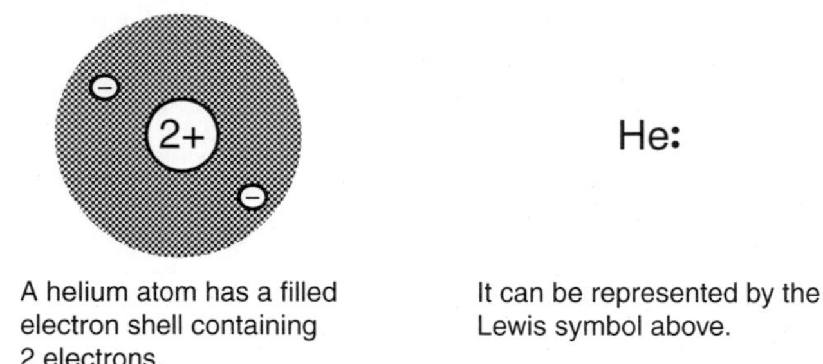

Figure 3.5. Two representations of the helium atom having a filled electron shell.

At this point, it is useful to introduce the concept of the **electron shell** to help explain electron energy levels and their influence on chemical behavior. Each electron shell can hold a maximum number of electrons. An atom with a **filled electron shell** is especially content in a chemical sense, with little or no tendency to lose, gain, or share electrons. Elements with these characteristics exist as gas-phase atoms and are called *noble gases*. The high stability of the noble gas electron configuration is explained in more detail in Section 3.9.

Examined in order of increasing atomic number, the first element consisting of atoms with filled electron shells is **helium**, He, atomic number 2. All helium atoms contain 2 protons and 2 electrons. Virtually all helium atoms are $^{4}_{2}\mathbf{He}$, containing 2 neutrons in their nuclei; the $^{3}_{2}\mathbf{He}$ isotope containing only 1 neutron in its nucleus occurs to a very small extent. The atomic mass of helium is 4.00260.

The two electrons in the helium atom are shown by the Lewis symbol illustrated in Figure 3.5. These electrons constitute a *filled electron shell*, so that helium is a *noble gas* composed of individual helium atoms that have no tendency to form chemical bonds with other atoms. Helium gas has a very low density of only 0.164 g/L at 25°C and 1 atm pressure.

Occurrence and Uses of Helium

Helium is extracted from some natural gas sources that contain up to 10% helium by volume. It has many uses that depend upon its unique properties. Because of its very low density compared with air, helium is used to fill weather balloons and airships. Helium is nontoxic, odorless, tasteless, and colorless. Because of these properties and its low solubility in blood, helium is mixed with oxygen for breathing by deep sea divers and persons with some respiratory ailments. Use of helium by divers avoids the very painful condition called "the bends" caused by bubbles of nitrogen forming from nitrogen gas dissolved in blood.

Liquid helium, which boils at a temperature of only 4.2 K above absolute zero, is especially useful in the growing science of **cryogenics**, which deals with very low temperatures. Some metals are superconductors at such temperatures, so that helium is used to cool electromagnets that develop very powerful magnetic fields for a relatively small magnet. Such magnets are employed in the very useful chemical technique known as nuclear magnetic resonance (NMR). A modification of this technique is used as a medical diagnostic tool called magnetic resonance imaging (MRI).

3.8. LITHIUM, THE FIRST ATOM WITH BOTH INNER AND OUTER ELECTRONS

The third element in the periodic table is **lithium**, Li, atomic number 3, atomic mass 6.941. The most abundant lithium isotope has 4 neutrons in its nucleus, so it has a mass number of 7 and is designated $^{7}_{3}\mathbf{Li}$. A less common isotope, $^{6}_{3}\mathbf{Li}$, has only 3 neutrons.

Lithium is the first element in the periodic table that is a *metal.* Lithium is the lightest metal, with a density of only 0.531 g/cm^3. As mentioned in Section 2.2, metals tend to have the following properties:

- Characteristic **luster** (like freshly polished silverware or a new penny)
- **Malleable** (can be pounded or pressed into various shapes without breaking)
- **Conduct electricity**
- Chemically, tend to **lose electrons** and form cations (see Section 1.6) with charges of +1 to +3

Uses of Lithium

Lithium compounds have a number of important uses in industry and medicine. Lithium carbonate, Li_2CO_3, is one of the most important lithium compounds and is used as the starting material for the manufacture of many other lithium compounds. It is an ingredient of specialty glasses, enamels, and specialty ceramic ware having low thermal expansion coefficients (minimum expansion when heated). Lithium carbonate is widely prescribed as a drug to treat acute mania in bipolar affective (manic-depressive) and schizoaffective mental disorders. Lithium hydroxide, LiOH, is an ingredient in the manufacture of lubricant greases and in some long-life alkaline storage batteries.

The lithium atom has three electrons. As shown in Figure 3.6, lithium has both **inner electrons**—in this case 2 contained in an **inner shell**—as in the immediately preceding noble gas helium, and an **outer electron** that is farther from, and less strongly attracted to, the nucleus. The outer electron is said to be in the atom's **outer shell**. The inner electrons are, on the average, closer to the nucleus than is the outer electron, are very difficult to remove from the atom, and do not become involved in chemical bonds. Lithium's outer electron is relatively easy to remove from the atom,

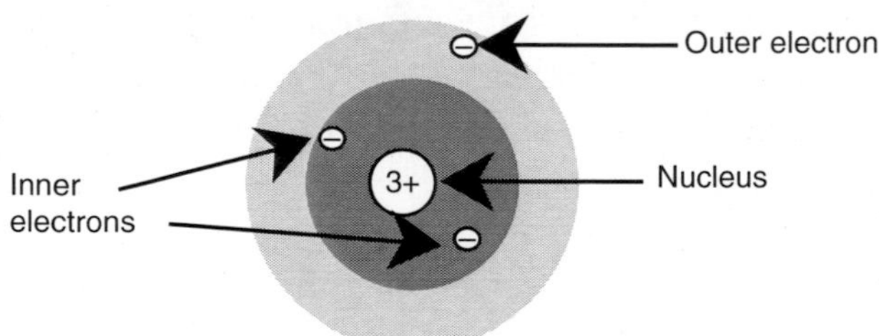

Figure 3.6. An atom of lithium, Li, has 2 inner electrons and 1 outer electron. The latter can be lost to another atom to produce the Li^+ ion, which is present in ionic compounds (see Section 1.6).

which is what happens when ionic bonds involving Li^+ ion are formed (see Section 1.4). The distinction between inner and outer electrons is developed to a greater extent later in this chapter.

In atoms such as lithium that have both outer and inner electrons, the Lewis symbol shows only the outer electrons. Therefore, the Lewis symbol of lithium is

$$Li\cdot$$

A lithium atom's loss of its single outer electron is shown by the **half-reaction** (one in which there is a net number of electrons on either the reactant or product side) in Figure 3.7. The Li^+ product of this reaction has the very stable helium core of 2 electrons. The Li^+ ion is a constituent of ionic lithium compounds, in which it is held by attraction for negatively charged anions (such as Cl^-) in the crystalline lattice of the ionic compound. The tendency to lose its outer electron and to be stabilized in ionic compounds determines lithium's chemical behavior.

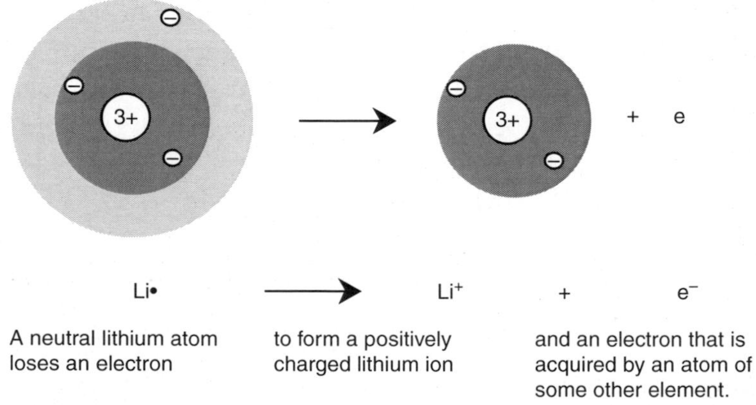

Figure 3.7. Half-reaction showing the formation of Li^+ from an Li atom. The Li^+ ion has the especially stable helium core of just 2 electrons. The atom to which the electron is lost is not shown, so this is a half-reaction.

Lithium, a Key Material in Green Technology

One of the greatest challenges in the efficient utilization of energy is electrical energy storage in efficient and compact devices. Electrical energy storage in batteries is required for portable electronic devices (especially laptop computers), in hybrid automobiles, and in many other applications. The low density of lithium and its ability to generate relatively high voltages makes batteries fabricated with lithium useful in a number of different applications. Cells containing lithium and SO_2 are used in submarines and space probes. Lithium–FeS_2 cells are used in small electronic devices, such as compact disk players. Lithium–iodine cells power heart pacemakers inserted surgically into the chests of heart patients and capable of lasting for several years before needing to be replaced.

One of the most promising kinds of batteries for storing electrical energy is the lithium-ion battery, which stores electrical energy during charging and liberates it during discharge through the movement of lithium ions, Li^+, between battery electrodes. When the battery is discharged, the lithium ions are bound to a thin electrode consisting of the cobalt compound $LiCoO_2$ separated from a carbon electrode by an electrolyte that allows movement of Li^+ ions between the electrodes. Charging the battery forces Li^+ ions to migrate through the electrolyte medium onto the carbon electrode, whereas during discharge the process is reversed.

Lithium-ion batteries have become quite popular because of their high energy-to-mass ratio, lack of memory effects, and slow loss of charge when not in use. They are favorable to the practice of green technology because they are made from relatively abundant and recyclable materials that are nontoxic, unlike the toxic heavy metals used in lead–acid batteries, nickel–cadmium batteries, and mercury batteries. Cases have been reported in which lithium-ion batteries have burned in laptop computers, but modern manufacturing practices have largely eliminated this hazard. The use of these batteries in hybrid automobiles would be advantageous because of their light mass and high storage capacities. In this application, the possibility of fire is a potentially serious problem because of the relatively large batteries required and the conflagration that could result from damage caused to the battery in an automobile accident. Huge banks of lithium-ion batteries may eventually prove to be useful in evening out the sporadic power output of renewable wind and solar electrical power sources.

3.9. THE SECOND PERIOD, ELEMENTS 4–10

In this section, elements 4–10 will be discussed and placed in the periodic table to complete a period in the table.

Beryllium, Atomic Number 4

Each atom of **beryllium**, Be—atomic number 4, atomic mass 9.01218—contains 4 protons and 5 neutrons in its nucleus. The beryllium atom has two inner electrons and two outer electrons, the latter designated by the two dots in the Lewis symbol below:

Be:

Beryllium can react chemically by losing 2 electrons from the beryllium atom. This occurs according to the half-reaction

$$\text{Be:} \rightarrow \text{Be}^{2+} + 2\text{e}^- \text{ (lost to another atom)} \qquad (3.9.1)$$

in which the beryllium atom, Lewis symbol Be:, loses two e^- to form a beryllium ion with a charge of +2. The loss of these two outer electrons gives the beryllium atom the same stable helium core as that of the Li^+ ion discussed in the preceding section.

Beryllium is melted together with certain other metals to give homogeneous mixtures of metals called **alloys**. The most important beryllium alloys are hard, corrosion-resistant, nonsparking, and good conductors of electricity. They are used to make such things as springs, switches, and small electrical contacts. A very high melting temperature of about 1290°C combined with good heat absorption and conduction properties has led to the use of beryllium metal in aircraft brake components.

Beryllium is an environmentally and toxicologically important element because it causes **berylliosis**, a disease marked by lung deterioration. Inhalation of beryllium is particularly hazardous, and atmospheric standards have been set at very low levels.

Boron, Atomic Number 5

Boron, B, has an atomic number of 5 and an atomic mass of 10.81. Most boron atoms have 6 neutrons in addition to 5 protons in their nuclei; a less common isotope has 5 protons. Two of boron's 5 electrons are in a helium core and 3 are outer electrons, as shown by the Lewis symbol

$\dot{\text{B}}$:

Boron—along with silicon, germanium, arsenic, antimony, and tellurium—is one of a few elements, called **metalloids**, with properties intermediate between those of metals and nonmetals. Although they have a luster like metals, metalloids do not form positively charged ions (cations). The melting temperature of boron is very high, 2190°C. Boron is added to copper, aluminum, and steel to improve their properties. It is used in control rods of nuclear reactors because of the good neutron-absorbing properties of the $^{10}_{5}\mathbf{B}$ isotope. Some chemical compounds of boron, especially boron nitride, BN, are noted for their hardness. Boric acid, H_3BO_3, is used as a flame retardant in cellulose insulation in houses. The oxide of boron, B_2O_3, is an ingredient of fiberglass, used in textiles and insulation.

Carbon, Atomic Number 6

Atoms of **carbon**, C, have 2 inner and 4 outer electrons, the latter shown by the Lewis symbol

$$\cdot\dot{\text{C}}:$$

The carbon-12 isotope with 6 protons and 6 neutrons in its nucleus, $^{12}_{6}C$, constitutes 98.9% of all naturally occurring carbon. The $^{13}_{6}C$ isotope makes up 1.1% of all carbon atoms. Radioactive carbon-14, $^{14}_{6}C$, is present in some carbon sources.

Carbon is an extremely important element with unique chemical properties without which life could not exist. All of organic chemistry (Chapter 9) is based upon compounds of carbon, and it is an essential element in life molecules (studied as part of biochemistry, Chapter 10). Carbon atoms are able to bond to each other to form long straight chains, branched chains, rings, and three-dimensional structures. As a result of its self-bonding abilities, carbon exists in several elemental forms. These include powdery carbon black; very hard, clear diamonds; and graphite so soft that it is used as a lubricant. Activated carbon prepared by treating carbon with air, carbon dioxide, or steam at high temperatures is widely used to absorb undesirable pollutant substances from air and water. Carbon fiber has been developed as a structural material in the form of composites consisting of strong strands of carbon bonded together with special plastics and epoxy resins.

Nitrogen, Atomic Number 7

Nitrogen, N, composes 78% by volume of air in the form of diatomic N_2 molecules. The atomic mass of nitrogen is 14.0067, and the nuclei of nitrogen atoms contain 7 protons and 7 neutrons. Nitrogen has 5 outer electrons, so its Lewis symbol is

$$\cdot\dot{\underset{\cdot}{\text{N}}}:$$

Like carbon, nitrogen is a nonmetal. Pure N_2 is prepared by distilling liquified air, and it has a number of uses. Since nitrogen gas is not very chemically reactive, it is used as an inert atmosphere in some industrial applications, particularly where fire or chemical reactivity may be a hazard. People have been killed by accidentally entering chambers filled with nitrogen gas, which acts as a simple asphyxiant with no odor to warn of its presence. Liquid nitrogen boils at a very cold −190°C. It is widely used to maintain very low temperatures in the laboratory, for quick-freezing foods, and in freeze-drying processes. Freeze-drying is used to isolate fragile biochemical compounds from water solution, for the concentration of environmental samples to be analyzed for pollutants, and for the preparation of instant coffee and other dehydrated foods. It has potential applications in the concentration and isolation of hazardous waste substances.

Like carbon, nitrogen is an essential element for life processes. Nitrogen is an ingredient of all of the amino acids found in proteins. Nitrogen compounds are fertilizers essential for the growth of plants. The **nitrogen cycle**, which involves incorporation of N_2 from the atmosphere into living matter and chemically bound nitrogen in soil and water, then back into the atmosphere again, is one of nature's fundamental cycles. Nitrogen compounds, particularly ammonia (NH_3) and nitric acid (HNO_3), are widely used industrial chemicals.

Oxygen, Atomic Number 8

Like carbon and nitrogen, **oxygen**, O, atomic number 8, is a major component of living organisms. Oxygen is a nonmetal existing as molecules of O_2 in the elemental gas state, and air is 21% oxygen by volume. Like all animals, humans require oxygen to breathe and to maintain their life processes. The nuclei of oxygen atoms contain 8 protons and 8 neutrons, and the atomic mass of oxygen is 15.9994. The oxygen atom has 6 outer electrons, as shown by its Lewis symbol:

In addition to O_2, there are two other important elemental oxygen species in the atmosphere. These are atomic oxygen, O, and ozone, O_3. These species are normal constituents of the stratosphere, a region of the atmosphere that extends from about 11 km to about 50 km in altitude. Oxygen atoms are formed when high-energy ultraviolet radiation strikes oxygen molecules high in the stratosphere:

$$O_2 \xrightarrow[\text{radiation}]{\text{Ultraviolet}} O + O \tag{3.9.2}$$

The oxygen atoms formed by the above reaction combine with O_2 molecules,

$$O + O_2 \rightarrow O_3 \tag{3.9.3}$$

to form ozone molecules. These molecules make up the **ozone layer** in the stratosphere and effectively absorb additional high-energy ultraviolet radiation. If it were not for this phenomenon, the ultraviolet radiation would reach the Earth's surface and cause painful sunburn and skin cancer in exposed people. However, ozone produced in photochemical smog at ground level is toxic to animals and plants.

The most notable chemical characteristic of elemental oxygen is its tendency to combine with other elements in energy-yielding reactions. Such reactions provide the energy that propels automobiles, heats buildings, and keeps body processes going. One of the most widely used chemical reactions of oxygen is that with hydrocarbons, particularly those from petroleum and natural gas. For example, butane, C_4H_{10}, a liquifiable gaseous hydrocarbon fuel, burns in oxygen from the atmosphere, a reaction that provides heat in home furnaces, water heaters, and other applications:

$$2C_4H_{10} + 13O_2 \rightarrow 8CO_2 + 10H_2O \quad (3.9.4)$$

Fluorine, Atomic Number 9

Fluorine, F, has 7 outer electrons, so its Lewis symbol is

Under ordinary conditions, elemental fluorine is a greenish-yellow gas consisting of F_2 molecules.

Fluorine compounds have many uses. One of the most notable of these is the manufacture of chlorofluorocarbon compounds known by the trade name Freon. These are chemical combinations of chlorine, fluorine, and carbon, an example of which is dichlorodifluoromethane, Cl_2CF_2. These compounds used to be widely employed as refrigerant fluids and blowing agents to make foam plastics; they were also once widely used as propellants in aerosol spray cans. Uses of chlorofluorocarbons have now been phased out because of their role in destroying stratospheric ozone (discussed with oxygen, above).

Neon, Atomic Number 10

The last element in the period of the periodic table under discussion is **neon**, Ne. Air is about 2 parts per thousand neon by volume, and neon is obtained by the distillation of liquid air. Neon is especially noted for its use in illuminated signs that consist of glass tubes containing neon, through which an electrical current is passed, causing the neon gas to emit a characteristic glow.

In addition to 10 protons, most neon atoms have 10 neutrons in their nuclei, although some have 12, and a very small percentage have 11. As shown by its Lewis symbol,

:N̈e:

the neon atom has 8 outer electrons, which constitute a *filled electron shell*, just as the 2 electrons in helium give it a filled electron shell. Because of this “satisfied” outer shell, the neon atom has no tendency to acquire, give away, or share electrons. Therefore, neon is a *noble gas*, like helium, and consists of individual neon atoms.

Stability of the Neon Noble Gas Electron Octet

In going through the rest of the periodic table, it can be seen that all other atoms with 8 outer electrons, like neon, are also noted for a high degree of chemical stability. In addition to neon, these noble gases are argon (atomic number 18), krypton (atomic

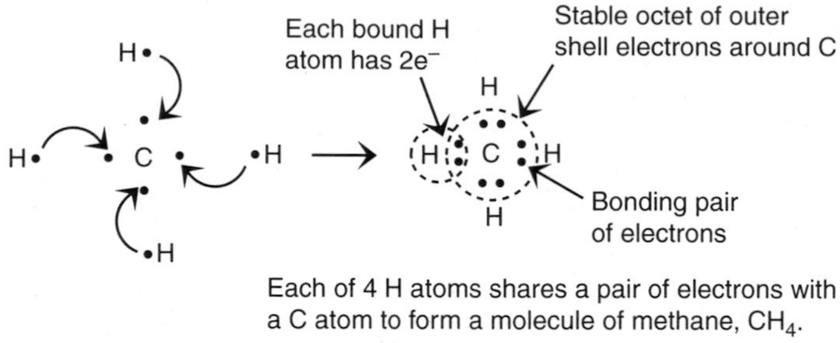

Figure 3.8. Illustration of the octet rule in methane.

number 36), xenon (atomic number 54), and radon (atomic number 86). Each of these may be represented by the Lewis symbol

where X is the chemical symbol of the noble gas. It is seen that these atoms each have 8 outer electrons, a group known as an **octet** of electrons. In many cases, atoms that do not have an octet of outer electrons acquire one by losing, gaining, or sharing electrons in chemical combination with other atoms; that is, they acquire a **noble gas outer electron configuration**. For all noble gases except helium, which has only 2 electrons, the noble gas outer electron configuration consists of 8 electrons. The tendency of elements to acquire an 8-electron outer electron configuration, which is very useful in predicting the nature of chemical bonding and the formulas of compounds that result, is called the **octet rule**. Although the use of the octet rule to explain and predict bonding is discussed in some detail in Chapter 4, at this point it is useful to show how it explains bonding between hydrogen and carbon in methane, as illustrated in Figure 3.8.

3.10. ELEMENTS 11–20, AND BEYOND

The abbreviated version of the periodic table will be finished with elements 11–20. The names, symbols, electron configurations, and other pertinent information about these elements are given in Table 3.2. An abbreviated periodic table with these elements in place is shown in Figure 3.9. This table shows the Lewis symbols of all of the elements to emphasize their orderly variation across periods and similarity in groups of the periodic table.

The first 20 elements in the periodic table are very important. They include the three most abundant elements on Earth's surface (oxygen, silicon, and aluminum); all elements of any appreciable significance in the atmosphere (hydrogen in H_2O vapor, N_2, O_2, carbon in CO_2, argon, and neon); the elements making up most of living plant and animal matter (hydrogen, oxygen, carbon, nitrogen, phosphorus, and sulfur); and elements such as sodium, magnesium, potassium, calcium, and chlorine that are essential for life processes. The chemistry of these elements is relatively straightforward

Table 3.2. Elements 11–20

Atomic Number	Name and Lewis Symbol	Atomic Mass	Number of Outer e^-	Major Properties and Uses
11	Sodium, Na·	22.9898	1	Soft, chemically very reactive metal. Nuclei contain 11 p and 12 n
12	Magnesium, Mg:	24.312	2	Lightweight metal used in aircraft components, extension ladders, portable tools. Chemically very reactive. Three isotopes with 12, 13, 14 n
13	Aluminum, Al:	26.9815	3	Lightweight metal used in aircraft, automobiles, electrical transmission line. Chemically reactive, but forms self-protective coating
14	Silicon, ·Si:	28.086	4	Nonmetal, 2nd most abundant metal in Earth's crust. Rock constituent. Used in semiconductors
15	Phosphorus, ·P:	30.9738	5	Chemically very reactive nonmetal. Highly toxic as elemental white phosphorus. Component of bones and teeth, genetic material (DNA), fertilizers, insecticides
16	Sulfur, ·S:	32.064	6	Brittle, generally yellow nonmetal. Essential nutrient for plants and animals, occurring in amino acids. Used to manufacture sulfuric acid. Present in pollutant sulfur dioxide, SO_2
17	Chlorine, ·Cl:	35.453	7	Greenish-yellow toxic gas composed of molecules of Cl_2. Manufactured in large quantities to disinfect water and to manufacture plastics and solvents
18	Argon, :Ar:	39.948	8	Noble gas used to fill light bulbs and as a plasma medium in inductively coupled plasma atomic emission analysis of elemental pollutants
19	Potassium, K·	39.098	1	Chemically reactive alkali metal very similar to sodium in chemical and physical properties. Essential fertilizer for plant growth as K^+ ion
20	Calcium, Ca:	40.078	2	Chemically reactive alkaline earth metal with properties similar to those of magnesium

First period	1 H 1.0							2 He 4.0
Second period	3 Li 6.9	4 Be 9.0	5 B 10.8	6 C 12.0	7 N 14.0	8 O 16.0	9 F 19.0	10 Ne 20.1
Third period	11 Na 23.0	12 Mg 24.3	13 Al 27.0	14 Si 28.1	15 P 31.0	16 S 32.1	17 Cl 35.5	18 Ar 39.9
Fourth period	19 K 39.1	20 Ca 40.1						

Figure 3.9. Abbreviated 20-element version of the periodic table showing Lewis symbols of the elements.

and easy to relate to their atomic structures. Therefore, emphasis is placed on them in the earlier chapters of this book. It is helpful to remember their names, symbols, atomic numbers, atomic masses, and Lewis symbols.

As mentioned in Section 1.5, the vertical columns of the table contain groups of elements that have similar chemical structures. Hydrogen, H, is an exception and is not regarded as belonging to any particular group because of its unique chemical properties. All elements other than hydrogen in the first column of the abbreviated table are **alkali metals**—lithium, sodium, and potassium. These are generally soft silvery-white metals of low density that react violently with water to produce hydroxides (LiOH, NaOH, and KOH) and with chlorine to produce chlorides (LiCl, NaCl, and KCl). The **alkaline earth** metals—beryllium, magnesium, calcium—are in the second column of the table. When freshly cut, these metals have a grayish-white luster. They are chemically reactive and have a strong tendency to form doubly charged cations (Be^{2+}, Mg^{2+}, and Ca^{2+}) by losing two electrons from each atom. Another group notable for the very close similarities of the elements in it consists of the **noble gases** in the far right column of the table. Each of these—helium, neon, argon—is a monatomic gas that does not react chemically.

The Elements Beyond Calcium

The electron structures of elements beyond atomic number 20 are more complicated than those of the lighter elements. The complete periodic table in Figure 1.4 shows, among the heavier elements, the transition metals, including chromium, manganese, iron, cobalt, nickel, and copper; the lanthanides; and the actinides, including thorium, uranium, and plutonium. The transition metals include a number of metals that are important in industry and in life processes. The actinides contain elements familiar to those concerned with nuclear energy, nuclear warfare, and related issues. A list of the known elements through atomic number 109 is given at the end of this chapter.

3.11. A MORE DETAILED LOOK AT ATOMIC STRUCTURE

So far, this chapter has covered some important aspects of atoms. These include the facts that an atom is made of three major subatomic particles, and consists of a very small, very dense, positively charged nucleus surrounded by a cloud of negatively charged electrons in constant, rapid motion. The first 20 elements have been discussed in some detail and placed in an abbreviated version of the periodic table. Important concepts introduced so far in this chapter include:

- Dalton's atomic theory
- Lewis symbols to represent outer e^-
- Electron shells
- Significance of filled electron shells
- Inner shell electrons
- Outer shell electrons
- Octet rule
- Abbreviated periodic table

The information presented about atoms so far in this chapter is adequate to meet the needs of many readers. These readers may choose to forgo the details of atomic structure presented in the rest of this chapter without major harm to their understanding of chemistry. However, for those who wish to go into more detail (or who do not have a choice), the remainder of this chapter discusses in more detail the electronic structures of atoms as related to their chemical behavior and introduces the quantum theory of electrons in atoms.

Electromagnetic Radiation

The **quantum theory** explains the unique behavior of charged particles that are as small and move as rapidly as electrons. Because of its close relationship to electromagnetic radiation, an appreciation of quantum theory requires an understanding of the following important points related to electromagnetic radiation:

- Energy can be carried through space at the speed of light, 3.00×10^9 m/s in a vacuum, by **electromagnetic radiation,** which includes visible light, ultraviolet radiation, infrared radiation, microwaves, and radio waves.
- Electromagnetic radiation has a **wave character**. The waves move at the speed of light, *c*, and have characteristics of **wavelength** λ (Greek "lambda"), amplitude, and **frequency** ν (Greek "nu") as illustrated below:

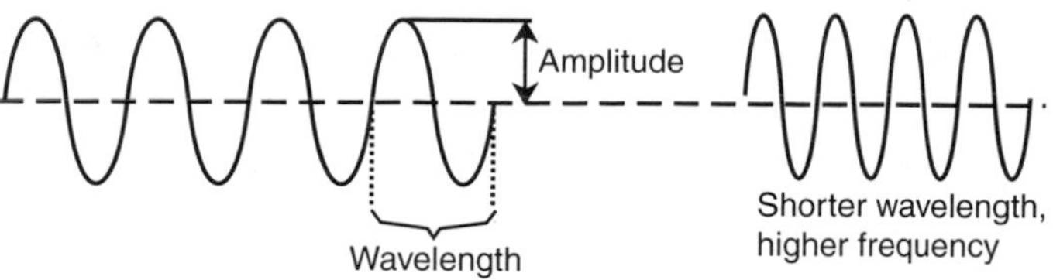

- The wavelength is the distance required for one complete cycle and the frequency is the number of cycles per unit time. They are related by the following equation:

$$\nu\lambda = c$$

where ν is in units of cycles per second (s^{-1}, a unit called the hertz, Hz) and λ is in meters (m).

- In addition to behaving as a wave, electromagnetic radiation also has characteristics of particles.
- This dual wave/particle nature is the basis of the quantum theory of electromagnetic radiation, which states that radiant energy can be absorbed or emitted only in discrete packets called quanta or photons. The energy E of each photon is given by

$$E = h\nu$$

where h is Planck's constant, 6.63×10^{-34} J-s (joule $\times$ second).

- From the preceding, it can be seen that the energy of a photon is higher when the frequency of the associated wave is higher (and the wavelength shorter).

3.12. QUANTUM AND WAVE MECHANICAL MODELS OF ELECTRONS IN ATOMS

The *quantum theory* introduced in the preceding section provided the key concepts needed to explain the energies and behavior of electrons in atoms. One of the best clues to this behavior, and one that ties the nature of electrons in atoms to the properties of electromagnetic radiation, is the emission of light by energized atoms. This is easiest to explain for the simplest atom of all, that of hydrogen, which consists of only one electron moving around a nucleus with a single positive charge. Energy added to hydrogen atoms, such as by an electrical discharge through hydrogen gas, is re-emitted in the form of light at very specific wavelengths (656, 486, 434, and 410 nm in the visible region). The highly energized atoms that can emit this light are said to be "excited" by the excess energy originally put into them and to be in an **excited state**. The reason for this is that the electrons in the excited atoms are forced farther from the nuclei of the atoms and, when they return to a lower-energy state, energy is emitted in the form of light. The fact that very specific wavelengths of light are emitted in this process means that electrons can be present only in specified states at highly specific energy levels. Therefore, the transition from one energy state to a lower one involves the emission of a specific energy of electromagnetic radiation (light). Consider the equation

$$E = h\nu \tag{3.12.1}$$

that relates energy E to frequency ν of electromagnetic radiation. If a transition of an electron from one excited state to a lower one involves a specific amount of energy E, then a corresponding value of ν is observed. This corresponds to a specific wavelength of light according to the following relationship:

$$\lambda = c/\nu \tag{3.12.2}$$

The first accepted explanation of the behavior outlined above was the **Bohr theory** advanced by the Danish physicist Neils Bohr in 1913. Although this theory has been shown to be too simplistic, it had some features that are still pertinent to atomic structure. The Bohr theory visualized an electron orbiting the nucleus (a proton) of the hydrogen atom in orbits. Only specific orbits called **quantum states** were allowed. When energy was added to a hydrogen atom, its electron could jump to a higher orbit. When the atom lost its energy as the electron returned to a lower orbit, the energy lost was emitted in the form of electromagnetic radiation as shown in Figure 3.10. Because the two energy levels are of a definite magnitude according to quantum theory, the energy lost by the electron must also be of a definite energy, $E = h\nu$. Therefore, the electromagnetic radiation (light) emitted is of a specific frequency and wavelength.

The Wave Mechanical Model of Atomic Structure

Although shown to have some serious flaws and long since abandoned, the Bohr model laid the groundwork for the more sophisticated theories of atomic structure that are accepted today and introduced the all-important concept that *only specific energy states are allowed for an electron in an atom*. Like electromagnetic radiation, electrons in atoms are now visualized as having a dual wave/particle nature.

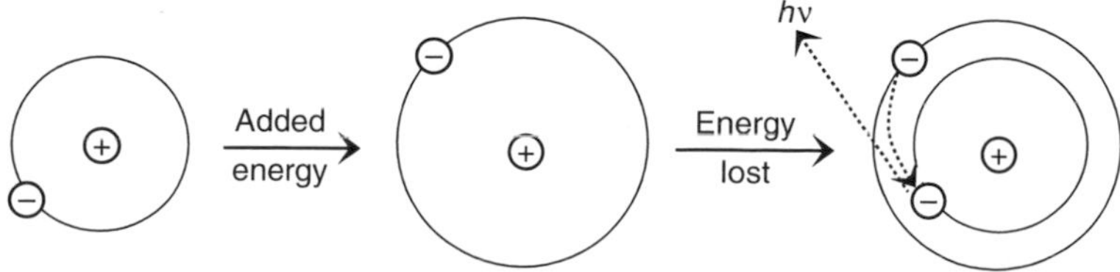

Figure 3.10. According to the Bohr model, adding energy to the hydrogen atom promotes an electron to a higher energy level. When the electron falls back to a lower energy level, excess energy is emitted in the form of electromagnetic radiation of a specific energy $E = h\nu$.

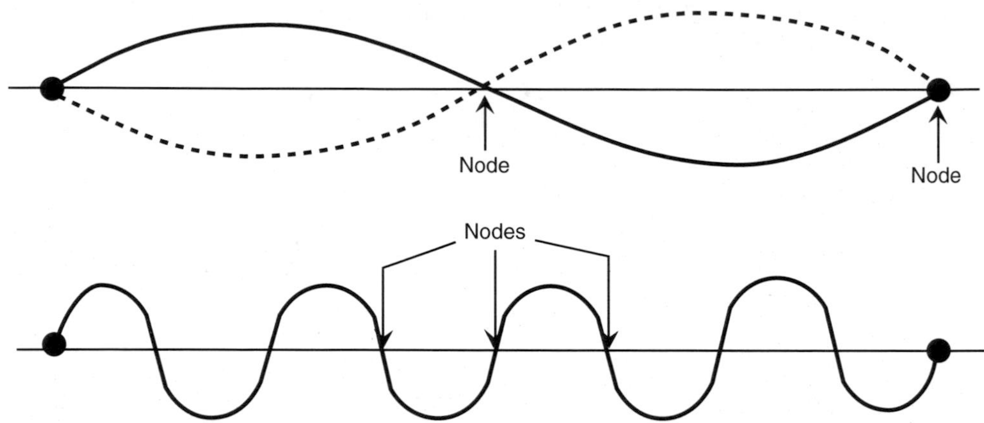

Figure 3.11. A string anchored at both ends (such as on a stringed musical instrument) can only vibrate at multiples of waves or half-waves. The illustration above shows one wave (top) and four waves (bottom). Each point at which a wave intersects the horizontal lines is called a node.

They are treated theoretically by the **wave mechanical model** as **standing waves** around the nucleus of an atom. The idea of a standing wave can be visualized for the string of a musical instrument as represented in Figure 3.11. Such a wave does not move along the length of a string, because both ends are anchored, which is why it is called a *standing* wave. Each wave has **nodes**, which are points of zero displacement. Because there must be a node on each end where the string is anchored, the standing waves can only exist as multiples of *half-wavelengths.*

According to the *wave mechanical* or *quantum mechanical* model of electrons in atoms, the known *quantization* of electron energy in atoms occurs because only specific multiples of the standing wave associated with an electron's movement are allowed. Such a phenomenon is treated mathematically with the **Schrödinger equation,** resulting from the work of Erwin Schrödinger, first published in 1926. Even for the one-electron hydrogen atom, the mathematics is quite complicated, and no attempt will be made to go into it here. The Schrödinger equation is represented as

$$H\psi = E\psi \tag{3.12.3}$$

Here, ψ (Greek "psi") is the **wave function**, a function of the electron's energy and the coordinates in space where it may be found; H is an **operator** composed of mathematical operations acting on ψ, and E is the sum of the kinetic energy due to the motion of the electron and the potential energy of the mutual attraction between the electron and the nucleus.

Solutions to the Schrödinger Equation, Electrons in Orbitals

With its motion governed by the laws of quantum mechanics, it is not possible to know exactly where an electron is relative to an atom's nucleus at any specific

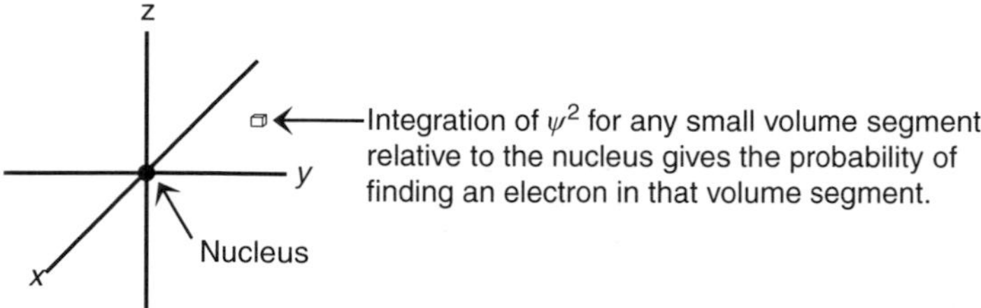

Figure 3.12. The *xyz* coordinate system used to describe the orientations in space of orbitals around the nucleus of an atom.

instant. However, the Schrödinger equation permits calculation of the probability of an electron being in a specified region. Solution of the equation gives numerous wave functions, each of which corresponds to a definite energy level and to the probability of finding an electron at various locations and distances relative to the nucleus (Figure 3.12). The wave function describes **orbitals**, *each of which has a characteristic energy and region around the nucleus where the electron has certain probabilities of being found.* The term "orbital" is used because quantum mechanical calculations do not give specific orbits consisting of defined paths for electrons, like planets going around the Sun.

The square of the wave function, ψ^2, calculated and integrated over a small segment of volume, is proportional to the probability of finding an electron in that volume, which can be regarded as part of an *electron cloud.* With this view, the electron is more likely to be found in a region where the cloud is relatively denser. The cloud has no definite outer limits, but fades away with increasing distance from the nucleus to regions in which there is essentially no probability of finding the electron.

Multielectron Atoms and Quantum Numbers

Wave mechanical calculations describe various **energy levels** for electrons in an atom. Each energy level has at least one orbital. *A single orbital may contain a maximum of only two electrons.* Although the Schrödinger approach was originally applied to the simplest atom, hydrogen, it has been extended to atoms with many electrons (multielectron atoms). Electrons in atoms are described by four **quantum numbers**, which are defined and described briefly below. With these quantum numbers and knowledge of the rules governing their use, it is possible to specify the orbitals allowed in a particular atom. *An electron in an atom has its own unique set of quantum numbers; no two electrons can have exactly identical quantum numbers.*

The Principal Quantum Number n

Main energy levels corresponding to *electron shells* discussed earlier in this chapter are designated by a **principal quantum number** *n*. Both the size of orbitals and the magnitude of the average energy of electrons contained therein increase with increasing n. Permitted values of n are 1, 2, 3,..., extending through 7 for the known elements.

The Azimuthal Quantum Number l

Within each main energy level (shell) represented by a principal quantum number, there are **sublevels** (subshells). Each sublevel is denoted by an **azimuthal quantum number** l. For any shell with a principal quantum number n, the possible values of l are $0, 1, 2, 3, \ldots, n-1$. This gives the following:

- For $n = 1$, there is only 1 possible subshell, $l = 0$.
- For $n = 2$, there are 2 possible subshells, $l = 0, 1$.
- For $n = 3$, there are 3 possible subshells, $l = 0, 1, 2$.
- For $n = 4$, there are 4 possible subshells, $l = 0, 1, 2, 3$.

From the above, it can be seen that the maximum number of sublevels within a principal energy level designated by n is equal to n. Within a main energy level, sublevels denoted by different values of l have slightly different energies. Furthermore, l designates *different shapes of orbitals.*

The letters *s*, *p*, *d*, and *f*, corresponding to values of $l = 0, 1, 2$, and 3, respectively, are conventionally used to designate sublevels. The number of electrons present in a given sublevel is limited to 2, 6, 10, and 14 for *s*, *p*, *d*, and *f*, sublevels, respectively. It follows that, since each orbital can be occupied by a maximum of 2 electrons, there is only 1 orbital in the *s* sublevel, 3 orbitals in the *p* sublevel, 5 in the *d* and 7 in the *f*. The value of the principal quantum number and the letter designating the azimuthal quantum number are written in sequence to designate both the shell and subshell. For example, 4*d* represents the *d* subshell of the fourth shell.

The Magnetic Quantum Number $\mathbf{m_l}$

The **magnetic quantum number** m_l is also known as the **orientational quantum number**. It designates the orientation of orbitals in space relative to each other and distinguishes orbitals within a subshell from each other. It is called the magnetic quantum number because the presence of a magnetic field can result in the appearance of additional lines among those emitted by electronically excited atoms (i.e., in atomic emission spectra). The possible values of m_l in a subshell with azimuthal quantum number l are given by $m_l = +l, +(l-1), \ldots, 0, \ldots, -(l-1), -l$. As examples, for $l = 0$, the only possible value of m_l is 0, and for $l = 3$, m_l may have values of 3, 2, 1, 0, −1, −2, −3.

Spin Quantum Number m_s

The fourth and final quantum number to be considered for electrons in atoms is the **spin quantum number** m_s, which can only have values of +½ or −½. This results from the fact that an electron spins in either of two directions and generates a tiny magnetic field with an associated magnetic moment. *Two electrons can occupy the same orbital only if they have opposite spins so that their magnetic moments cancel each other.*

Quantum Numbers Summarized

The information given by each quantum number is summarized below:

- The value of the principal quantum number *n* gives the shell and main energy level of the electron.
- The value of the azimuthal quantum number *l* specifies sublevels with somewhat different energies within the main energy levels and describes the shapes of orbitals.
- The magnetic quantum *number* m_l distinguishes the orientations in space of orbitals in a subshell and may provide additional distinctions of orbital shapes.
- The spin quantum number m_s, with possible values of only +½ and −½, accounts for the fact that each orbital can be occupied by a maximum number of only 2 electrons with opposing spins.

Specification of the values of four quantum numbers describes each electron in an atom, as shown in Table 3.3.

For each electron in an atom, the orbital that it occupies and the direction of its spin are specified by values of *n*, *l*, m_l, and m_s assigned to it. These values are unique for each electron; *no two electrons in the same atom can have identical values of all four quantum numbers.* This rule is known as the **Pauli exclusion principle**.

3.13. ENERGY LEVELS OF ATOMIC ORBITALS

Electrons in atoms occupy the lowest energy levels available to them. Figure 3.13 is an energy level diagram that shows the relative energy levels of electrons in various atomic orbitals. Each dash, —, in the figure represents *one* orbital that is a potential slot for *two* electrons with opposing spins.

The electron energies represented in Figure 3.13 can be visualized as those required to remove an electron from a particular orbital to a location completely away from the atom. As indicated by its lowest position in the diagram, the most

Table 3.3. Quantum Numbers for Electrons in Atoms[a]

n	l	m_l	m_s	
1	0 (1*s*)	0	(m_s = +½ or −½)	1 orbital
2	0 (2*s*)	0	(m_s = +½ or −½)	1 orbital
	1 (2*p*)	−1	(m_s = +½ or −½)	3 orbitals
		0	(m_s = +½ or −½)	
		+1	(m_s = +½ or −½)	
3	0 (3*s*)	0	(m_s = +½ or −½)	1 orbital
	1 (3*p*)	−1	(m_s = +½ or −½)	3 orbitals
		0	(m_s = +½ or −½)	
		+1	(m_s = +½ or −½)	
	2 (3*d*)	−2	(m_s = +½ or −½)	5 orbitals
		−1	(m_s = +½ or −½)	
		0	(m_s = +½ or −½)	
		+1	(m_s = +½ or −½)	
		+2	(m_s = +½ or −½)	
4	0 (4*s*)	0	(m_s = +½ or −½)	1 orbital
	1 (4*p*)	−1	(m_s = +½ or −½)	3 orbitals
		0	(m_s = +½ or −½)	
		+1	(m_s = +½ or −½)	
	2 (4*d*)	−2	(m_s = +½ or −½)	5 orbitals
		−1	(m_s = +½ or −½)	
		0	(m_s = +½ or −½)	
		+1	(m_s = +½ or −½)	
		+2	(m_s = +½ or −½)	
	3 (4*f*)	−3	(m_s = +½ or −½)	3 orbitals
		−2	(m_s = +½ or −½)	
		−1	(m_s = +½ or −½)	
		0	(m_s = +½ or −½)	7 orbitals
		+1	(m_s = +½ or −½)	
		+2	(m_s = +½ or −½)	
		+3	(m_s = +½ or −½)	

[a] The four quantum numbers are as follows:
n, principal quantum number denoting energy level, or shell
l, azimuthal quantum number of a sublevel or subshell
m_l, magnetic quantum number of designating an orbital; each entry in this column designates an orbital capable of holding 2 electrons with opposing spins
m_s, spin quantum number, which may have a value of +½ or −½

energy would be needed to remove an electron in the 1*s* orbital; comparatively little energy is required to remove electrons from orbitals having higher values of n, such as 5, 6, or 7. Furthermore, the diagram shows decreasing separation of the energy levels (values of n) with increasing n. It also shows that some sublevels with a particular value of n have lower energies than sublevels for which the principal quantum number

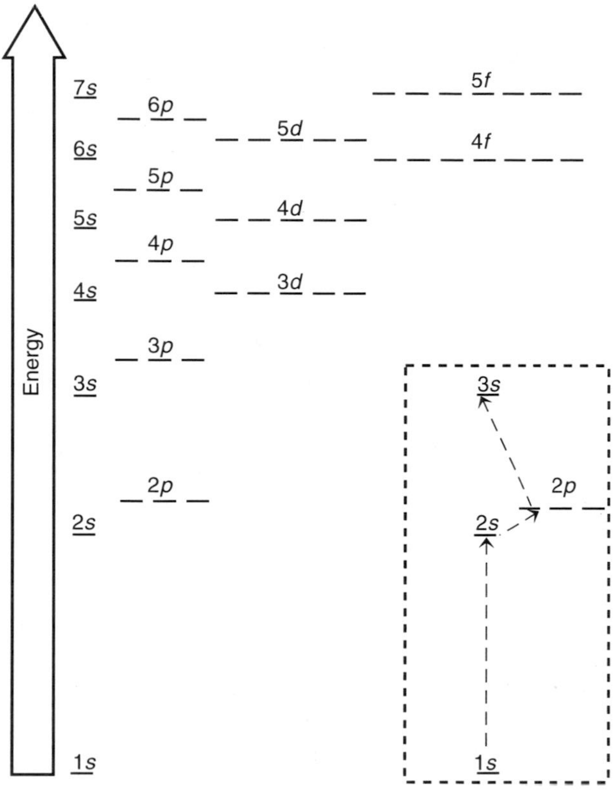

Figure 3.13. Energy levels of atomic orbitals. Each orbital capable of containing 2 electrons is shown with a dash, —. The order of placing electrons in orbitals is from the lowest-lying orbitals up, as shown in the inset.

is $n - 1$. This is first seen from the placement of the 4*s* sublevel below that of the 3*d* sublevel. As a consequence, with increasing atomic number, the 4*s* orbital acquires 2 electrons before any electrons are placed in the 3*d* orbitals.

The value of the principal quantum number *n* of an orbital is a measure of the relative distance of the maximum electron density from the nucleus. An electron with a lower value of *n*, being on the average closer to the nucleus, is more strongly attracted to the nucleus than an electron with a higher value of *n*, which is at a relatively greater distance from the nucleus.

Hund's Rule of Maximum Multiplicity

Orbitals that are in the same sublevel, but that have different values of m_l, have the same energy. This is first seen in the 2*p* sublevel, where the three orbitals with m_l of −1, 0, +1 (shown as three dashes on the same level in Figure 3.13) all have the same energies. The order in which electrons go into such a sublevel follows **Hund's rule of maximum multiplicity**, which states that *electrons in a sublevel are* *distributed to give the maximum number of unpaired electrons* (i.e., those with parallel spins having the same sign of m_s). Therefore, the first three electrons to be placed in the $2p$ sublevel would occupy the three separate available orbitals and would have the same spins. Not until the fourth electron out of a maximum number of 6 is added to this sublevel are two electrons placed in the same orbital.

3.14. SHAPES OF ATOMIC ORBITALS

In trying to represent different shapes of orbitals, it is important to keep in mind that they do not contain electrons within finite volumes, because there is no outer boundary of an orbital at which the probability of finding an electron drops to exactly 0. However, an orbital can be drawn as a figure ("fuzzy cloud") around the atom nucleus within which there is a relatively high probability (typically 90%) of finding an electron within the orbital. These figures are called **contour representations of orbitals** and are as close as one can come to visualizing the shapes of orbitals. Each point on the surface of a contour representation of an orbital has the same value of ψ^2 (the square of the wave function of the Schrödinger equation; see Section 3.13), and the entire surface encloses the volume within which an electron spends 90% of its time. The two most important aspects of an orbital are its size and shape, both of which are rather well illustrated by a contour representation.

Figure 3.14 shows the contour representations of the first three *s* orbitals. These are spherically shaped and increase markedly in size with increasing principal quantum number, reflecting increased average distance of the electron from the nucleus.

Figure 3.15 shows contour representations of the three $2p$ orbitals. These are seen to have different orientations in space; they have directional properties. In fact, *s* orbitals are the only ones that are spherically symmetrical. The shapes of orbitals are involved in molecular geometry and help to determine the shapes of molecules. The shapes of *d* and *f* orbitals are more complex and varied than those of *p* orbitals and are not discussed in this book.

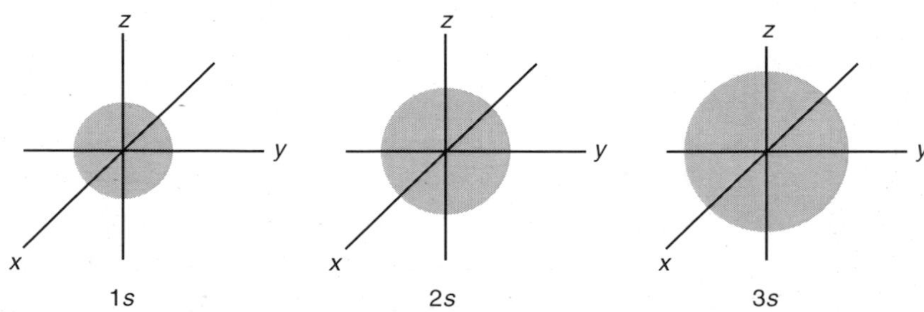

Figure 3.14. Contour representations of *s* orbitals for the first three main energy levels.

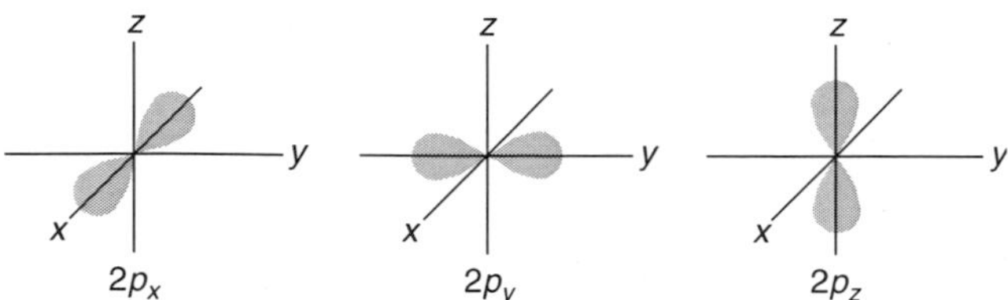

Figure 3.15. Contour representations of 2*p* orbitals.

3.15. ELECTRON CONFIGURATION

Electron configuration is a means of stating which kinds of orbitals contain electrons and the numbers of electrons in each kind of orbital of an atom. It is expressed by the number and letter representing each kind of orbital, together with superscript numbers indicating how many electrons are in each sublevel. The hydrogen atom's 1 electron in the 1*s* orbital is shown by the following notation:

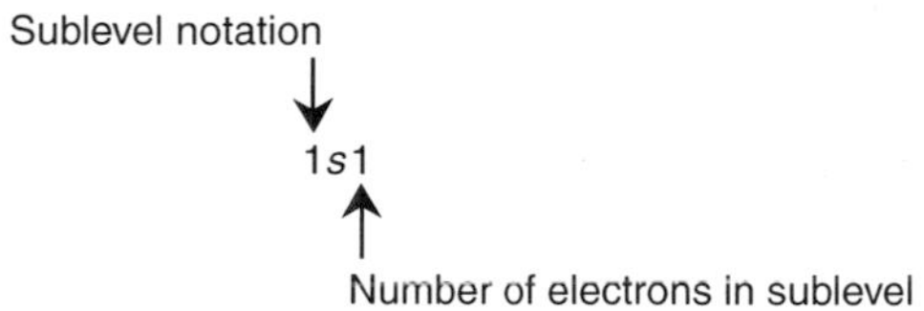

Nitrogen, atomic number 7, has 7 electrons, of which 2 are paired in the 1*s* orbital, 2 are paired in the 2*s* orbital, and 3 occupy singly each of the 3 available 2*p* orbitals. This electron configuration is designated as $1s^22s^22p^3$.

The orbital diagram is an alternative way of expressing electron configurations in which each separate orbital is represented by a box. Individual electrons in the orbitals are shown as arrows pointing up or down to represent opposing spins (m_s = +½ or –½). The orbital diagram for nitrogen is the following:

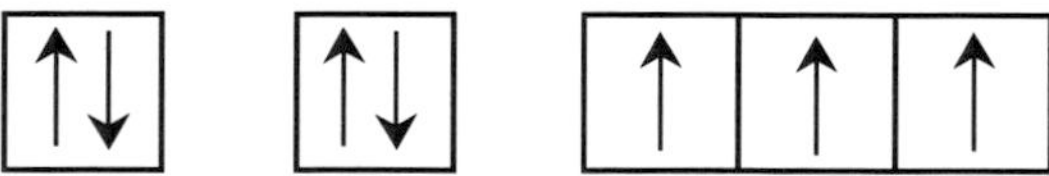

This gives all the information contained in the notation $1s^22s^22p^3$, but emphasizes that the three electrons in the three available 2*p* orbitals each occupy separate orbitals, a condition that is a consequence of Hund's rule of maximum multiplicity (Section 3.13).

Most of the remainder of this chapter is devoted to a discussion of the placement of electrons in atoms with increasing atomic number, its effects upon the chemical behavior of atoms, and how it leads to a systematic organization of the elements in the periodic table. This placement of electrons is in the order shown in increasing energy levels from bottom to top in Figure 3.13, as illustrated for nitrogen in Figure 3.16.

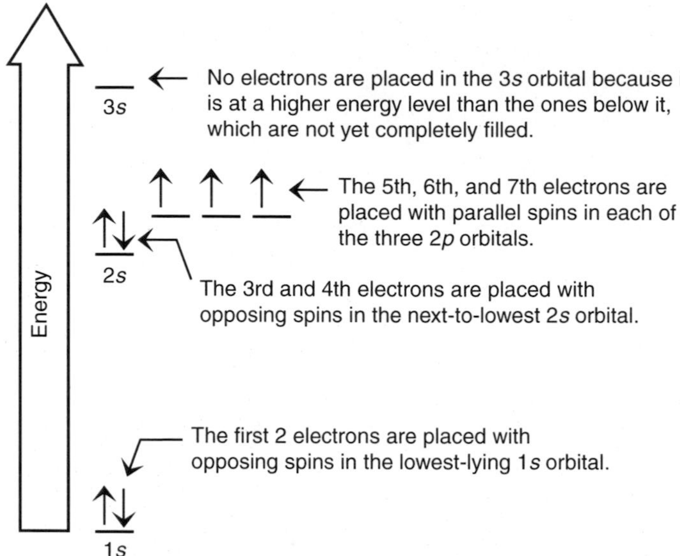

Figure 3.16. Placement of electrons in orbitals for the 7-electron nitrogen atom according to the energy level diagram.

3.16. ELECTRONS IN THE FIRST 20 ELEMENTS

The most meaningful way to place electrons in the orbitals of atoms is on the basis of the periodic table. This enables relating electron configurations to chemical properties and the properties of elements in groups and periods of the periodic table. In this section, electron configurations are deduced for the first 20 elements and given in an abbreviated version of the periodic table.

Electron Configuration of Hydrogen

The 1 electron in the hydrogen atom goes into its lowest-lying 1*s* orbital. Figure 3.17 summarizes all the information available about this electron and its electron configuration.

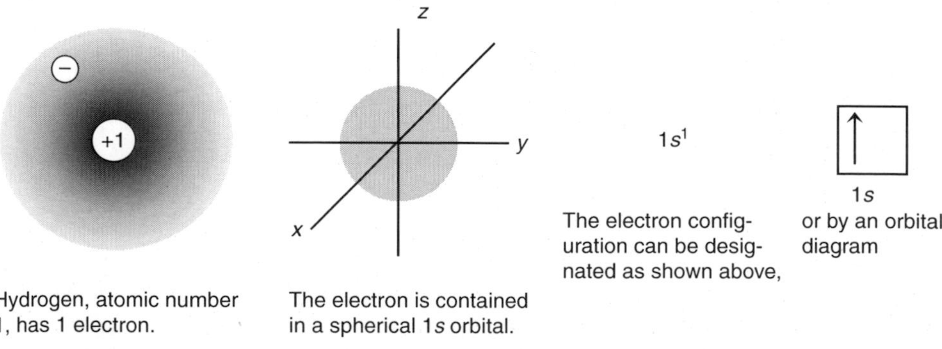

Figure 3.17. Representation of the electron on the hydrogen atom.

Electron Configuration of Helium

An atom of helium, atomic number 2, has 2 electrons, both contained in the 1*s* orbital and having quantum numbers $n = 1$, $l = 0$, $m_l = 0$, and $m_s = +½$ and $-½$. Both electrons have the same set of quantum numbers except for m_s. The electron configuration of helium can be represented as $1s^2$, showing that there are 2 electrons in the 1*s* orbital. Two is the maximum number of electrons that can be contained in the first principal energy level; additional electrons in atoms with atomic number greater than 2 must go into principal energy levels with *n* greater than 1. There are several other noble gas elements in the periodic table, of which neon and argon have already been mentioned, but helium is the only one with a filled shell of 2 electrons—the rest have stable outer shells of 8 electrons.

Electron Configurations of Elements 2–20

The electron configurations of elements with atomic number through 20 are very straightforward. Electrons are placed in order of orbitals with increasing energy as shown in Figure 3.13. This order is

$$1s^2 2s^2 2p^6 3s^2 3p^6 4s^2$$

Note that in this configuration the electrons go into the 4*s* orbital before the 3*d* orbital, which lies at a slightly higher energy level. In filling the *p* orbitals, it should also be kept in mind that 1 electron goes into each of three *p* orbitals before pairing occurs. In order to follow the discussion of electron configurations for elements through 20, it is useful to refer to the abbreviated periodic table in Figure 3.20.

Lithium

For lithium, atomic number 3, two electrons are placed in the 1*s* orbital, leaving the third electron for the 2*s* orbital. This gives an electron configuration of $1s^2 2s^1$. The two 1*s* electrons in lithium are in the stable noble gas electron configuration of helium and are very difficult to remove. These are lithium's *inner electrons* and, along with the nucleus, constitute the **core** of the lithium atom. The electron configuration *of the core* of the lithium atom is $1s^2$, the same as that of helium. Therefore, the lithium atom is said to have a *helium core. The core of any atom consists of its nucleus plus its inner electrons, those with the same electron configuration as the noble gas immediately preceding the element in the periodic table.*

Valence Electrons

Lithium's lone 1*s* electron is an outer electron contained in the *outer shell* of the atom. Outer shell electrons are also called **valence electrons**, and are the electrons

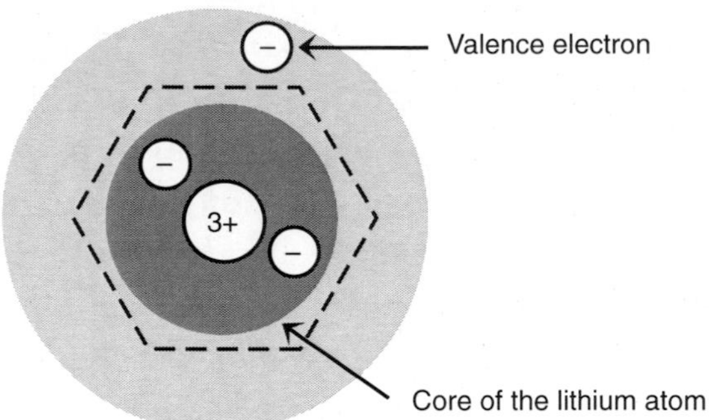

Figure 3.18. Core (kernel) and valence electrons of the lithium atom.

that can be shared in covalent bonding or lost to form cations in ionic compounds. Lithium's valence electron is illustrated in Figure 3.18.

Beryllium

Beryllium, atomic number 4, has 2 inner electrons in the 1*s* orbital and 2 outer electrons in the 2*s* orbital. Therefore, beryllium has a *helium core*, plus 2 *valence electrons*. Its electron configuration is $1s^22s^2$.

Filling the 2*p* Orbitals

Boron, atomic number 5, is the first element containing an electron in a 2*p* orbital. Its electron configuration is $1s^22s^22p^1$. Two of the five electrons in the boron atom are contained within the spherical orbital closest to the nucleus, and 2 more are in the larger spherical 2*s* orbital. The lone 2*p* electron is in an approximately dumbbell-shaped orbital in which the average distance of the electron from the nucleus is about the same as that of the 2*s* electrons. This electron could have any one of the three orientations in space shown for *p* orbitals in Figure 3.15.

The electron configuration of carbon, atomic number 6, is $1s^22s^22p^2$. The four outer (valence) electrons are shown by the Lewis symbol

$$\cdot\dot{\mathrm{C}}:$$

Two of the four valence electrons are represented by a pair of dots, :, to indicate that these electrons are paired in the same orbital. These are the two 2*s* electrons. The other two are shown as individual dots to represent two unpaired 2*p* electron in separate orbitals.

The 7 electrons in nitrogen are in a $1s^22s^22p^3$ electron configuration. Nitrogen is the first element to have at least one electron in each of 3 possible *p* orbitals (Figure 3.19).

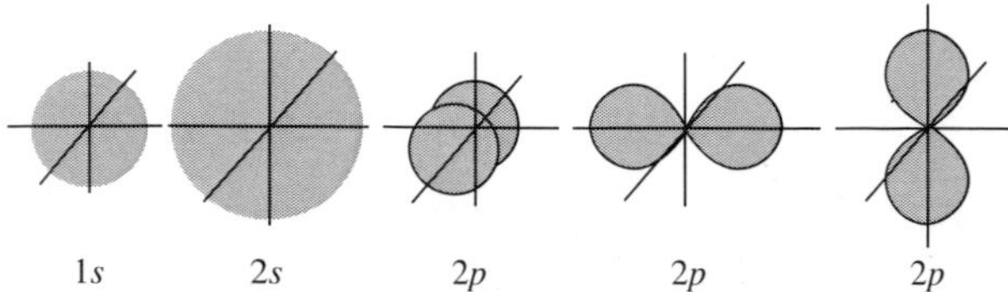

Figure 3.19. The nitrogen atom contains 2 electrons in the 1*s* orbital, 2 in the 2*s* orbital, and 1 in each of three separate 2*p* orbitals oriented in different directions in space.

The next element to be considered is oxygen, which has an atomic number of 8. Its electron configuration is $1s^22s^22p^4$. It is the first element in which it is necessary for 2 electrons to occupy the same *p* orbital, as shown by the following orbital diagrams:

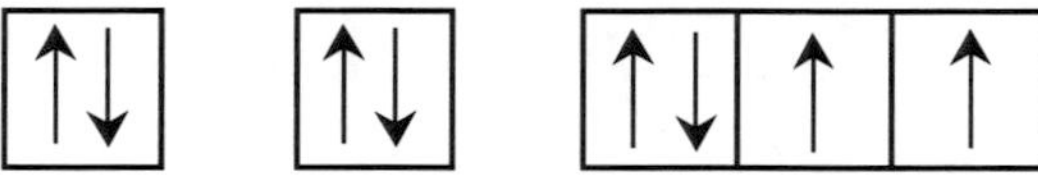

The electron configuration of fluorine, atomic number 9, is $1s^22s^22p^5$. Its Lewis symbol

shows that the fluorine atom has only one unpaired electron of the 7 electrons in its valence shell. In its chemical reactions, fluorine seeks to obtain another electron to give a stable *octet.*

Neon, atomic number 10, is at the end of the second period of the periodic table and is a noble gas, as shown by its Lewis symbol

denoting a filled octet of electrons. Its electron configuration is $1s^22s^22p^6$. In this configuration, the $2s^22p^6$ portion stands for the outer electrons. Like neon, all other elements with the outer electron configuration $\boldsymbol{ns^2np^6}$ are noble gases located in the far right of the periodic table (e.g., argon, with the electron configuration).

Filling the 3*s*, 3*p*, and 4*s* Orbitals

The 3*s* and 3*p* orbitals are filled in going across the third period of the periodic table from sodium through argon. Atoms of these elements, like all atoms beyond neon, have their 10 innermost electrons in the neon electron configuration of $1s^22s^22p^6$. Therefore, these atoms have a *neon core*, which may be designated {Ne}:

{Ne} stands for $1s^22s^22p^6$

1 H $1s^1$							2 He $1s^2$
3 Li {He}$2s^1$	4 Be {He}$2s^2$	5 B {He}$2s^22p^1$	6 C {He}$2s^22p^2$	7 N {He}$2s^22p^3$	8 O {He}$2s^22p^4$	9 F {He}$2s^22p^5$	10 Ne {He}$2s^22p^6$
11 Na {Ne}$3s^1$	12 Mg {Ne}$3s^2$	13 Al {Ne}$3s^23p^1$	14 Si {Ne}$3s^23p^2$	15 P {Ne}$3s^23p^3$	16 S {Ne}$3s^23p^4$	17 Cl {Ne}$3s^23p^5$	18 Ar {Ne}$3s^23p^6$
19 K {Ar}$4s^1$	20 Ca {Ar}$4s^2$						

Figure 3.20. Abbreviated periodic table showing the electron configurations of the first 20 elements.

With this notation, the electron configuration of element number 11, sodium, may be shown as {Ne}$3s^1$, which is an abbreviation for $1s^22s^22p^63s^1$. The former notation has some advantage in simplicity, while showing the outer electrons specifically. In the example just cited, it is easy to see that sodium has 1 outer-shell $3s$ electron, which it can lose to form the Na^+ ion with its stable noble gas neon electron configuration.

At the end of the third period is located the noble gas argon, atomic number 18, with the electron configuration $1s^22s^22p^63s^23p^6$. For elements beyond argon, that portion of the electron configuration identical to argon's may be represented simply as {Ar}. Therefore, the electron configuration of potassium, atomic number 19, is $1s^22s^22p^63s^23p^64s^1$, abbreviated {Ar}$4s^1$, and that of calcium, atomic number 20, is $1s^22s^22p^63s^23p^64s^2$, abbreviated {Ar}$4s^2$. Calcium completes the periodic table through element number 20, as shown in Figure 3.20.

3.17. ELECTRON CONFIGURATIONS AND THE PERIODIC TABLE

There are several learning devices to assist expression of the order in which atomic orbitals are filled (electron configuration for each element). However, it is of little significance to express electron configurations without an understanding of their meaning. By far the most meaningful way to understand this important aspect of chemistry is within the context of the periodic table, as outlined in Figure 3.21. To avoid clutter, only atomic numbers of key elements are shown in this table. The double-pointed arrows drawn horizontally across the periods are labeled with the kind of orbital being filled in that period.

The first step in using the periodic table to figure out electron configurations is to note that the periods are numbered 1 through 7 from top to bottom along the left side of the table. These numbers correspond to the principal quantum numbers (values of n) for the orbitals that become filled across the period for both s and

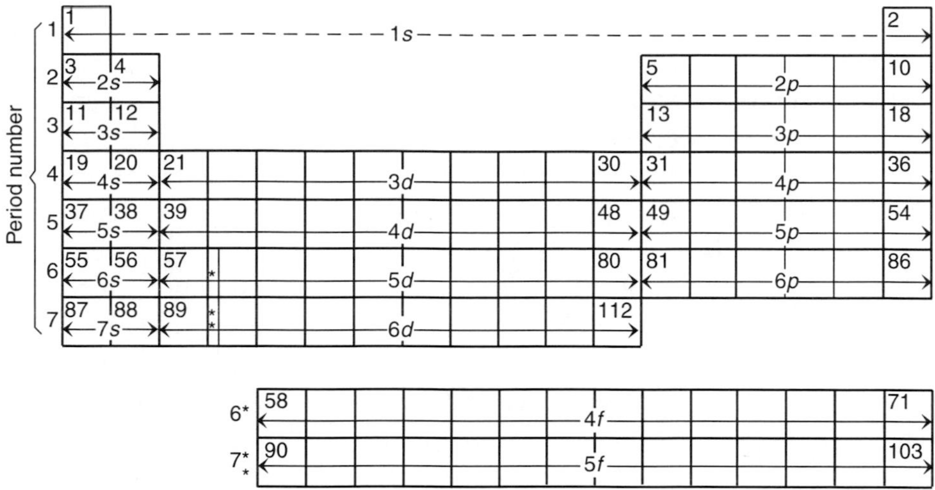

Figure 3.21. Outline of the periodic table labeled to show the filling of atomic orbitals. The type of element being filled with increasing atomic number is shown by the labeled horizontal lines across the periods. The left point of each arrow, ⟵, marks an element in which the filling of a new kind of orbital starts, and the right point of each arrow, ⟶, marks an element for which the filling of a new kind of orbital is completed. The atomic numbers are given for each element at which the filling of a kind of orbital begins or is completed.

p orbitals. Therefore, in the first group of elements—those with atomic numbers 1, 3, 11, 19, 37, 55, and 87 located in the left vertical column—the last electron added is in the *ns* orbital, for example, 5*s* for Rb, atomic number 37. The last electron added to each of the second group of elements—those with atomic numbers 4, 12, 20, 38, 56, and 88—is the second electron going into the *ns* orbital. For example, in the third period, Mg, atomic number 12, has a filled 3*s* orbital containing 2 electrons. For the group of elements in which the *p* orbitals start to be filled—those in the column with atomic numbers 5, 13, 31, 49, 81—the last electron added to each atom is the first one to enter an *np* orbital. For example for element 31, Ga, which is contained in the 4th period, the outermost electron is in the 4*p* orbital. In going to the right across each period in this part of the periodic table, the outermost electrons are, successively, np^1, np^2, np^3, np^4, np^5, and np^6. Therefore, for elements with atomic numbers 13, 14, 15, 16, 17, and 18 in the 3rd period, the outermost electrons are $3p^1$, $3p^2$, $3p^3$, $3p^4$, $3p^5$, and $3p^6$. Each of the noble gases beyond helium has filled *np* orbitals with a total of 6 outermost *p* electrons in the three *np* orbitals.

Each set of five *d* orbitals becomes filled for the *transition metals* in the three horizontal periods beginning with atomic numbers 21, 39, and 57 and ending with, successively, atomic numbers 30, 48, and 80. *For each of these orbital s, the value of n is **1** less than the period number in which the orbitals become filled.* The first *d* orbitals to become filled are the lowest-lying ones possible, the 3*d* orbitals, which become filled in the fourth period. Across the 5th period, where the 5*s* and 5*p* orbitals become filled for elements 37 and 38 and for elements 49–54, respectively, the 4*d* orbitals (n is 1 less than the period number) become filled for the transition metals, atomic numbers 39–48.

Below the main body of the periodic table are the inner transition elements consisting of two rows of elements that are actually parts of the 6th and 7th periods, respectively. The first f orbitals begin to fill with the first of the *lanthanides*, element number 58 (Figure 3.21). The principal quantum numbers of their f orbitals are *2 less than* their period numbers. These are $4f$ orbitals, of which there are 7 that are filled completely with element number 71. The $4f$ orbitals are filled in the *6th* period. The $5f$ orbitals are filled with the actinide elements, atomic numbers 90–103.

With Figure 3.21 in mind, it is possible to write the expected electron configurations of any of the elements; these are given for all elements in Table 3.4. Consider the following examples:

- Atomic No. 16, S: 3rd period, 16 electrons, Ne core, outermost electrons $3p$, electron configuration {Ne}$3s^2 3p^4$
- Atomic No. 23, V: 4th period, 23 electrons, Ar core, outermost electrons $3d$, electron configuration {Ar}$4s^2 3d^3$
- Atomic No. 35, Br: 4th period, 35 electrons, Ar core, all $3d$ orbitals filled, outermost electrons $4p$, electron configuration {Ne}$4s^2 3d^{10} 4p^5$
- Atomic No. 38, Sr: 5th period, 38 electrons, Kr core, outermost electrons $5s$, electron configuration {Kr}$5s^2$
- Atomic No. 46, Pd: 5th period, 46 electrons, Kr core, outermost electrons $4d$, electron configuration {Kr}$5s^2 4d^8$
- Atomic No. 77, Ir: 6th period, 77 electrons, Xe core, all $4f$ orbitals filled in the 6th period, outermost electrons $5d$, electron configuration {Xe}$6s^2 4f^{14} 5d^7$

Considering the above, it is possible to write the expected electron configurations of any of the elements. The actual electron configurations are given in Table 3.4. In some cases, these vary slightly from those calculated according to the rules outlined above. These exceptions occur because of the relatively higher stabilities of two half-filled sets of outermost orbitals, or orbitals in which one is half-filled and one entirely filled. The following examples illustrate this point:

- Cr, atomic number 24. Rules predict {Ar}$4s^2 3d^4$. However the actual electron configuration is {Ar}$4s^1 3d^5$ because this gives the slightly more stable electron configuration with *half-filled* 4*s and* 3*d* orbitals.
- Cu, atomic number 29. Rules predict {Ar}$4s^2 3d^9$. However the actual electron configuration is {Ar}$4s^1 3d^{10}$ because this gives *half-filled* 4*s* and *filled* 3*d* orbitals.

Table 3.4. Electron Configurations of the Elements

Atomic Number	Symbol	Electron Configuration	Atomic Number	Symbol	Electron Configuration
1	H	$1s^1$	55	Cs	$\{Xe\}6s^1$
2	He	$1s^2$	56	Ba	$\{Xe\}6s^2$
3	Li	$\{He\}2s^1$	57	La	$\{Xe\}6s^25d^1$
4	Be	$\{He\}2s^2$	58	Ce	$\{Xe\}6s^24f^15d^1$
5	Be	$\{He\}2s^22p^1$	59	Pr	$\{Xe\}6s^24f^3$
6	C	$\{He\}2s^22p^2$	60	Nd	$\{Xe\}6s^24f^4$
7	N	$\{He\}2s^22p^3$	61	Pm	$\{Xe\}6s^24f^5$
8	O	$\{He\}2s^22p^4$	62	Sm	$\{Xe\}6s^24f^6$
9	F	$\{He\}2s^22p^5$	63	Eu	$\{Xe\}6s^24f^7$
10	Ne	$\{He\}2s^22p^6$	64	Gd	$\{Xe\}6s^24f^75d^1$
11	Na	$\{Ne\}3s^1$	65	Tb	$\{Xe\}6s^24f^9$
12	Mg	$\{Ne\}3s^2$	66	Dy	$\{Xe\}6s^24f^{10}$
13	Al	$\{Ne\}3s^23p^1$	67	Ho	$\{Xe\}6s^24f^{11}$
14	Si	$\{Ne\}3s^23p^2$	68	Er	$\{Xe\}6s^24f^{12}$
15	P	$\{Ne\}3s^23p^3$	69	Tm	$\{Xe\}6s^24f^{13}$
16	Si	$\{Ne\}3s^23p^4$	70	Yb	$\{Xe\}6s^24f^{14}$
17	Cl	$\{Ne\}3s^23p^5$	71	Lu	$\{Xe\}6s^24f^{14}5d^1$
18	Ar	$\{Ne\}3s^23p^6$	72	Hf	$\{Xe\}6s^24f^{14}5d^2$
19	K	$\{Ar\}4s^1$	73	Ta	$\{Xe\}6s^24f^{14}5d^3$
20	Ca	$\{Ar\}4s^2$	74	W	$\{Xe\}6s^24f^{14}5d^4$
21	Sc	$\{Ar\}4s^23d^1$	75	Re	$\{Xe\}6s^24f^{14}5d^5$
22	Ti	$\{Ar\}4s^23d^2$	76	Os	$\{Xe\}6s^24f^{14}5d^6$
23	V	$\{Ar\}4s^23d^3$	77	Ir	$\{Xe\}6s^24f^{14}5d^7$
24	Cr	$\{Ar\}4s^13d^5$	78	Pt	$\{Xe\}6s^24f^{14}5d^9$
25	Mn	$\{Ar\}4s^23d^5$	79	Au	$\{Xe\}6s^14f^{14}5d^{10}$
26	Fe	$\{Ar\}4s^23d^6$	80	Hg	$\{Xe\}6s^24f^{14}5d^{10}$
27	Co	$\{Ar\}4s^23d^7$	81	Tl	$\{Xe\}6s^24f^{14}5d^{10}6p^1$
28	Ni	$\{Ar\}4s^23d^8$	82	Pb	$\{Xe\}6s^24f^{14}5d^{10}6p^2$
29	Cu	$\{Ar\}4s^13d^{10}$	83	Bi	$\{Xe\}6s^24f^{14}5d^{10}6p^3$
30	Zn	$\{Ar\}4s^23d^{10}$	84	Po	$\{Xe\}6s^24f^{14}5d^{10}6p^4$
31	Ga	$\{Ar\}4s^23d^{10}4p^1$	85	At	$\{Xe\}6s^24f^{14}5d^{10}6p^5$

Continued

Table 3.4. (continued)

Atomic Number	Symbol	Electron Configuration	Atomic Number	Symbol	Electron Configuration
32	Ge	{Ar}$4s^23d^{10}4p^2$	86	Rn	{Xe}$6s^24f^{14}5d^{10}6p^6$
33	As	{Ar}$4s^23d^{10}4p^3$	87	Fr	{Rn}$7s^1$
34	Se	{Ar}$4s^23d^{10}4p^4$	88	Ra	{Rn}$7s^2$
35	Br	{Ar}$4s^23d^{10}4p^5$	89	Ac	{Rn}$7s^16d^1$
36	Kr	{Ar}$4s^23d^{10}4p^6$	90	Th	{Rn}$7s^16d^2$
37	Rb	{Kr}$5s^1$	91	Pa	{Rn}$7s^25p^26d^1$
38	Sr	{Kr}$5s^2$	92	U	{Rn}$7s^25p^36d^1$
39	Y	{Kr}$5s^24d^1$	93	Np	{Rn}$7s^25p^46d^1$
40	Zr	{Kr}$5s^24d^2$	94	Pu	{Rn}$7s^25f^6$
41	Nb	{Kr}$5s^14d^4$	95	Am	{Rn}$7s^25f^7$
42	Mo	{Kr}$5s^14d^5$	96	Cm	{Rn}$7s^25f^76d^1$
43	Tc	{Kr}$5s^24d^5$	97	Bk	{Rn}$7s^25f^9$
44	Ru	{Kr}$5s^14d^7$	98	Cf	{Rn}$7s^25f^{10}$
45	Rh	{Kr}$5s^14d^8$	99	Es	{Rn}$7s^25f^{11}$
46	Pd	{Kr}$4d^{10}$	100	Fm	{Rn}$7s^25f^{12}$
47	Ag	{Kr}$5s^14d^{10}$	101	Md	{Rn}$7s^25f^{13}$
48	Cd	{Kr}$5s^24d^{10}$	102	No	{Rn}$7s^25f^{14}$
49	In	{Kr}$5s^24d^{10}5p^1$	103	Lr	{Rn}$7s^2f^{14}6d^1$
50	Sn	{Kr}$5s^24d^{10}5p^2$	104	Rf	{Rn}$7s^2f^{14}6d^2$
51	Sb	{Kr}$5s^24d^{10}5p^3$	105	Ha	{Rn}$7s^2f^{14}6d^3$
52	Te	{Kr}$5s^24d^{10}5p^4$	106	Unh	{Rn}$7s^2f^{14}6d^4$
53	I	{Kr}$5s^24d^{10}5p^5$	107	Uns	{Rn}$7s^2f^{14}6d^5$
54	Xe	{Kr}$5s^24d^{10}5p^6$	108	—	{Rn}$7s^2f^{14}6d^6$
			109	Une	{Rn}$7s^2f^{14}6d^7$

CHAPTER SUMMARY

The chapter summary below is presented in a programmed format to review the main points covered in this chapter. It is used most effectively by filling in the blanks, referring back to the chapter as necessary. The correct answers are given at the end of the summary.

Briefly, the basic parts of the atomic theory are [1]____________________, [2]____________________, [3]____________________

______, [4] ______________________________, and [5] ______________ ______________________. Three fundamental laws that are explained by the atomic theory are [6] ______________, [7] ____________, and [8] ____________. The atomic mass unit is used to [9] __________________ and is defined as [10] ____________________. Most of the volume of an atom is composed of [11] ____________. The three subatomic particles of concern to chemists, their charges, and mass numbers are [12] __________________. The atomic number and mass number of $^{14}_{7}N$ are [13] ____________. A systematic arrangement of elements that places those with similar chemical properties and electron configurations in the same groups is the [14] __________. Most hydrogen atoms have a nucleus consisting of [15] __________. The notation Ca: is an example of [16] ________________ and H:H is an example of [17] ________________ Helium is the first element with a [18] _______________. Lithium is the first element having both [19] __________ and [20] ________ electrons. Four general characteristics of metals are [21] ____________ ______________________________. The Lewis symbols of elements 3–10 are [22] ____________________________. The octet rule is [23] ______________ ____________________. The names of the elements in the third period of the periodic table are [24] ____________________________. The noble gases in the first 20 elements are [25] ____________. The unique behavior of charged particles that are as small and move as rapidly as electrons is explained by [26] __________. Electromagnetic radiation has a characteristic [27] _________ and [28] _________ related by the equation [29] ________. According to the quantum theory, radiant energy may be absorbed or emitted only in discrete packets called [30] _____________, the energy of which is given by the equation [31] __________. The Bohr model introduced the all-important concept that [32] ________________________. The wave mechanical model of electrons treats them as [33] ___________ around the nucleus of an atom. The wave mechanical model of electrons in atoms is treated by the [34] __________ equation expressed mathematically as [35] ________. In this equation ψ is the [36] ______________ and is a function of the electron's [37] __________________. According to the wave mechanical model, electrons occupy [38] ______________, each of which has [39] ______________________________. A single orbital can contain a maximum of [40] __________ electrons. An electron in an atom is described by four [41] ____________, which may not be [42] ______________ for any two electrons in an atom. The symbol n represents the [43] ____________ which may have values of [44] ______________. The symbol l represents the [45] _________ quantum number, which can have values of [46] _____________. The symbol m_l represents [47] ______________________, with possible values of [48] ____________. The symbol m_s is the [49] ________________, which may have values of only [50] ____________. Using standard notation for electron configuration (starting $1s^2 2s^2 2p^6$), the order of filling of orbitals and the maximum number of electrons in each is [51] ______________. The orbital diagram for the p electrons in nitrogen,

↑	↑	↑

illustrates the rule that [52] ____________________. The

entire surface of a contour representation of an orbital encloses [53]__________________________________. The contour representation of an *s* orbital is that of [54]______ whereas that of a *p* orbital is shaped like a [55]________________________. [56]_______________ is a means of stating which kinds of orbitals contain electrons and the numbers of electrons in each kind of orbital of an atom. It is expressed by the number and letter representing [57]________ and superscript numbers telling [58]________________. The electron configurations of phosphorus (P), potassium (K), and arsenic (As) are, respectively, [59]___________. The part of an atom consisting of its nucleus and the electrons in it equivalent to those of the noble gas immediately preceding the element in the periodic table is the [60]___________ of the atom. The electronic configuration of the core of the lithium atom is $1s^2$, the same as that of helium. Therefore, the lithium atom is said to have a [61]__________. Electrons that can be shared in covalent bonding or lost to form cations in ionic compounds are called [62]____________. All elements with the outer electron configuration ns^2np^6 are [63]____________________. In respect to period number in the periodic table, the principal quantum number of *s* electrons is [64]____________, that of *p* electrons is [65]_______________, that of *d* electrons is [66]______________, and that of *f* electrons is [67]____________________. The types of elements in which *d* orbitals become filled are [68]_______________. The electron configuration of Cr, atomic number 24, is {Ar}$4s^13d^5$, which appears to deviate slightly from the rules because it gives a [69]______________________________________.

Answers to Chapter Summary

1. Elements are composed of small objects called atoms.
2. Atoms of different elements do not have identical chemical properties.
3. Chemical compounds are formed by combination of atoms of different elements in definite ratios.
4. Chemical reactions involve the separation and combination of atoms.
5. During the course of ordinary chemical reactions, atoms are not created, destroyed, or changed to atoms of other elements.
6. law of conservation of mass
7. law of multiple proportions
8. law of constant composition
9. express masses of atoms
10. exactly 1/12 mass of carbon-12 isotope

11. electrons around the nucleus
12. proton (+1, 1), electron (−1, 0), neutron (0, 1)
13. 7 and 14
14. periodic table
15. 1 proton
16. an electron-dot symbol or Lewis symbol
17. an electron-dot formula or Lewis formula
18. filled electron shell
19. inner
20. outer
21. luster, malleable, conduct electricity, tend to lose electrons to form cations
22. 3 Li· 4 Be: 5 Ḃ: 6 ·Ċ: 7 ·N̈: 8 ·Ö: 9 ·F̈: 10 :N̈e:
23. the tendency of elements to acquire an 8-electron outer electron configuration in chemical compounds
24. sodium, magnesium, aluminum, silicon, phosphorus, sulfur, chlorine, and argon
25. helium, neon, argon
26. quantum theory
27. wavelength
28. frequency
29. $\nu\lambda = c$
30. quanta or photons
31. $E = h\nu$
32. only specific energy states are allowed for an electron in an atom
33. standing waves
34. Schrödinger equation
35. $H\psi = E\psi$
36. wave function

37. energy and the coordinates in space where it may be found
38. orbitals
39. characteristic energy and region around the nucleus where the electron has certain probabilities of being found
40. two
41. quantum numbers
42. identical
43. principal
44. 1, 2, 3, 4, 5, 6, 7,...
45. azimuthal
46. 0, 1, 2, 3,..., $n - 1$
47. magnetic quantum number
48. $+l, +(l-1), \ldots, 0, \ldots, -(l-1), -l$
49. spin quantum number
50. +½ or −½
51. $1s^22s^22p^63s^23p^64s^23d^{10}4p^65s^24d^{10}5p^66s^24f^{14}5d^{10}6p^67s^2$
52. electrons in a sublevel are distributed to give the maximum number of unpaired electrons
53. the volume within which an electron spends 90% of its time
54. a sphere
55. dumbbell, or two "pointed" spheres touching at the nucleus
56. Electron configuration
57. each kind of orbital
58. how many electrons are in each sublevel
59. $1s^22s^22p^63s^23p^3$, $1s^22s^22p^63s^23p^64s^1$, $1s^22s^22p^63s^23p^64s^23d^{10}4p^3$
60. core
61. helium core
62. valence electrons

63. noble gases located at the far right of the periodic table

64. the same as the period number

65. the same as the period number

66. one less than the period number

67. two less than the period number

68. transition metals

69. slightly more-stable electron configuration with half-filled 4*s* and 3*d* orbitals

QUESTIONS AND PROBLEMS

1. Match the law or observation denoted by letters below with the portion of Dalton's atomic theory that explains it denoted by numbers:

   A. Law of Conservation of Mass

   B. Law of Constant Composition

   C. Law of Multiple Proportions

   D. The reaction of C with O_2 does not produce SO_2.

   (1) Chemical compounds are formed by the combination of atoms of different elements in definite constant ratios that usually can be expressed as integers or simple fractions.

   (2) During the course of ordinary chemical reactions, atoms are not created or destroyed

   (3) During the course of ordinary chemical reactions, atoms are not changed to atoms of other elements.

   (4) Illustrated by groups of compounds such as $CHCl_3$, CH_2Cl_2, or CH_3Cl.

2. Particles of pollutant fly ash may be very small. Estimate the number of atoms in such a small particle assumed to have the shape of a cube that is 1 micrometer (μm) to the side. Assume also that an atom is shaped like a cube 100 picometers (pm) on a side.

3. Explain why it is incorrect to say that atomic mass is the mass of any atom of an element. How is atomic mass defined?

4. The ${}^{12}_{6}C$ isotope has a mass of exactly 12 u. Compare this with the sum of the masses of the subatomic particles that compose this isotope. Is it correct to say that the mass of an isotope is exactly equal to the sum of the masses of its subatomic particles? Is it close to the sum?

5. What is the distinction between the mass of a subatomic particle and its mass number?

6. Look up the isotopic composition of fluorine, atomic number 9, on the internet. Add the masses of the subatomic particles composing the flourine atom. Is the sum exactly that of the atomic mass given for F? Should it be?

7. Fill in the blanks in the following table:

Subatomic Particle	Symbol	Unit Charge	Mass Number	Mass (u)	Mass (g)
Proton	(a) _____	(b) _____	(c) _____	(d) _____	(e) _____
(f) _____	n	0	1	(g) _____	(h) _____
Electron	e	(i) _____	(j) _____	(k) _____	9.1096×10^{-28}

8. Define what is meant by *x*, *y*, and A in the notation ${}^{y}_{x}A$.

9. Describe what happens to the magnitude and direction of the forces between charged particles (electrons, protons, and nuclei) of (a) like charge and (b) unlike charge with distance and magnitude of charge.

10. What is the Lewis symbol of hydrogen and what does it show? What is the Lewis formula of H_2 and what does it show?

11. In many respects, the properties of elemental hydrogen are unique. List some of these properties and some of the major uses of H_2.

12. Give the Lewis symbol of helium and explain what it has to do with (a) electron shell, (b) filled electron shell, and (c) noble gases.

13. Where is helium found, and for what purpose is it used?

14. Using dots to show *all* of its electrons, give the Lewis symbol of Li. Explain how this symbol shows (a) inner and outer electrons and electron shells, (b) valence electrons, (c) and how the Li^+ ion is formed.

15. Discuss the chemical and physical properties of lithium that indicate that it is a metal.

16. What is a particular health concern with beryllium?

17. Based upon its electronic structure, suggest why boron behaves like a metalloid, showing properties of both metals and nonmetals.

18. Carbon has two isotopes that are of particular importance. What are they and why are they important?

19. Why might carbon be classified as a "life element"?

20. What two species other than O_2 are possible for elemental oxygen, particularly in the stratosphere?

21. What do particular kinds of fluorine compounds have to do with atmospheric ozone?

22. In reference to neon, define and explain the significance of (a) noble gas, (b) octet of outer shell electrons, (c) noble gas outer electron configuration, and (d) octet rule.

23. To which class of elements do lithium, sodium, and potassium belong? What are their elemental properties?

24. Define and explain (a) electromagnetic radiation, (b) wave character of electromagnetic radiation, and (c) quanta (photons).

25. What is the significance of the fact that very specific wavelengths of light are emitted when atoms in an excited state revert back to a lower-energy state (ground state)?

26. What are the major accomplishments and shortcomings of the Bohr theory?

27. How are electrons visualized in the wave mechanical model of the atom?

28. What is the fundamental equation for the wave mechanical or quantum mechanical model of electrons in atoms? What is the significance of ψ in this equation?

29. Why are orbitals not simply called "orbits" in the quantum mechanical model of atoms?

30. Name and define the four quantum numbers used to describe electrons in atoms.

31. What does the Pauli exclusion principle say about electrons in atoms?

32. Use the notation employed for electron configurations to denote the order in which electrons are placed in orbitals through the 4*f* orbitals.

33. What is a statement and significance of Hund's rule of maximum multiplicity?

34. Discuss the shapes and sizes of *s* orbitals with increasing principal quantum number.

35. Complete the following figure for contour representations of the three 2*p* orbitals:

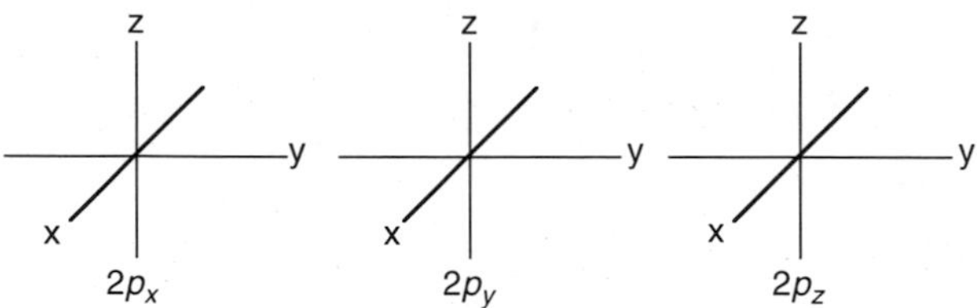

36. An atom of nitrogen has an electron configuration of $1s^22s^22p^3$. Give the electron configuration of an atom with 5 more electrons and also give the orbital diagram of such an atom.

37. Give the electron configurations of carbon, neon, aluminum, phosphorus, and calcium.

38. Give the symbols of the elements with the following electron configurations:

    (a) $1s^2$-$2s^2$ (b) $1s^2$-$2s^2$-$2p^6$-$3s^1$ (c) $1s^2$-$2s^2$-$2p^6$-$3s^2$-$3p^4$ (d) $1s^2$-$2s^2$-$2p^6$-$3s^2$-$3p^6$-$4s^1$

39. What is the core of an atom? Specifically, what is the neon core?

40. What are valence electrons? What are the electron configurations of the valence electrons in aluminum?

41. What is wrong with the following orbital diagram?

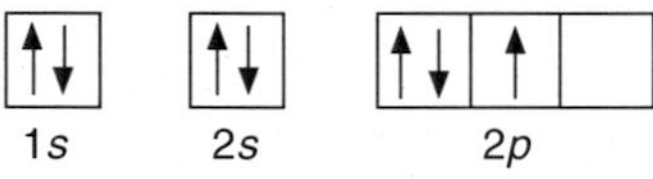

42. State the rule that gives which orbitals are filled for transition elements in relation to the period number of the elements involved. Do the same for the inner transition elements.

43. Explain why the electron configuration of Cr is {Ar}$4s^13d^5$ and why that of Cu is {Ar}$4s^13d^{10}$.

44. The isotope,

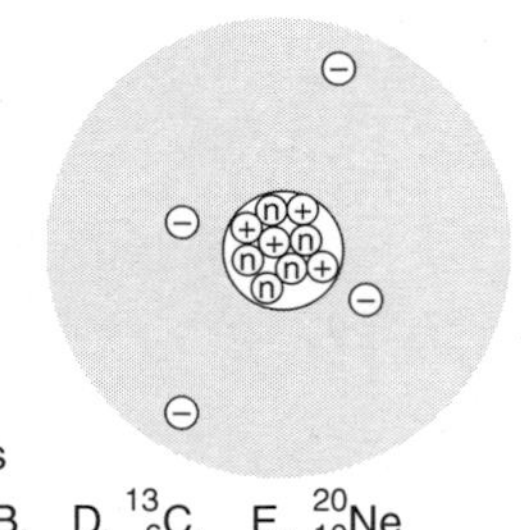

could be designated as
A. $^{7}_{3}Li$, B. $^{9}_{4}Be$, C. $^{11}_{5}B$, D. $^{13}_{6}C$, E. $^{20}_{10}Ne$

45. Of the following, the statement that is part of Dalton's atomic theory is

A. Atoms of different elements are likely to have identical chemical properties.

B. Two or more different elements combine in definite, constant ratios to form chemical compounds.

C. Atoms of one element may be converted to atoms of other elements during chemical reactions.

D. The total mass of the products of chemical reactions may be different from the total mass of the reactants.

E. Each element is composed of diatomic molecules, such as O_2.

46. Four of the following elements share a common characteristic insofar as they relate to the periodic table. The one that does not share this characteristic is

A. Neon B. Carbon C. Calcium D. Oxygen E. Nitrogen

47. Of the following, the statement that is **untrue** pertaining to atoms is

A. About half of the volume is contained in the nucleus.

B. Essentially all of the mass is in the nucleus.

C. Essentially all the volume is composed of a cloud of electrons.

D. An atom of ^1_1H has no neutrons.

E. A neutral atom has equal numbers of protons and electrons.

List of the Elements[a]

Atomic Number	Name	Symbol	Atomic Mass	Atomic Number	Name	Symbol	Atomic Mass
1	Hydrogen	H	1.00794	56	Barium	Ba	137.327
2	Helium	He	4.0026	57	Lanthanum	La	138.9055
3	Lithium	Li	6.941	58	Cerium	Ce	140.115
4	Beryllium	Be	9.01218	59	Praseodymium	Pr	140.9077
5	Boron	B	10.811	60	Neodymium	Nd	144.24
6	Carbon	C	12.011	61	Promethium	Pm	145
7	Nitrogen	N	14.0067	62	Samarium	Sm	150.36
8	Oxygen	O	15.9994	63	Europium	Eu	151.965

Continued

List of the Elements[a] **(continued)**

Atomic Number	Name	Symbol	Atomic Mass	Atomic Number	Name	Symbol	Atomic Mass
9	Fluorine	F	18.9984	64	Gadolinium	Gd	157.25
10	Neon	Ne	20.1797	65	Terbium	Tb	158.925
11	Sodium	Na	22.9898	66	Dysprosium	Dy	162.50
12	Magnesium	Mg	24.305	67	Holmium	Ho	164.9303
13	Aluminum	Al	26.98154	68	Erbium	Er	167.26
14	Silicon	Si	28.0855	69	Thulium	Tm	168.9342
15	Phosphorus	P	30.9738	70	Ytterbium	Yb	173.04
16	Sulfur	S	32.066	71	Lutetium	Lu	174.967
17	Chlorine	Cl	35.4527	72	Hafnium	Hf	178.49
18	Argon	Ar	39.948	73	Tantalum	Ta	180.9497
19	Potassium	K	39.0983	74	Tungsten	W	183.85
20	Calcium	Ca	40.078	75	Rhenium	Re	186.207
21	Scandium	Sc	44.9559	76	Osmium	Os	190.2
22	Titanium	Ti	47.88	77	Iridium	Ir	192.22
23	Vanadium	V	50.9415	78	Platinum	Pt	195.08
24	Chromium	Cr	51.9961	79	Gold	Au	196.9665
25	Manganese	Mn	54.938	80	Mercury	Hg	200.59
26	Iron	Fe	55.847	81	Thallium	Tl	204.383
27	Cobalt	Co	58.9332	82	Lead	Pb	207.2
28	Nickel	Ni	58.6934	83	Bismuth	Bi	208.98
29	Copper	Cu	63.546	84	Polonium	Po	209
30	Zinc	Zn	65.39	85	Astatine	At	210
31	Gallium	Ga	69.723	86	Radon	Rn	222
32	Germanium	Ge	72.61	87	Francium	Fr	223
33	Arsenic	As	74.9216	88	Radium	Ra	226.0254
34	Selenium	Se	78.96	89	Actinium	Ac	227.0278
35	Bromine	Br	79.904	90	Thorium	Th	232.038
36	Krypton	Kr	83.8	91	Protactinium	Pa	231.0359
37	Rubidium	Rb	85.4678	92	Uranium	U	238.0289
38	Strontium	Sr	87.62	93	Neptunium	Np	237.048

Continued

List of the Elements[a] (continued)

Atomic Number	Name	Symbol	Atomic Mass	Atomic Number	Name	Symbol	Atomic Mass
39	Yttrium	Y	88.9056	94	Plutonium	Pu	244
40	Zirconium	Zr	91.224	95	Americium	Am	243
41	Niobium	Nb	92.9064	96	Curium	Cm	247
42	Molybdenum	Mo	95.94	97	Berkelium	Bk	247
43	Technetium	Tc	98	98	Californium	Cf	251
44	Ruthenium	Ru	101.07	99	Einsteinium	Es	252
45	Rhodium	Rh	102.9055	100	Fermium	Fm	257.1
46	Palladium	Pd	106.42	101	Mendelevium	Md	258.1
47	Silver	Ag	107.8682	102	Nobelium	No	255
48	Cadmium	Cd	112.411	103	Lawrencium	Lr	260
49	Indium	In	114.82	104	Rutherfordium	Rf	261.11
50	Tin	Sn	118.710	105	Dubnium	Db	262.11
51	Antimony	Sb	121.757	106	Seaborgium	Sg	263.12
52	Tellurium	Te	127.60	107	Bohrium	Bh	262.12
53	Iodine	I	126.9045	108	Hassium	Hs	265
54	Xenon	Xe	131.29	109	Meitnerium	Mt	266
55	Cesium	Cs	132.9054				

[a] Elements above atomic number 92 have been made artificially.

4. CHEMICAL BONDS, MOLECULES, AND COMPOUNDS

4.1. CHEMICAL BONDS AND COMPOUND FORMATION

Chemical compounds and chemical bonds have already been mentioned several times in this book. It is not possible to even begin the study of chemistry without some knowledge of compounds and bonding, because they are the essence of chemical science. The first three chapters have provided an overview of chemistry, a discussion of the properties of matter, and an explanation of atoms and elements. With this background, it is now possible to discuss chemical bonds, molecules, and compounds in more detail.

By now, the reader should have an appreciation of the diversity of types of chemical compounds, all formed from just a few elements. This variety is possible because atoms of the elements are joined together through chemical bonds. The chemical behavior of the elements can only be understood with knowledge of their abilities to form chemical bonds. Shapes and structures of molecules are determined by the orientation of chemical bonds attached to the atoms composing the elements. The strengths of chemical bonds in a compound largely determine its stability and resistance to chemical change. For example, photons (as represented by the symbol $h\nu$; see Section 3.11) of sunlight in the ultraviolet and shorter-wavelength regions of the visible light spectrum can be absorbed by a molecule of nitrogen dioxide, NO_2, and break a chemical bond between an N and O atom,

$$NO_2 \xrightarrow{h\nu} NO + O \qquad (4.1.1)$$

yielding very reactive atoms of oxygen. These atoms attack hydrocarbons (compounds composed of C and H) introduced as pollutants into the atmosphere from incompletely burned automobile exhaust gases and other sources. This sets off a series of reactions—chain reactions—that lead to the production of atmospheric ozone, formaldehyde, additional NO_2, and other products known as photochemical smog (Figure 4.1).

Figure 4.1. Breakage of a chemical bond between N and O in NO_2 starts the process involving a series of many reactions leading to the formation of unhealthy pollutant photochemical smog.

Chemical Bonds and Valence Electrons

Chemical bonds are normally formed by the transfer or sharing of *valence electrons*, that is, those that are in the *outermost shell* of the atom (see Section 3.8). To visualize how this is done, it is helpful to consider the first 20 elements in the periodic table and their valence electrons represented by Lewis symbols, as shown in Figure 4.2.

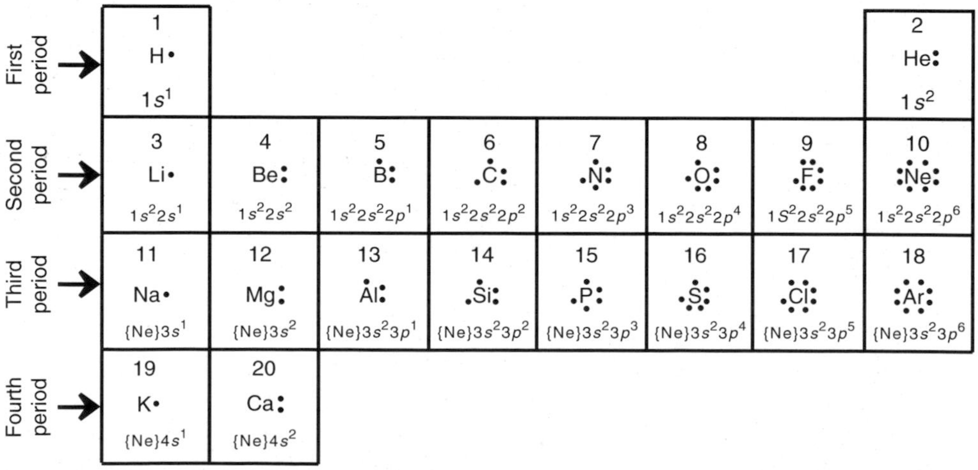

First period →	1 H• $1s^1$							2 He: $1s^2$
Second period →	3 Li• $1s^22s^1$	4 Be: $1s^22s^2$	5 ·B: $1s^22s^22p^1$	6 ·C: $1s^22s^22p^2$	7 ·N: $1s^22s^22p^3$	8 ·O: $1s^22s^22p^4$	9 ·F: $1s^22s^22p^5$	10 :Ne: $1s^22s^22p^6$
Third period →	11 Na• $\{Ne\}3s^1$	12 Mg: $\{Ne\}3s^2$	13 ·Al: $\{Ne\}3s^23p^1$	14 ·Si: $\{Ne\}3s^23p^2$	15 ·P: $\{Ne\}3s^23p^3$	16 ·S: $\{Ne\}3s^23p^4$	17 ·Cl: $\{Ne\}3s^23p^5$	18 :Ar: $\{Ne\}3s^23p^6$
Fourth period →	19 K• $\{Ne\}4s^1$	20 Ca: $\{Ne\}4s^2$						

Figure 4.2. Abbreviated version of the periodic table showing outer shell (valence) electrons as dots. Electron configurations are also given.

4.2. CHEMICAL BONDING AND THE OCTET RULE

The octet rule for chemical bonding was mentioned briefly in Chapter 3. At this point, it will be useful to examine this rule in more detail. Although there are many exceptions to it, the octet rule remains a valuable concept for an introduction to chemical bonding.

The Octet Rule for Some Diatomic Gases

Recall that some elemental gases—hydrogen, nitrogen, oxygen, and fluorine—do not consist of individual atoms, but of *diatomic molecules* of H_2, N_2, O_2, and F_2, respectively. This can be explained by the tendencies of these elements to attain a noble gas outer electron configuration. The bonds in molecules of F_2 are shown by the following:

$$:\ddot{\underset{..}{F}}\cdot + \cdot\ddot{\underset{..}{F}}: \rightarrow :\ddot{\underset{..}{F}}:\ddot{\underset{..}{F}}: \qquad (4.2.1)$$

Octets

It can be seen that the F_2 molecule is held together by a *single covalent bond* consisting of 2 shared electrons. As shown by the circles around them, each atom in the molecules has 8 outer electrons, some of them shared. These 8 electrons constitute an *octet* of outer electrons. Such an octet is possessed by the neon atom, which is the noble gas nearest to F in the periodic table.

The Octet Rule for Chemical Compounds

The octet rule illustrated for molecules of diatomic elemental fluorine above applies to molecules of chemical compounds. Recall from Section 1.6 that a *chemical compound* is formed from atoms or ions of two or more elements joined by chemical bonds. In many cases, compound formation enables elements to attain a noble gas electron configuration, usually an octet of outer shell electrons.

The most straightforward way for atoms to gain a stable octet is through loss and gain of electrons to form ions. This may be illustrated by the ionic compound sodium chloride as shown in Figure 4.3. By considering the Lewis symbols of Na

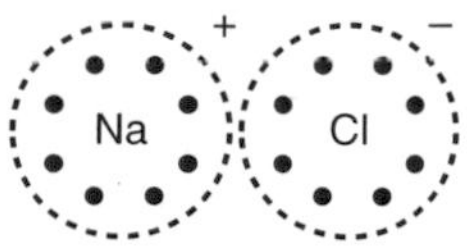

Figure 4.3. In the ionic compound NaCl, both the Na^+ ion and the Cl^- ion have octets of outer shell electrons. The Na^+ ion is formed by the loss of 1 electron from an Na atom (see Figure 4.2) and the Cl^- ion is formed by a Cl atom gaining an electron.

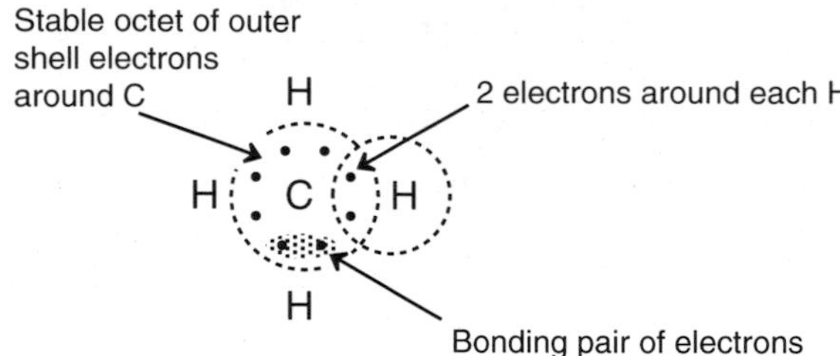

Figure 4.4. Stable outer electron shells from covalent bonding in compounds.

and Cl atoms in Figure 4.2, it is easy to see that the octet of the Na^+ ion is formed by the loss of an electron from an Na atom and that of the Cl^- ion is produced by a Cl atom gaining 1 electron.

Normally, atoms of elements in the middle of a period of the periodic table, such as carbon or nitrogen atoms, share electrons to form *covalent bonds*. In methane, CH_4, each of 4 H atoms shares an electron with 1 C atom, giving each C atom an 8-electron outer shell (octet) like neon (Figure 4.4). Each H atom has 2 electrons, both shared with C, which provides a shell of 2 electrons like that in the noble gas helium.

4.3. IONIC BONDING

An **ion** consists of an atom or group of atoms having an unequal number of electrons and protons and, therefore, a net electrical charge. A **cation** has a positive charge and an **anion** has a negative charge. The simplest kind of ion to visualize is one formed by the loss or gain by an atom of one or more electrons. As examples, loss of 2 electrons from a calcium atom gives a Ca^{2+} cation, and the gain of an electron by a Cl atom gives a Cl^- anion. An **ionic compound** is one that contains cations and anions. Such a compound is held together by **ionic bonds** that result from the mutual attraction of positively charged cations and negatively charged anions.

Electron Configurations of Ions from a Single Atom

Electron configurations of atoms were discussed in some detail in Chapter 3. Emphasis has been placed on the stability of a stable *octet* of outer shell electrons, which is characteristic of a *noble gas*. Such an octet is in the $\boldsymbol{ns^2np^6}$ electron configuration. The first element to have this configuration is neon, electron configuration $1s^22s^22p^6$. (Recall that helium is also a noble gas, but with only 2 electrons, its electron configuration is simply $1s^2$.) The next noble gas beyond neon is argon, which has the electron configuration $1s^22s^22p^63s^23p^6$, commonly abbreviated {Ne}$3s^23p^6$.

The electron configurations of ions consisting of only one atom can be shown in the same way as illustrated for atoms above. For the lighter elements, the electron

Table 4.1. Ions with Neon and Argon Electron Configurations

Element[a]	Electron Configuration	Electrons Gained/Lost	Ion	Ion Electron Configuration	
N (7)	$1s^22s^22p^3$	Gain 3 e^-	N^{3-}	$1s^22s^22p^3$	
O (8)	$1s^22s^22p^4$	Gain 2 e^-	O^{2-}	$1s^22s^22p^6$	
F (9)	$1s^22s^22p^5$	Gain 1 e^-	F^-	$1s^22s^22p^6$	Neon
Na (11)	$1s^22s^22p^63s^1$	Lose 1 e^-	Na^+	$1s^22s^22p^6$	
Mg (12)	$1s^22s^22p^63s^2$	Lose 2 e^-	Mg^{2+}	$1s^22s^22p^6$	
S (16)	$1s^22s^22p^63s^23p^4$	Gain 2 e^-	S^{2-}	$1s^22s^22p^63s^23p^6$	
Cl (17)	$1s^22s^22p^63s^23p^5$	Gain 1 e^-	Cl^-	$1s^22s^22p^63s^23p^6$	
K (19)	$1s^22s^22p^63s^23p^64s^1$	Lose 1 e^-	K^+	$1s^22s^22p^63s^23p^6$	Argon
Ca (20)	$1s^22s^22p^63s^23p^64s^2$	Lose 2 e^-	Ca^{2+}	$1s^22s^22p^63s^23p^6$	

[a] Atomic number in parentheses.

configurations are those of the nearest noble gas. Therefore, in forming ions, an atom of an element that is just before a particular noble gas in the periodic table *gains* enough electrons to get the configuration of that noble gas, whereas an atom of an element just beyond a noble gas *loses* enough electrons to attain the electron configuration of the nearest noble gas. This is best seen in reference to the periodic table showing electron configurations as seen in Figure 4.2.

As an example of electron configurations of ions, consider an anion, F^-, and a cation, Mg^{2+}, both near neon in the periodic table. From Figure 4.2, it is seen that the electron configuration of the neutral fluorine atom is $1s^22s^22p^5$. To get the stable noble gas electron configuration of neon, the fluorine atom may gain 1 electron to become an F^- ion, electron configuration $1s^22s^22p^6$. The electron configuration of magnesium, atomic number 12, is $1s^22s^22p^63s^2$. A magnesium atom can attain the stable electron configuration of neon by losing its 2 outer shell 3*s* electrons to become a Mg^{2+} ion, electron configuration $1s^22s^22p^6$. The electron configurations of some other ions near neon and argon in the periodic table are given in Table 4.1.

Sodium Chloride as an Ionic Compound

The formation of sodium chloride, NaCl, is often cited as an example of the production of an ionic compound and its constituent ions, and the stable octets of Na^+ and Cl^- ions were shown in Figure 4.3. Here the nature of sodium chloride as an ionic compound is considered in more detail. The electron configurations of both the Na^+ ion and the Cl^- ion are given in Table 4.1. The reaction between a neutral elemental sodium atom and a neutral elemental chlorine atom to form NaCl consisting of Na^+ and Cl^- ions is as follows:

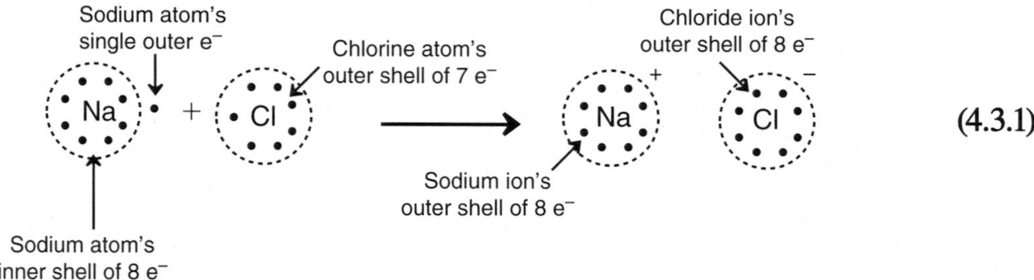

(4.3.1)

A compound such as NaCl that is composed of ions is an ionic compound. The ionic bonds in such a compound exist because of the electrostatic attraction between ions of opposite charge. An ionic compound possesses a crystal structure such that a particular ion is located as close as possible to attracting ions of opposite charge and as far as possible from repelling ions with the same charge. Figure 4.5 shows such a structure for NaCl. Visualized in three dimensions, this figure shows how the six nearest neighbors of each ion are ions of opposite charge. Consider the Cl^- ion with the arrow pointing to it in the center of the structure. Every line intersecting the sphere that represents the Cl^- ion leads to a nearest-neighbor ion, each of which is an Na^+ ion. The closest Cl^- ions are actually farther away than any of the six nearest-neighbor Na^+ ions. The crystal structure is such that each Na^+ ion is similarly surrounded by six nearest-neighbor Cl^- ions. Therefore, every ion in the crystal is closest to ions of opposite charge, resulting in forces of attraction that account for the stability of ionic bonds. Depending upon the relative numbers of cations and

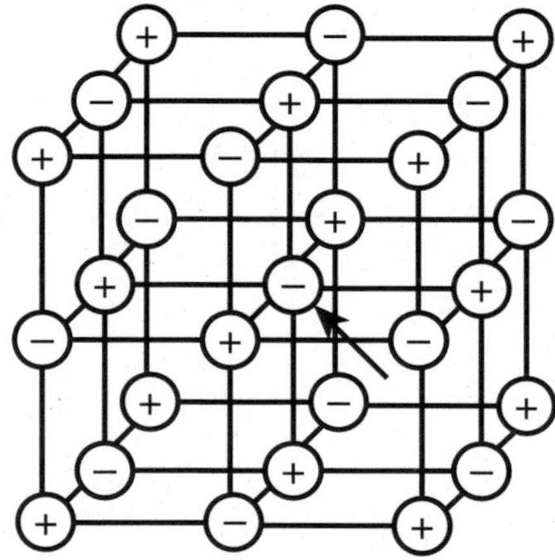

Figure 4.5. Representation of the structure of NaCl. This structure can be visualized in three dimensions as "balls" with interconnecting "sticks." The balls are Na^+ and Cl^- ions with the charges designated.

anions in a compound and upon their sizes, there are various crystalline structures that maximize the closeness of ions of opposite charge and, therefore, the strengths of ionic bonds. Although the ball-and-stick model in Figure 4.5 shows the relative positions of the ions, the ions are actually considered to be of a size such that they fill the spaces in the structure and touch each other (the Cl^- ion is larger than the Na^+ ion, as shown in Figure 4.7).

Energetics in Ionic Bonding

The stability of the ionic bond is the result of the interaction of several factors involving the energetics of the attraction of atoms for electrons and of oppositely charged ions for each other. To understand some of the factors involved, consider the reaction of isolated gas-phase atoms of Na and Cl to form solid NaCl. This is a somewhat simplified view in that it does not take account of the fact that sodium is a solid under normal conditions and that elemental chlorine exists as diatomic molecules of Cl_2; when solid Na and gaseous Cl_2 react, energy is involved in removing Na atoms from the elemental solid, as well as in breaking the covalent bond between Cl atoms in Cl_2. The reaction to be considered is the following:

$$\mathrm{Na}(g) + \cdot\ddot{\underset{\cdot\cdot}{\mathrm{Cl}}}\!:(g) \rightarrow \mathrm{Na}^+(g) \quad + \quad :\ddot{\underset{\cdot\cdot}{\mathrm{Cl}}}{}^-\!:(g) \quad \rightarrow \quad \mathrm{Na}^+\,:\ddot{\underset{\cdot\cdot}{\mathrm{Cl}}}{}^-\!:(s)$$

When gas-phase sodium and chlorine atoms react,	*ionization energy* is involved in the removal of an electron from a sodium atom to form Na^+,	*electron affinity* is involved in the addition of an electron to a chlorine atom to produce Cl^-,	and *lattice energy* is released by oppositely charged ions coming together in the sodium chloride crystal.

(4.3.2)

Three energy changes are listed for the process outlined in the above equation. These are the following:

- Even though the sodium ion has been represented as being "willing" to lose its valence electron to attain a noble gas electron configuration, some energy is required. This is called the **ionization energy**. In the case of sodium, the ionization energy for the removal of the single outer shell electron is +490 kJ/mol. The positive value indicates that energy must be put into the system to remove the electron from Na. For each mole of Na atoms, the amount of energy is 490 kJ.
- When an electron is added to a Cl atom, the energy is expressed as **electron affinity**, with a value of −349 kJ/mol. The negative sign indicates that energy is *released*; for the addition of 1 electron to each atom in a mole of Cl atoms, 349 kJ of energy is evolved.

- A very large amount of energy is released when a mole of solid NaCl is formed by a mole of Na^+ and a mole of Cl^- ions coming together to form a **crystal lattice** of ions arranged in a crystalline structure. The energy change for this process is −785 kJ/mole.

The net energy change for Reaction 4.3.2 is given by the following:

$$\begin{aligned}\text{Energy change} &= \text{Ionization energy for Na}(g) + \text{Electron affinity of Cl}(g) \\ &\quad + \text{Energy released by Na}^+ \text{ and Cl}^- \text{ coming together} \\ &\quad\ \ \text{to form NaCl}(s) \qquad (4.3.3)\\ &= 490\text{ kJ} - 349\text{ kJ} - 785\text{ kJ} \\ &= -644\text{ kJ}\end{aligned}$$

This calculation shows that when 1 mole of solid NaCl is formed from 1 mole each of gas-phase Na and Cl atoms, the energy change is −644 kJ; the negative sign shows that energy is released, so that the process is energetically favored. Under normal conditions of temperature and pressure, however, sodium exists as the solid metal, Na(*s*), and chlorine is the diatomic gas, $Cl_2(s)$. The reaction of 1 mole of solid sodium with one mole of chlorine atoms contained in gaseous Cl_2 to yield 1 mole of NaCl is represented by

$$\text{Na}(s) + \tfrac{1}{2}\,\text{Cl}_2(g) \rightarrow \text{NaCl}(g) \qquad (4.3.4)$$

The energy change for this reaction is −411 kJ/mol, representing a significantly lower release of energy from that calculated in Equation 4.3.3. This difference is due to the energy used to vaporize solid Na and to break the Cl–Cl bonds in diatomic Cl_2. The total process for the formation of solid NaCl from solid Na and gaseous Cl_2 is outlined in Figure 4.6, a form of the Born–Haber cycle.

Energy of Ion Attraction

In the case of NaCl, it is seen that the largest component of the energy change that occurs in the formation of the ionic solid is that released when the ions come together to form the solid. This may be understood in light of the potential energy of interaction, *E*, between two charged bodies, given by the equation

$$E = k\frac{Q_1Q_2}{d} \qquad (4.3.5)$$

where Q_1 and Q_2 are the charges on the two bodies in coulombs (abbreviated C), *d* is the distance between their centers in meters, and *k* is a constant with a value of 8.99×10^9 J⁻m/C². If one of the bodies has a positive and the other a negative charge, as is the case with oppositely charged ions, the sign of *E* is negative, consistent with

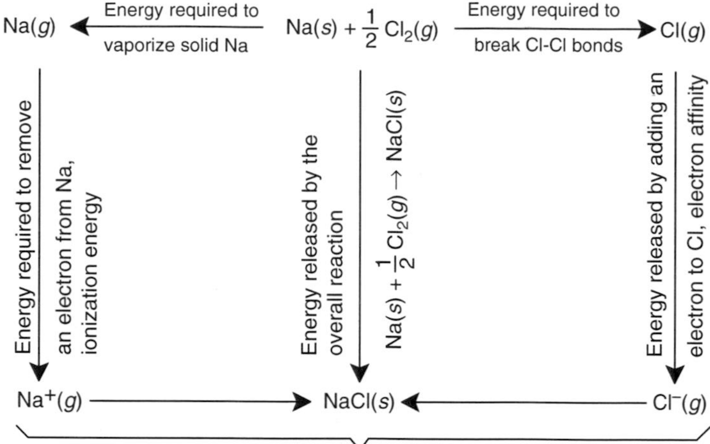

Figure 4.6. Energy changes involved in the formation of an ionic compound from its constituent elements. The overall process is shown by the vertical arrow in the middle and the constituent parts by the arrows around the periphery.

the fact that energy is released when oppositely charged bodies come together. The magnitude of E is greater for larger values of Q_1 and Q_2 (higher ionic charges, such as the 2+ for Ca^{2+} ion compared with the 1+ for the Na^+ ion) and greater for smaller values of d (smaller radii of ions allowing them to approach more closely).

Lattice Energy

Having considered the energy of interaction between charged particles, next consider the large amount of energy involved with the packing of ions into a crystalline lattice. The lattice energy for a crystalline ionic compound is that required to separate all of the ions in the compound and remove them a sufficient distance from each other so that there is no interaction between them. Since energy input is required to do this for ionic compounds, lattice energy has a positive value; it is conventionally expressed in kilojoules per mole of compound. For the example being considered, the lattice energy of NaCl is +785 kJ/mol. However, in considering the production of an ionic compound, as shown in Reaction 4.3.3, the opposite process was considered, which is the assembling of a mole of Na^+ ions and a mole of Cl^- ions to produce a mole of solid NaCl. This releases the lattice energy, so that its contribution to the energy change involved was given a minus sign, that is, −785 kJ/mol.

Ion Size

Equation 4.3.5 shows that the distance between the centers of charged bodies, d, affects their energy of interaction. For ionic compounds, this is a function of ionic size; smaller ions can get closer, and therefore their energies of interaction as measured

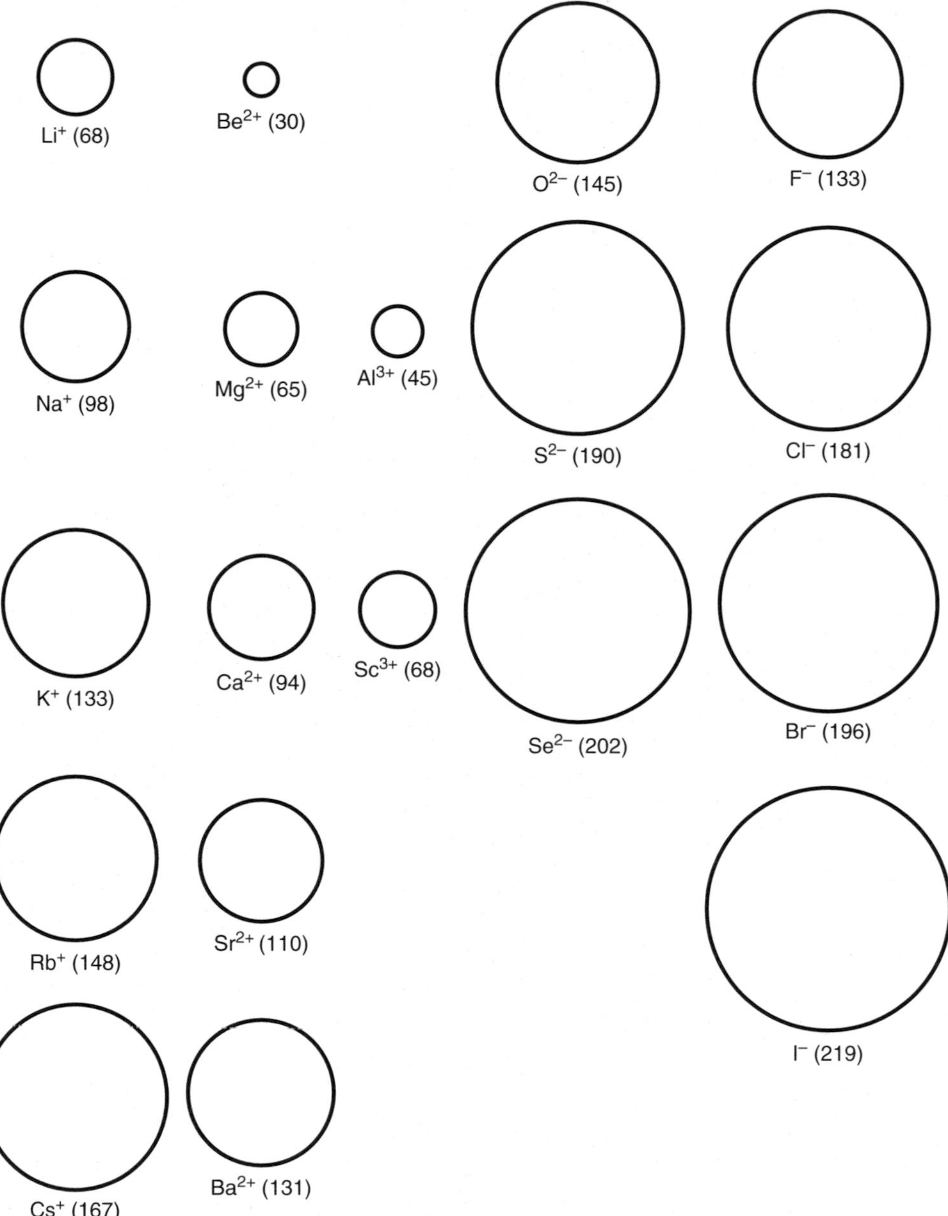

Figure 4.7. Diameters of some ions formed from single atoms, in units of picometers (1 Ångstrom = 100 pm).

by the lattice energy are greater. Because of the removal of their outer shell electrons, cations are smaller than the atoms from which they are formed; because of the addition of outer shell electrons, anions are larger than their corresponding neutral atoms. Some ion sizes are shown in Figure 4.7. This figure shows some trends in ion size across periods of the periodic table and within vertical groups in the table. In going from left to right across a period, it is seen that the cations decrease in size with increasing ionic charge. The size of cations within a period decreases significantly in the order of 1+, 2+, and 3+ ions. There is a large increase in ion size as anions are formed. The doubly charged anions within a period are larger than the singly charged ones. For example, the O^{2-} ion is larger than the F^-, ion as shown in Figure 4.7.

Formation of Some Example Ionic Compounds

Next consider several other ionic compounds that contain some of the ions shown in Figure 4.7. To better illustrate the transfer of electrons, these are shown for the formation of ions from single atoms, even in cases where diatomic gases are involved. Calcium and chlorine react,

$$\mathrm{Ca{:}} + 2\,\mathrm{\cdot\ddot{\underset{..}{Cl}}{:}} \rightarrow \mathrm{{:}\ddot{\underset{..}{Cl}}{:}^-\ Ca^{2+}\ {:}\ddot{\underset{..}{Cl}}{:}^-} \quad (4.3.6)$$

to form calcium chloride, $CaCl_2$. This compound is a byproduct of some industrial processes and is used to salt "icy" streets, causing the ice to melt and, unfortunately, automobiles to rust. The reaction of magnesium and oxygen,

$$\mathrm{Mg{:}} + \mathrm{\cdot\dot{\underset{..}{O}}{:}} \rightarrow \mathrm{Mg^{2+}\,{:}\ddot{\underset{..}{O}}{:}^{2-}} \quad (4.3.7)$$

yields magnesium oxide, MgO. This compound is used in heat-resistant refractory materials (such as those used to line industrial furnaces), insulation, cement, and paper manufacture. Sodium and sulfur react,

$$\mathrm{Na\cdot} + \mathrm{Na\cdot} + \mathrm{\cdot\dot{\underset{..}{S}}{:}} \rightarrow \mathrm{Na^+\ {:}\ddot{\underset{..}{S}}{:}^{2-}\ Na^+} \quad (4.3.8)$$

to yield sodium sulfide, Na_2S. Among its other uses, this ionic compound is an intermediate in one of the processes used in wood pulping for paper manufacture. As a final example, consider the following reaction of aluminum and oxygen:

$$2\,\mathrm{\dot{Al}{\cdot}} + 3\,\mathrm{\cdot\dot{\underset{..}{O}}{:}} \rightarrow \mathrm{{:}\ddot{\underset{..}{O}}{:}^{2-}Al^{3+}{:}\ddot{\underset{..}{O}}{:}^{2-}Al^{3+}{:}\ddot{\underset{..}{O}}{:}^{2-}} \quad (4.3.9)$$

The product, aluminum oxide, $A1_2O_3$, is the most widely used aluminum compound. It is commonly called alumina. Among the products that it is used to make are abrasives, ceramics, antacids, and antiperspirants.

Exercise: Fill in the following table for the ionic products of the reaction of the metal and nonmetal indicated:

Metal	Nonmetal	Cation formed	Anion formed	Ionic compound
Li	Cl	(a) ________	(b) ________	(c) ________
K	S	(d) ________	(e) ________	(f) ________
Ca	F	(g) ________	(h) ________	(i) ________
Ba	O	(j) ________	(k) ________	(l) ________
Ca	N	(m) ________	(n) ________	(o) ________

Answers: (a) Li^+; (b) Cl^-; (c) LiCl; (d) K^+; (e) S^{2-}; (f) K_2S; (g) Ca^{2+}; (h) F^-; (i) CaF^{2+}; (j) Ba^{2+}; (k) O^{2-}; (l) BaO; (m) Ca^{2+}; (n) N^{3-}; (o) Ca_3N_2

4.4. FUNDAMENTALS OF COVALENT BONDING

Chemical Bonds and Energy

The energy changes involved in the formation of ions and ionic compounds were mentioned prominently in the preceding discussion of ionic compounds. In fact, bonds form largely as the result of the tendency of atoms to attain minimum energy. Several kinds of bonds are possible between atoms, and energy considerations largely determine the type of bonds that join atoms together. It was shown in the preceding section that *ionization energy, electron affinity*, and *lattice energy* are all strongly involved in forming ionic bonds. Ionic compounds form when the sum of the energy released in forming anions (electron affinity) and in the cations and anions coming together (lattice energy) is sufficient to remove electrons from a neutral atom to produce cations (ionization energy). It was seen that, for simple monatomic ions, conditions are most likely to be favorable for the formation of ionic bonds when the cations are derived from elements on the left side of the periodic table and the anions from those near the right side. In addition, the transition elements can lose 1–3 electrons to form simple monatomic cations, such as Cu^{2+}, Zn^{2+}, and Fe^{3+}.

In many cases, however, the criteria outlined above are not energetically favorable for the formation of ionic bonds. In such cases, bonding can be accomplished by the sharing of electrons in covalent bonds. Whereas oppositely charged ions attract each other, neutral atoms have a tendency to repel each other because of the forces of repulsion between their positively charged nuclei and the negatively charged clouds of electrons around the nuclei. Sharing electrons in covalent bonds between atoms enables these forces of repulsion to be overcome and results in energetically favored bonding between atoms. Covalent bonding is discussed in this and the following sections.

Covalent Bonding

A **covalent bond** is one that joins 2 atoms through the sharing of 1 or more pairs of electrons between them. Covalent bonds were discussed briefly in Chapter 1.

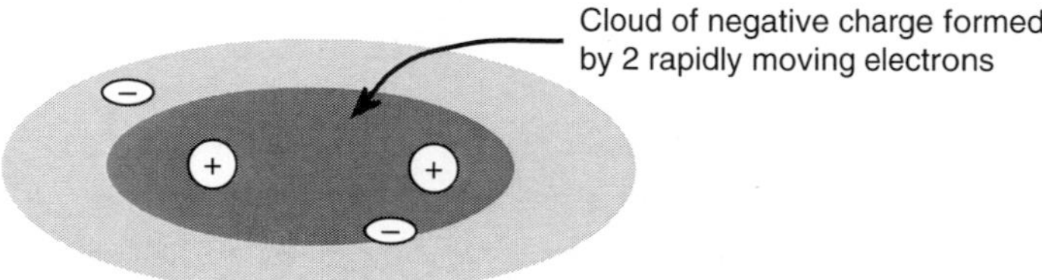

Figure 4.8. Two hydrogen nuclei, +, covalently bonded together by sharing 2 electrons, −. The negatively charged cloud of electrons is concentrated between the 2 nuclei, reducing their natural tendency to repel each other.

The simplest such bond to visualize is that which forms between 2 hydrogen atoms, to yield a diatomic molecule of hydrogen, H_2:

$$H\cdot + \cdot H \rightarrow (H:H) \quad (4.4.1)$$

Figures 4.8 and 4.9 illustrate some important characteristics of the H–H bond. As shown in Figure 4.8, there is an electron cloud between the 2 positively charged H atom nuclei. This is the covalent bond consisting of a pair of shared electrons that holds the 2 atoms together. Figure 4.9 is an energy diagram for the H_2 molecule. The 2 hydrogen nuclei are shown at the distance that results in minimum energy, the bond length. Forcing the nuclei close together results in rapidly increasing energy because of the repulsive forces between the nuclei. The energy increases more gradually with distances farther apart than the bond length, approaching zero energy as complete separation is approached. The bond energy of 435 kJ/mol means that a total energy of 435 kJ is required to break all the bonds in a mole of H_2 molecules and totally separate the resulting H atoms.

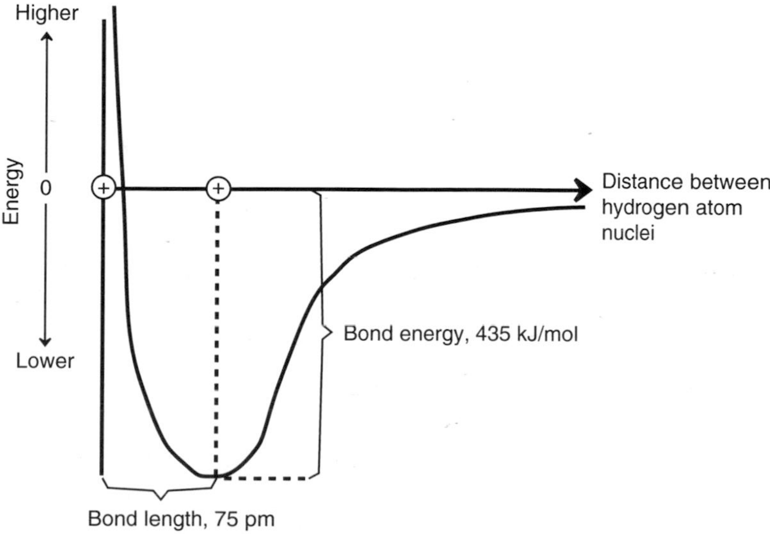

Figure 4.9. Relationship between energy and distance between hydrogen nuclei (energy diagram of H_2). The nuclei are shown at the most energetically favored distance apart, that is, the bond length of the H_2 atom.

4.5. COVALENT BONDS IN COMPOUNDS

Consider next some example covalent bonds between atoms of some of the lighter elements. These are best understood in reference to Figure 4.2, the abbreviated version of the periodic table showing the Lewis symbols (outer shell valence electrons) of the first 20 elements. As is the case with ions, atoms that are covalently bonded in molecules often have an arrangement of outer shell electrons like that of the noble gas with an atomic number closest to the element in question. It was just seen that covalently bonded H atoms in molecules of H_2 have 2 outer shell electrons like the nearby noble gas helium. For atoms of many other elements, the tendency is to acquire 8 outer shell electrons—an octet—in sharing electrons through covalent bonds. This tendency forms the basis of the *octet rule* discussed in Section 4.2. In illustrating the application of the octet rule to covalent bonding, consider first the bonding of atoms of hydrogen to atoms of elements with atomic numbers 6–9 in the second period of the periodic table. These elements are close to the noble gas neon and tend to attain a "neon-like" octet of outer shell electrons when they form covalently bonded molecules.

As shown in Figure 4.10, when each of the 4 H atoms shares an electron with 1 C atom to form a molecule of methane, CH_4, each C atom attains an 8-electron outer shell (octet) like neon. Every hydrogen atom attains an outer shell of 2 electrons through the arrangement of shared electrons. There are a total of 4 covalent bonds, 1 per H attached to the C atom, and each composed of a shared pair of electrons. Each of the hydrogen atoms has 2 electrons like the noble gas helium and carbon has an octet of outer shell electrons like the noble gas neon.

As also shown in Figure 4.10, nitrogen bonds with hydrogen such that 1 N atom shares 2 electrons with each of 3 H atoms to form a molecule of ammonia, NH_3.

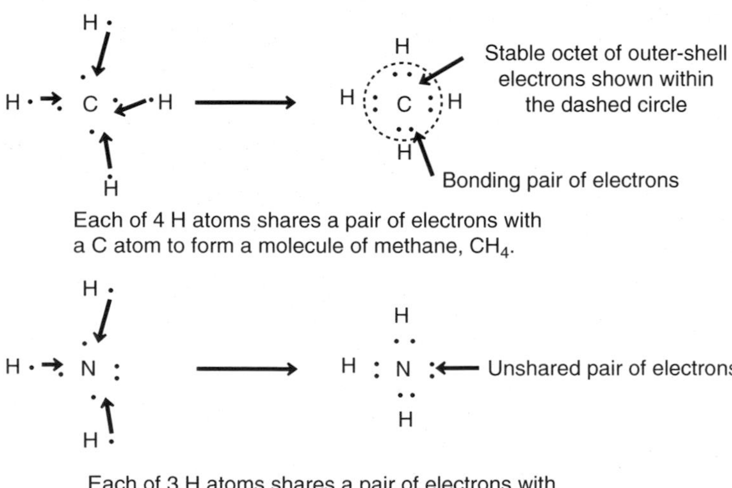

Figure 4.10. Formation of stable outer electron shells by covalent bonding in compounds.

This arrangement gives the N atom 8 outer shell electrons, of which 6 are shared with H atoms and 2 constitute an **unshared pair** of electrons. In both CH_4 and NH_3, each H atom has 2 shared electrons, which provides a shell of 2 electrons like that in the noble gas helium.

In discussing covalent bonds and molecules, it is sometimes convenient to use the term **central atom** in reference to an atom to which several other atoms are bonded. In the case of CH_4, carbon, C, is the central atom. For NH_3, nitrogen, N, is the central atom.

The use of Lewis symbols and formulas in the preceding examples readily shows how many atoms are bound together in each of the compounds and the types of bonds in each. The products of each reaction are shown in two ways: one in which all the valence electrons are represented as dots; and the other in which each pair of valence electrons in a chemical bond is shown as a dash, and the unbonded valence electrons as dots. In cases where all that is needed is to show which atoms are bonded together and the types of bonds (one dash for a single covalent bond, two for a double bond of 2 shared electron pairs, and three for a triple bond of 3 shared electron pairs), the dots representing unshared electrons may be omitted.

Figure 4.11 shows covalent bonding in the formation of hydrogen compounds of O and F. In the case of oxygen, 2 H atoms are combined with 1 O atom having 6 valence electrons, sharing electrons such that each H atom has 2 electrons and the O atom has 8 outer shell electrons, 4 of which are in 2 shared pairs. To form HF, only 1 H atom is required to share its electron with an atom of F having 7 outer shell electrons, leading to a compound in which the F atom has 8 outer shell electrons, 2 of which are shared with H.

Of the compounds whose formation is shown above, CH_4 is methane, the simplest of the hydrocarbon compounds and the major component of natural gas. The compound formed with nitrogen and hydrogen is ammonia, NH_3. It is the second most widely produced synthetic chemical, a pungent-smelling gas with many uses in the manufacture of fertilizer, explosives, and synthetic chemicals. The product of the reaction between hydrogen and oxygen is, of course, water, H_2O. (Another compound composed only of hydrogen and oxygen is hydrogen peroxide, H_2O_2, a reactive compound widely used as a bleaching agent.) Hydrogen fluoride, HF, is the product of the chemical combination of H and F. It is a toxic corrosive gas (boiling point 19.5°C) used as a raw material to manufacture organofluorine compounds and Teflon plastics.

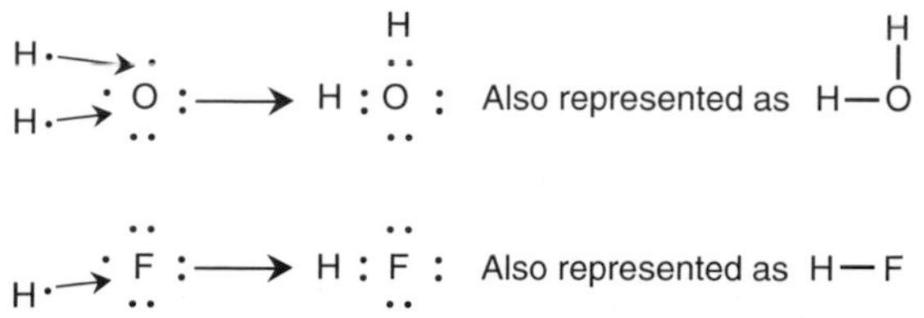

H· + H· + :Ö: ⟶ H:Ö: (with second H above O) Also represented as H—O (with H above O)

H· + :F̤: ⟶ H:F̤: Also represented as H—F

Figure 4.11. Covalent bonding of H to O and F to produce H_2O and HF, respectively.

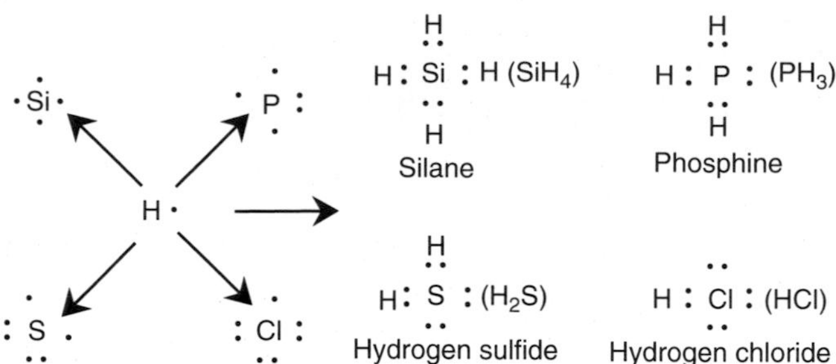

Figure 4.12. Chemical compounds of Si, P, S, and Cl with H.

Having seen how hydrogen combines with carbon, nitrogen, oxygen, and fluorine, it is easy to predict the hydrogen compounds with the elements directly below C, N, O, and F in the third period of the periodic table. These elements are silicon, phosphorus, sulfur, and chlorine, atomic numbers 14–17. As shown in Figure 4.12, these elements combined with hydrogen have covalent bonds such that every hydrogen atom has 2 outer shell electrons and each of the other elements has an octet of outer shell electrons.

Silane, SiH_4, is a colorless gas. **Phosphine**, PH_3, is also a gas; it catches fire spontaneously when exposed to air, is very toxic, and has a bad odor. **Hydrogen sulfide**, H_2S, is a toxic gas with a foul, rotten-egg odor. Natural gas contaminated with hydrogen sulfide is said to be sour, and the hydrogen sulfide must be removed before the natural gas can be used as a fuel. Fortunately, H_2S is readily converted to elemental sulfur or to sulfuric acid, for which there are ready markets. **Hydrogen chloride**, HCl, gas has a sharp odor and is very soluble in water. Solutions of hydrogen chloride in water are called **hydrochloric acid**. About 2.5 million tons of hydrochloric acid are produced for industrial applications in the U.S. each year.

4.6. SOME OTHER ASPECTS OF COVALENT BONDING

Multiple Bonds and Bond Order

Several examples of single bonds consisting of a pair of shared electrons have just been seen. Two other types of covalent bonds are the *double bond* consisting of two shared pairs of electrons (4 electrons total) and the *triple bond* made up of three shared pairs of electrons (6 electrons total). These are both examples of **multiple bonds**. Examples of a double bond in ethylene and of a triple bond in acetylene are shown in Figure 4.13.

Atoms of carbon, oxygen, and nitrogen bonded to each other are the most likely to form multiple bonds. In some cases, sulfur forms multiple bonds. Multiple bonds are especially abundant in the functional groups of organic compounds (Chapter 9).

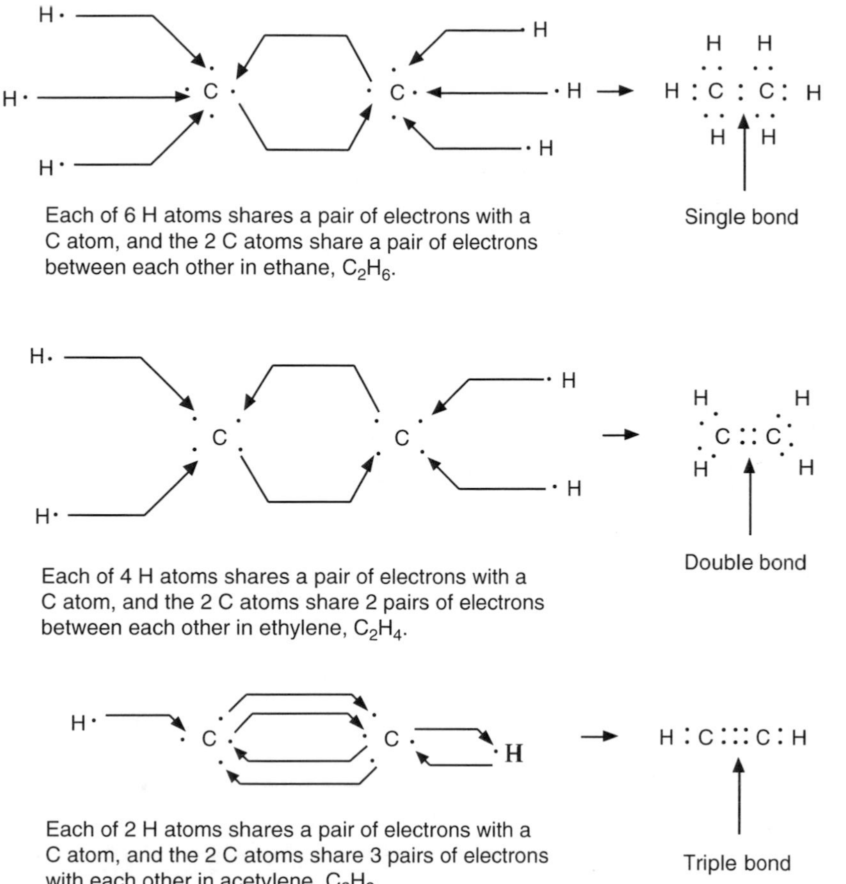

Figure 4.13. Single, double, and triple bonds between C atoms in three different hydrocarbon molecules, each of which contains 2 C atoms.

Lengths and Strengths of Multiple Bonds

In various compounds, bonds between the same two atoms (either of the same or different elements) may be single, double, or triple. This is seen below for the three types of carbon–carbon bonds illustrated in Figure 4.13:

Bond type:	Single in C_2H_6	Double in C_2H_2	Triple in C_2H_4
Bond length:	154 pm	134 pm	120 pm
Bond strength:	348 kJ/mol	614 kJ/mol	839 kJ/mol

The preceding examples of bond type, length, and strength show that as the bond multiplicity increases, the bonds become shorter. Furthermore, the bond strength, expressed as the energy in kilojoules needed to break a mole of the bonds, increases. These trends are reasonable if a covalent bond is viewed as consisting of electrons located between the two atoms that are bonded together, such that the positively charged nuclei of the atoms are attracted to the electrons in the bond. Increasing multiplicity gives a denser cloud of electrons, enabling the atoms to approach closer and resulting in shorter bond length. The same reasoning explains the greater strength of multiple bonds. One of the strongest common bonds is the triple N≡N bond in molecular nitrogen, N_2:

$$\cdot\dot{\underset{\cdot}{N}}\!: \quad \cdot\dot{\underset{\cdot}{N}}\!: \;\rightarrow\; :N\!\vdots\vdots\!N: \tag{4.6.1}$$

The number of bonding electron pairs that make up the bond between two atoms is called the **bond order**. The single, double, and triple bonds described above have bond orders of 1, 2, and 3, respectively. It has just been shown that bond strength increases and bond length decreases with increasing numbers of electrons in the bonds connecting two atom nuclei. Therefore, increased bond order is associated with increased bond energy and decreased bond length. These relationships were illustrated above for C—C, C=C, and C≡C bonds.

The absorption of infrared radiation by bonds in molecules provides the chemist with a powerful probe to determine bond orders, lengths, and energies. Infrared radiation was mentioned as a form of electromagnetic radiation in Chapter 3. When absorbed by molecules, infrared energy causes atoms to vibrate and (when more than 2 atoms are present) bend relative to each other. The chemical bond can be compared to a spring connecting 2 atomic nuclei (Figure 4.14). A short, strong bond acts like a short, strong spring, so that the atoms connected by the bond vibrate rapidly relative to each other. The infrared radiation absorbed by the bond causes it to vibrate. For a particular mode of bond vibration, the infrared radiation absorbed must be of a specific energy, which results in specific frequencies (ν) and wavelengths (λ) for the infrared radiation absorbed (see Section 3.11 in Chapter 3 for the relationships among energies, frequencies, and wavelengths of electromagnetic radiation).

Electronegativity and Covalent Bonding

Electronegativity refers to the ability of a bonded atom to attract electrons to itself. Whereas the previously discussed ionization energy and electron affinity deal

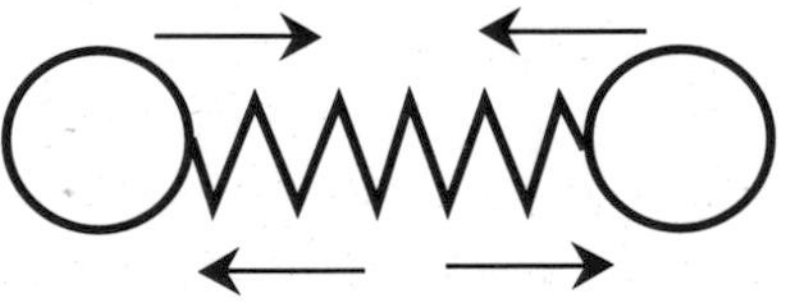

Figure 4.14. Representation of vibration of a chemical bond as a spring connecting the nuclei of 2 atoms (circles).

1 H 2.2							2 He -
3 Li 1.0	4 Be 1.6	5 B 1.8	6 C 2.5	7 N 3.0	8 O 3.4	9 F 4.0	10 Ne -
11 Na 0.93	12 Mg 1.3	13 Al 1.6	14 Si 1.9	15 P 2.2	16 S 2.6	17 Cl 3.2	18 Ar -
19 K 0.82	20 Ca 1.0						

Figure 4.15. Electronegativities of the first 20 elements in the periodic table.

with the attraction for electrons of isolated gas-phase atoms, electronegativity applies to bonded atoms. Tabulated values of electronegativity range from 0.79 for cesium to 4.0 for fluorine. Unlike ionization energies and electron affinities, which are directly measurable intrinsic properties of atoms, electronegativities are numerical estimates calculated from several elemental properties. Electronegativity values for the first 20 elements are shown in Figure 4.15.

Examination of Figure 4.15 shows that electronegativity values increase from left to right across a period of the periodic table (there is little discernible trend for the transition elements). Within a group of elements, electronegativity decreases with increasing atomic number. For the elements in the abbreviated version of the periodic table shown, this is most obvious for the active metals in the first two groups.

As will be seen in the next section, electrons are often shared unequally between atoms in chemical bonds. The electronegativity concept is useful in dealing with this important phenomenon in chemical bonding.

Sharing Electrons—Unequally

The Lewis formula for hydrogen chloride, HCl, is as follows:

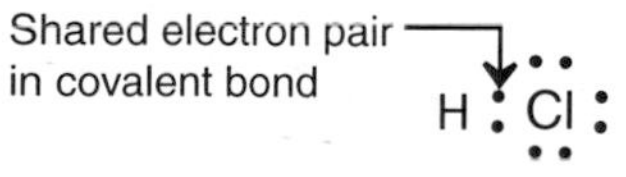

Shared electron pair in covalent bond

H : C̈l :

This formula might leave the impression that the two electrons in the covalent bond between H and Cl are shared equally between the two atoms. However, the numbers in Figure 4.15 show that the electronegativity of Cl, 3.2, is significantly higher than that of H, 2.2. This indicates that the Cl atom has a greater attraction for electrons than does the H atom. Therefore, the 2 electrons shared in the bond spend a greater fraction of time in the vicinity of the Cl atom than they do around the H atom. The result of this unequal sharing of electrons is that the chloride end of the molecule acquires a partial negative charge and the hydrogen end a partial positive charge. Although the overall charge of the molecule is zero, it is distributed unequally over the molecule. A body with an unequal electrical charge distribution is said to be **polar**. A covalent bond in which the electrons are not shared equally is called a **polar covalent bond**. A covalent bond in which the sharing of electrons is exactly equal is a **nonpolar covalent bond**:

H : H   :N⋮⋮N:

Shared electron pairs between two identical atoms are shared equally and the covalent bond is nonpolar

The ultimate in unequal sharing of electrons is the ionic bond, in which there is a complete transfer of electrons.

The two common ways of showing a polar bond are illustrated below. The δs represent partial positive and partial negative charge; the point of the arrow is toward the more electronegative atom, which attracts electrons more strongly than the other atom:

$$\overset{\delta+}{H}\text{———}\overset{\delta-}{Cl} \qquad \overset{\longrightarrow}{H\text{———}Cl}$$

Coordinate Covalent Bonds

The covalent bonds examined so far have consisted of electrons contributed equally from both of the atoms involved in the bond. It is possible to have covalent bonds in which only 1 of the 2 atoms that are joined together contributes both of the electrons in the bond. The bond so formed is called a **coordinate covalent bond** or a **dative bond**. A typical coordinate covalent bond forms when ammonia gas and hydrogen chloride gas react. This sometimes happens accidentally in the laboratory when these gases are evolved from beakers of concentrated ammonia solution and concentrated hydrochloric acid (a solution of HCl gas) that are accidentally left uncovered. When these two gases meet, a white chemical fog is formed. It is ammonium chloride, NH_4Cl, a salt that is produced by the reaction

Site of coordinate covalent bond formation (after the bond has formed, the four N-H bonds are indistinguishable)

$$H:\underset{\cdot\cdot}{\overset{H}{\overset{\cdot\cdot}{N}}}: + H:\underset{\cdot\cdot}{\overset{\cdot\cdot}{Cl}}: \longrightarrow \left[H:\underset{\underset{H}{\cdot\cdot}}{\overset{H}{\overset{\cdot\cdot}{N}}}:H\right]^+ :\underset{\cdot\cdot}{\overset{\cdot\cdot}{Cl}}:^- \qquad (4.6.2)$$

White solid ammonium chloride

in which the unshared pair of electrons from NH_3 forms a coordinate covalent bond with an H^+ ion released from HCl. The product is an NH_4^+ ion. After the coordinate covalent bond has formed, the ammonium ion has four equivalent N—H bonds;

the one formed by coordinate covalent bonding is indistinguishable from the other three bonds.

Compounds That Do Not Conform to the Octet Rule

By now, a number of examples of bonding have been shown that are explained very well by the octet rule. There are numerous exceptions to the octet rule, however. These fall into the three following major categories:

1. *Molecules with an uneven number of valence electrons.* A typical example is nitric oxide, NO. Recall that the nitrogen atom has 5 valence electrons and the oxygen atom 6, so that NO must have 11, an uneven number. The Lewis structure of NO may be represented as one of the two following forms (these are resonance structures, which are discussed later in this section):

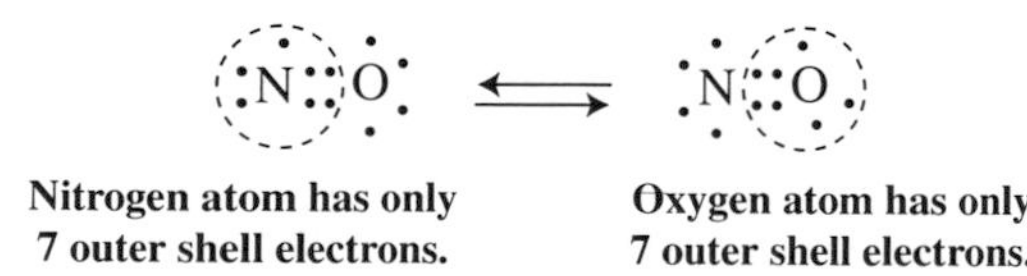

2. *Molecules in which an atom capable of forming an octet has fewer than 8 outer electrons.* A typical example is highly reactive, toxic, boron trichloride, BCl_3:

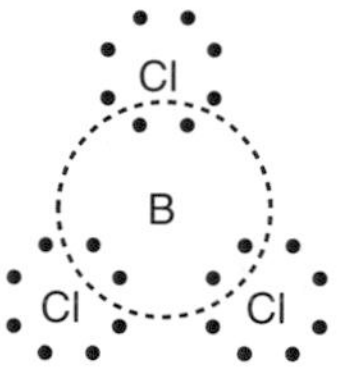

The dashed circle around the central B atom shows that it has only 6 outer shell electrons. However, it can accept 2 more from another compound such as NH_3 to fill a vacant orbital and complete an octet.

3. *Molecules in which an atom has more than 8 outer electrons.* This can occur with elements in the third and higher periods of the periodic table because their underlying *d* orbitals are capable of accepting electron pairs, so that the valence shell is no longer confined to one *s* orbital and three *p* orbitals that make up an octet of electrons. An example of a compound with more than 8 outer electrons is chlorine trifluoride,

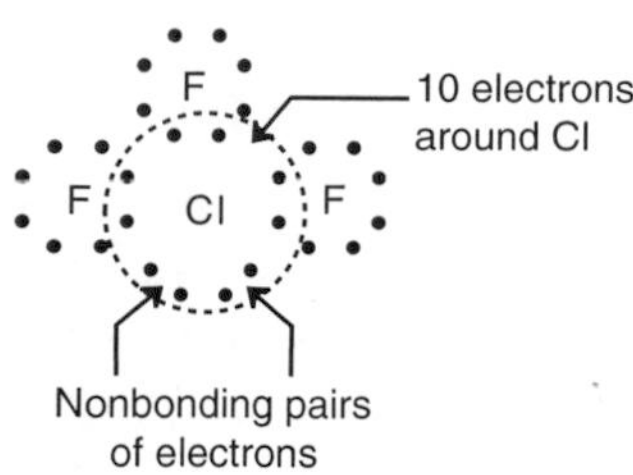

composed of 4 halide atoms (Cl and F), each of which has 7 valence electrons for a total of 28. These are accommodated by placing 5 electron pairs—2 of them nonbonding—around the central Cl atom.

Resonance Structures

For some compounds and multi-atom ions, it is possible to draw two or more equivalent arrangements of electrons. As an example, consider the following for sulfur dioxide:

An atom of sulfur and 2 atoms of oxygen compose a molecule of sulfur dioxide, SO_2, for which 2 equivalent electron structures may be drawn.

It can be seen that in order to have an octet of electrons around each atom, it is necessary to have a double bond between the S atom and one of the O atoms. However, either O atom may be chosen, so that there are two equivalent structures, which are called **resonance structures**. Resonance structures are conventionally shown with a double arrow between them, as in the example above. They differ only in the locations of bonds and unshared electrons, not in the positions of the atoms. Although the name and structures might lead one to believe that the molecule shifts back and forth between the structures, this is not the case, and the molecule is a hybrid of the two.

Another compound with different resonance forms is nitrogen dioxide:

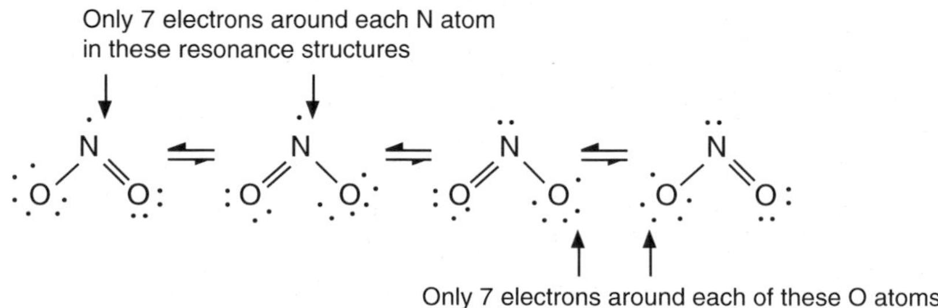

Four resonance structures of NO_2.

This compound is produced in smog-forming atmospheres from pollutant NO emitted by automobile engines, and is a major air pollutant. It can be seen from the above that there are two aspects of the electron structure of NO_2 relating to material that is under discussion. First, there are 4 resonance structures. Second, in each of these resonance structures, there is 1 atom with an incomplete octet of outer shell electrons. This is because NO_2 has an uneven number of electrons. This type of compound is, therefore, an exception to the octet rule. The molecule has an "extra" electron for which there is no other electron to form an electron pair. Compounds consisting of molecules with such an electron are attracted by a magnetic field and are said to be **paramagnetic**.

4.7. CHEMICAL FORMULAS OF COMPOUNDS

What a Chemical Formula States

Chemical formulas were introduced in Chapter 1, and a number of them have been given so far, such as H_2O for water and CO_2 for carbon dioxide. In this and later sections, chemical formulas and their meaning are discussed in more detail. A chemical formula contains a lot of information. It is one of the most important tools for communicating in chemical language. Therefore, it is important to understand what a chemical formula means, and how to use it properly.

To best illustrate all of the information contained in a chemical formula, consider the moderately complicated example of calcium phosphate, $Ca_3(PO_4)_2$. This compound is composed of Ca^{2+} ions, each of which has a +2 charge, and phosphate ions, PO_4^{3-}, each with a −3 charge. The 4 oxygen atoms and the phosphorus atom in the phosphate ion are held together with covalent bonds. The calcium and phosphate ions are bonded together ionically in a lattice composed of these two kinds of ions. Therefore, both ionic and covalent bonds are involved in calcium phosphate. Figure 4.16 summarizes the information contained in the calcium phosphate formula.

Exercise: In the formula for aluminum sulfate, $Al_2(SO_4)_3$.

(a) the elements present are __________, __________, and __________.

(b) In each formula unit there are _____ Al^{3+} ions and _____ SO_4^{2-} ions.

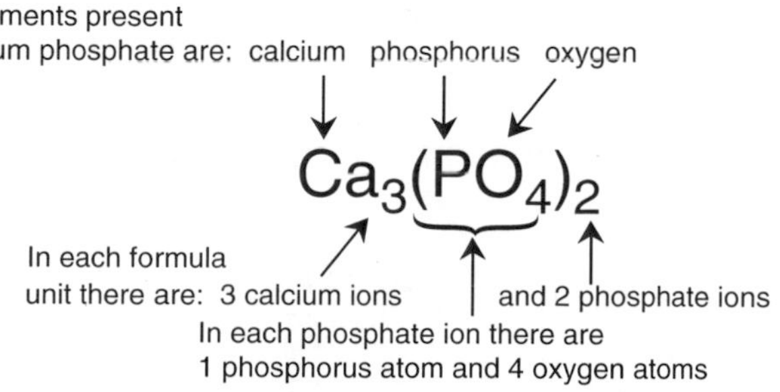

Figure 4.16. Summary of information contained in the chemical formula of calcium phosphate.

(c) Each sulfate ion contains _____ sulfur atoms and _____ oxygen atoms.

(d) The total numbers of each kind of atom in a formula unit of $Al_2(SO_4)_3$ are _____ Al, _____ S, and _____ O.

Answers: (a) aluminum, sulfur, and oxygen; (b) 2, 3; (c) 1, 4; (d) 2, 3, 12.

Percentage Composition from Chemical Formulas

The percentage elemental composition of a chemical compound is readily calculated from either its empirical or molecular formula. One way to calculate the percentage composition of a compound is to consider a mole of the compound, work out the number of grams of each element in 1 mole of the compound, and then calculate the percentages. This can be shown with glucose (blood sugar), molecular formula $C_6H_{12}O_6$. The subscripts indicate that each mole of glucose contains 6 mol C, 12 mol H, and 6 mol O. The mass of each element in a mole of glucose and the mass of a mole of glucose are given by the following:

$$\text{For C: } 6\text{ mol} \times \frac{12.01\text{ g}}{1\text{ mol}} = 72.06\text{ g}$$

$$\text{For H: } 12\text{ mol} \times \frac{1.01\text{ g}}{1\text{ mol}} = 12.12\text{ g}$$

$$\text{For O: } 6\text{ mol} \times \frac{16.00\text{ g}}{1\text{ mol}} = 96.00\text{ g}$$

$$\text{Molar mass of glucose} = 180.18\text{ g}$$

The percentage of each element in glucose is given as follows:

$$\text{Percent C: } \frac{72.06\text{ g}}{180.18\text{ g}} \times 100 = 40.0\%$$

$$\text{Percent H: } \frac{12.12\text{ g}}{180.18\text{ g}} \times 100 = 6.7\%$$

$$\text{Percent O: } \frac{96.0\text{ g}}{180.18\text{ g}} \times 100 = 53.3\%$$

Example: What is the percentage elemental composition of calcium oxalate, CaC_2O_4?

Answer: Exactly 1 mole of CaC_2O_4 contains 1 mol Ca, 2 mol C, and 4 mol O. Therefore,

$$\text{For Ca: } 1\text{ mol} \times \frac{40.08\text{ g}}{1\text{ mol}} = 40.08\text{ g} \quad \text{Percent Ca: } \frac{40.08}{128.10\text{ g}} \times 100 = 31.29\%$$

$$\text{For C: } 2 \text{ mol} \times \frac{12.01 \text{ g}}{1 \text{ mol}} = 24.02 \text{ g} \quad \text{Percent C: } \frac{24.02 \text{ g}}{128.10 \text{ g}} \times 100 = 18.75\%$$

$$\text{For O: } 4 \text{ mol} \times \frac{16.00 \text{ g}}{1 \text{ mol}} = 64.00 \text{ g} \quad \text{Percent O: } \frac{64.0 \text{ g}}{128.10 \text{ g}} \times 100 = 49.96\%$$

$$\text{Molar mass } CaC_2O_4 = 128.10 \text{ g/mol} \qquad \text{Total} = 100.0\%$$

Calculation of Chemical Formulas

In earlier sections of this chapter, it was shown how the chemical formulas of some compounds follow logically from the sharing or exchange of electrons in accordance with the octet rule. However, long before the nature of atoms or the existence of electrons was known, accurate chemical formulas were written for many common compounds. In this section, it is shown how chemical formulas are calculated from laboratory data.

First consider the calculation of the **empirical formula** of a chemical compound. The empirical formula gives the lowest whole number values of atoms in the formula. Empirical formulas may not be the same as the **molecular formulas** of substances that exist as molecules, which show the total number of each kind of atom in a molecule. For example, the empirical formula of benzene is CH. However, each molecule of benzene contains 6 C atoms and 6 H atoms, so the molecular formula is C_6H_6, illustrating the true number of each of the two kinds of atoms in the molecule.

Before calculating some empirical formulas, it is important to know what is meant by a mole of substance. The mole concept was discussed in detail in Section 2.3 and is summarized below:

1. The mole is based upon the number of particles or entities in a substance. These may be molecules of an element or compound, atoms of an element, atoms of a specified element in a particular mass of a compound, or ions in a compound.

2. In expressing moles, it is essential to state clearly the entity being described, such as atoms of Si, molecules of F_2, molecules of H_2O, formula units (see Section 2.3) of ionic Na_2SO_4, or ions of Na^+.

3. A mole of an entity contains Avogadro's number, 6.02×10^{23}, of that entity.

4. The mass of a mole (molar mass) of an element, compound, or portion of a compound (such as the SO_4^{2-} ion in Na_2SO_4) is the mass in grams equal numerically to the atomic mass, molecular mass, or formula mass. For example, given 23.0, 32.1, and 16.0 for the atomic masses of Na, S, and O,

respectively, the following can be stated: (a) A mole of Na atoms has a mass of 23.0 g, (b) a mole of Na_2SO_4 has a mass of $2 \times 23.0 + 32.1 + 4 \times 16.0 = 142.1$ g, and (c) a mole of SO_4^{2-} ions has a mass of $32.1 + 4 \times 16.0 = 96.1$ g.

5. Number of moles of substance $= \dfrac{\text{mass of substance in grams}}{\text{molar mass of substance in grams per mole}}$ (4.7.1)

Empirical Formula from Percentage Composition

Suppose that the only two elements in a compound are phosphorus and oxygen and that it is 43.64% P and 56.36% O by mass. The following steps can be used to find the empirical formula of the compound:

1. From the percentages, it follows that for each 100 g of compound there are 43.64 g of P and 56.36 g of O. The number of moles of each of these elements in exactly 100 g of the compound is calculated from the number of grams of each per mole of compound as follows:

$$43.64 \text{ g compound} \times \frac{1 \text{ mol P}}{30.97 \text{ g P}} = 1.409 \text{ mol P} \tag{4.7.2}$$

↑ In 100.0 g of the compound there are 43.64 g of P

↑ In 1 mol of P there are 30.97 g of P (from the atomic mass of P)

↑ Therefore, 100.0 g of the compound contains 1.409 mol of P

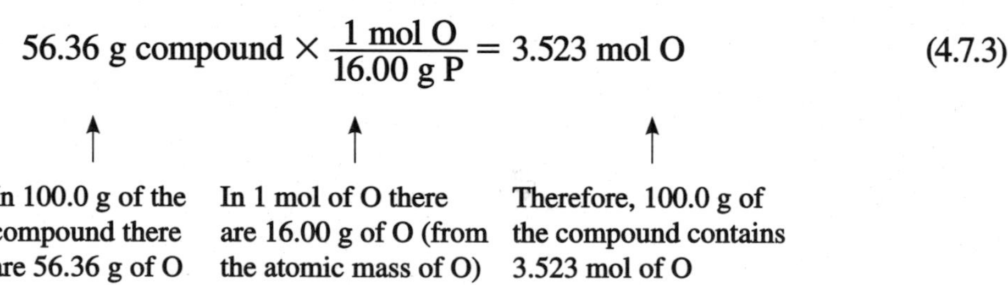

(4.7.3)

2. Calculate the relative number of moles of O per mole of P:

$$\frac{3.523 \text{ mol O}}{1.409 \text{ mol P}} = \frac{2.500 \text{ mol O}}{1.000 \text{ mol P}} \tag{4.7.4}$$

3. Calculate the ratio of moles of O to moles of P in terms of whole numbers. The denominator of the ratio 2.500 mol O/1.000 mol P is already a whole number, and the numerator can be converted to a whole number (5) if it is

multiplied by 2. Therefore, both the numerator and the denominator are multiplied by 2 to get the whole number ratio,

$$\frac{2.500 \text{ mol O}}{1.000 \text{ mol P}} \times \frac{2}{2} = \frac{5.000 \text{ mol O}}{2.000 \text{ mol P}} \tag{4.7.5}$$

4. Knowing the ratio of moles, write the empirical formula of the compound:

$$P_2O_5$$

The empirical formula does not necessarily give the actual number of atoms per molecule; it gives the smallest whole number ratio for the number of each kind of atom in the molecule. The molecular formula of this compound is in fact P_4O_{10}, indicating that a molecule of it contains twice the number of each kind of atom indicated by the empirical formula. The correct name of the compound is tetraphosphorus decoxide, although it is commonly known as "phosphorus pentoxide". It is made commercially by burning elemental phosphorus in dry air. It reacts violently with water to produce commercially important orthophosphoric acid, H_3PO_4. The strong affinity of P_4O_{10} for water makes it very useful as a drying and dehydrating agent.

Empirical formulas are actually determined in the laboratory by chemical analyses for each element in the compound. There is always some error involved in such an analysis, so the numbers obtained for the ratios of one element to another may not be as nicely rounded as in the example of the calculation of the empirical formula P_2O_5. As an example of this, suppose that a compound used as an inert filler in laundry detergent was found by chemical analysis to be 32.59% Na, 22.74% S, and 44.67% O. What is the empirical formula of the compound?

1. According to the percentages given, there are 32.59 g of Na, 22.74 g of S and 44.67 g of O in 100.0 g of the compound. Using the appropriate atomic masses gives the following number of moles of each element in 100 g of the compound:

$$32.59 \text{ g Na} \times \frac{1 \text{ mol Na}}{22.99 \text{ g Na}} = 1.418 \text{ mol g Na} \tag{4.7.6}$$

$$22.74 \text{ g S} \times \frac{1 \text{ mol S}}{32.06 \text{ g S}} = 0.7093 \text{ mol S} \tag{4.7.7}$$

$$44.67 \text{ g O} \times \frac{1 \text{ mol O}}{16.00 \text{ g O}} = 2.792 \text{ mol O} \tag{4.7.8}$$

2. Of the three elements present, the element with the least number of moles is sulfur. Therefore, the ratios of the number of moles of each of the other elements relative to a mole of S is calculated:

$$\frac{1.418 \text{ mol Na}}{0.7093 \text{ mol S}} = \frac{1.999 \text{ mol Na}}{1.000 \text{ mol S}} \tag{4.7.9}$$

$$\frac{2.792 \text{ mol O}}{0.7093 \text{ mol S}} = \frac{3.936 \text{ mol O}}{1.000 \text{ mol S}} \tag{4.7.10}$$

3. Examination of the results above shows that, rounding 1.999 to 2 and 3.936 to 4, there are 2 mol of Na and 4 mol of O per 1 mol of S, leading to the empirical formula

$$Na_2SO_4$$

This compound is called sodium sulfate.

Example: Analysis of a compound shows it to be 26.49% potassium, 35.58% Cr, and 37.93% O. What is the empirical formula of the compound?

Answer: In 100.0 g of the compound there are 26.49 g K, 35.58 g Cr, and 37.93 g O, and the following number of moles of each element:

$$26.49 \text{ g K} \times \frac{1 \text{ mol K}}{39.10 \text{ g K}} = 0.6775 \text{ mol K} \tag{4.7.11}$$

$$35.58 \text{ g Cr} \times \frac{1 \text{ mol Cr}}{52.00 \text{ g Cr}} = 0.6842 \text{ mol Cr} \tag{4.7.12}$$

$$37.93 \text{ g O} \times \frac{1 \text{ mol O}}{16.00 \text{ g O}} = 2.371 \text{ mol O} \tag{4.7.13}$$

Examination of the above figures reveals that there is essentially the same number of moles of K and Cr in a gram of the compound (rounding gives 0.68 mol of each per gram of compound), whereas there is a higher number of moles of O. Calculating the number of moles of K and O relative to the number of moles of Cr gives the following:

$$\frac{0.6775 \text{ mol K}}{0.6842 \text{ mol Cr}} \times 100.0 \text{ g compound} = 0.9902 \text{ mol K} \tag{4.7.14}$$

$$\frac{2.371 \text{ mol O}}{0.6842 \text{ mol Cr}} \times 100.0 \text{ g compound} = 3.465 \text{ mol K} \tag{4.7.15}$$

Rounding these numbers to the nearest 0.1 mol gives 1.0 mol of K and 3.5 mol of O per mol of Cr. One can write the empirical formula as

$$KCrO_{3.5}$$

but this does not give a whole number of O atoms in the formula unit. Multiplying the subscripts by 2 (in the formula above, there is understood to be a subscript 1 after K and after Cr) yields an empirical formula with whole numbers of atoms:

$$K_2Cr_2O_7$$

This is the formula of the compound potassium dichromate, each formula unit of which consists of 2 K^+ ions and a dichromate ion, $Cr_2O_7^{2-}$.

4.8. THE NAMES OF CHEMICAL COMPOUNDS

One of the more important aspects of chemical language is the correct use of names for chemical compounds—chemical nomenclature. So far in the text, the names of a number of compounds have been given. This section presents a systematic approach to the naming of inorganic compounds. These are all of the compounds that do not contain carbon plus a few—such as CO_2 and Na_2CO_3—that do contain carbon. Most carbon-containing compounds are organic compounds; organic compound nomenclature will not be considered here; it is discussed in Chapter 10. Inorganic nomenclature will be divided into several categories so that it can be approached in a systematic manner.

Binary Molecular Compounds

In this section, the rules for naming binary molecular compounds—those composed of only two elements bonded together as molecules—are examined. Most such compounds are composed of nonmetallic elements. In a binary molecular compound, one of the elements is usually regarded as being somewhat more positive than the other. This is a result of unequal sharing of electrons in covalent bonds, a characteristic of bonds considered briefly in Section 4.6. The symbol of the element with a more positive nature is given first in the chemical formula of a binary compound. For example, in HCl, hydrogen is regarded as having a somewhat more positive and Cl a somewhat more negative character.

In naming binary molecular compounds, the first part of the name is simply that of the first element in the chemical formula, and the second part of the name is that of the second element with an *-ide* ending. Therefore, the name of HCl is hydrogen chloride. In most cases, however, it is necessary to have prefixes to designate the relative number of atoms in the molecular formula. These prefixes are the following:

1 = mono	3 = tri	5 = penta	7 = hepta	9 = nano
2 = di	4 = tetra	6 = hexa	8 = octa	10 = deca

The uses of these prefixes are illustrated by the names of several oxygen compounds (oxides) in Table 4.2. Other examples of compounds in which prefixes are used for naming are $SiCl_4$, silicon tetrachloride; Si_2F_6, disilicon hexafluoride; PCl_5, phosphorus pentachloride; and SCl_2, sulfur dichloride.

It can be seen from the examples given that the prefix is omitted before the name of the first element in the chemical formula when there is only 1 atom of that element per molecule. In cases where the name of the second element in a binary

Table 4.2. Some Oxides and Their Names

Formula	Name	Formula	Name
CO	Carbon monoxide	SO_2	Sulfur dioxide
CO_2	Carbon dioxide	SO_3	Sulfur trioxide
N_2O_5	Dinitrogen pentoxide	SiO_2	Silicon dioxide
N_2O_4	Dinitrogen tetroxide	P_2O_4	Diphosphorus tetroxide
NO_2	Nitrogen dioxide	P_4O_7	Tetraphosphorus heptoxide
NO	Nitrogen monoxide	P_4O_{10}	Tetraphosphorus decoxide
N_2O	Dinitrogen oxide (nitrous oxide)	Cl_2O_7	Dichlorine heptoxide

molecular compound formula begins with a vowel, the "a" or "o" at the end of the prefix may be deleted. Thus, CO is called carbon monoxide, rather than carbon monooxide and Cl_2O_7 is called dichlorine heptoxide, not dichlorine heptaoxide.

A number of compounds, including those other than binary molecular compounds, have been known for so long and have been so widely used that they have acquired common names that do not describe their chemical formulas. The best example of these is H_2O, which is always known as water, not dihydrogen monoxide. Other examples are H_2O_2, hydrogen peroxide; NH_3, ammonia; and N_2O, nitrous oxide.

Names of Ionic Compounds

Recall that ionic compounds consist of aggregates of positively charged cations and negatively charged anions held together by the mutual attraction of their opposite electrical charges. The easiest of these compounds to visualize are those that are composed of only two elements, the cation formed by the loss of all its outer shell electrons and the anion formed by the gain of enough outer shell electrons to give it a stable octet of outer electrons (as shown for sodium and chlorine in Equation 4.3.1). The formation of such ions is readily shown by consideration of some of the elements in the abbreviated version of the periodic table in Figure 4.2, as illustrated by the following half-reactions (literally "half a reaction" in which electrons are reactants or products):

Half-reactions for cation formation

$$\mathrm{Li}\!:\!\cdot \longrightarrow \mathrm{Li}\!:^{+} + e^{-} \qquad \mathrm{Mg}\!: \longrightarrow \,:\!\underset{\cdot\cdot}{\ddot{\mathrm{Mg}}}\!:^{2+} + 2e^{-}$$

$$\mathrm{Na}\cdot \longrightarrow \,:\!\underset{\cdot\cdot}{\ddot{\mathrm{Na}}}\!:^{+} + e^{-} \qquad \mathrm{Ca}\!: \longrightarrow \,:\!\underset{\cdot\cdot}{\ddot{\mathrm{Ca}}}\!:^{2+} + 2e^{-}$$

$$\mathrm{K}\cdot \longrightarrow \,:\!\underset{\cdot\cdot}{\ddot{\mathrm{K}}}\!:^{+} + e^{-} \qquad \dot{\mathrm{Al}}\!: \longrightarrow \,:\!\underset{\cdot\cdot}{\ddot{\mathrm{Al}}}\!:^{3+} + 3e^{-}$$

Half-reactions for anion formation

$$\cdot\ddot{\underset{\cdot}{N}}: + 3e^- \longrightarrow :\ddot{\underset{\cdot\cdot}{N}}:^{3-} \qquad H\bullet + e^- \longrightarrow H:^-$$

$$\cdot\ddot{\underset{\cdot}{O}}: + 2e^- \longrightarrow :\ddot{\underset{\cdot\cdot}{O}}:^{2-} \qquad \cdot\ddot{\underset{\cdot\cdot}{F}}: + e^- \longrightarrow :\ddot{\underset{\cdot\cdot}{F}}:^-$$

$$\cdot\ddot{\underset{\cdot}{S}}: + 2e^- \longrightarrow :\ddot{\underset{\cdot\cdot}{S}}:^{2-} \qquad \cdot\ddot{\underset{\cdot\cdot}{Cl}}: + e^- \longrightarrow :\ddot{\underset{\cdot\cdot}{Cl}}:^-$$

The cations formed as shown by the half-reactions above are simply given the names of the metals that produced them, such as sodium for Na^+ and calcium for Ca^{2+}. Since they consist of only one element, the name for each anion has an -ide ending, that is, N^{3-}, nitride; O^{2-}, oxide; S^{2-}, sulfide; H^-, hydride; F^-, fluoride; and Cl^-, chloride. An advantage in the nomenclature of ionic compounds is that it is not usually necessary to use prefixes to specify the numbers of each kind of ion in a formula unit. This is because the charges on the ions determine the relative numbers of each, as shown by the examples in Table 4.3.

Exercise: For the compounds listed below, where x and y are unspecified subscripts, give the compound formula and name.

(a) K_xCl_y	(d) Li_xCl_y	(g) K_xN_y	(j) Mg_xO_y
(b) K_xS_y	(e) Li_xF_y	(h) Ca_xN_y	(k) Ca_xO_y
(c) Al_xF_y	(f) Li_xO_y	(i) Mg_xN_y	(l) Ca_xS_y

Answers: (a) KCl, potassium chloride; (b) K_2S, potassium sulfide; (c) AlF_3, aluminum fluoride; (d) LiCl, lithium chloride; (e) LiF, lithium fluoride; (f) Li_2O, lithium oxide, (g) K_3N, potassium nitride; (h) Ca_3N_2, calcium nitride; (i) Mg_3N_2, magnesium nitride; (j) MgO, magnesium oxide; (k) CaO, calcium oxide; (l) CaS, calcium sulfide.

Table 4.3. Formulas and Names of Ionic Compounds Formed from Two Elements

Compound Formula	Compound Name	Neutral Compound Formed from the Ions Below
NaF	Sodium fluoride	One +1 ion and one −1 ion
K_2O	Potassium oxide	Two +1 ions and one −2 ion
Na_3N	Sodium nitride	Three +1 ions and one −3 ion
$MgCl_2$	Magnesium chloride	One +2 ion and two −1 ions
CaH_2	Calcium hydride	One +2 ion and two −1 ions
$AlCl_3$	Aluminum chloride	One +3 ion and three −1 ions
Al_2O_3	Aluminum oxide	Two +3 ions and three −2 ions

4.9. ACIDS, BASES, AND SALTS

Most inorganic chemical compounds belong to one of three classes: acids, bases, or salts. These three categories of compounds are addressed briefly in this section and in more detail in Chapter 6.

Acids

For the present, an acid can be defined as a substance that dissolves in water to produce hydrogen ion, H^+ (*aq*). (Recall that *aq* in parentheses after a species formula shows that it is dissolved in water.) The most obviously acidic compounds are those that contain ionizable H in their formulas. Typically, hydrogen chloride gas, HCl(*g*), dissolves in water,

$$HCl(g) \xrightarrow{H_2O} H^+(aq) + Cl^-(aq) \qquad (4.9.1)$$

and completely dissociates (falls apart) to give an aqueous solution containing H^+ cation and Cl^- anion. Note that the H_2O above the arrow in this chemical equation shows that the reaction occurs in water. The presence of a substantial concentration of H^+ ions in water provides an acidic solution, in this case a solution of hydrochloric acid.

Some acids produce more than 1 hydrogen ion per molecule of acid. For example, sulfuric acid dissociates in two steps,

$$\underset{\text{Sulfuric acid}}{H_2SO_4(aq)} \longrightarrow H^+(aq) + \underset{\text{hydrogen sulfate ion}}{HSO_4^-(aq)} \qquad (4.9.2)$$

$$HSO_4^- \longrightarrow H^+(aq) + SO_4^{2-}(aq) \qquad (4.9.3)$$

to yield 2 H^+ per H_2SO_4 molecule; the second of the 2 hydrogen ions comes off much less readily than the first. Some acids have hydrogen atoms that do not produce H^+ ion in water. For example, each molecule of acetic acid produces only one H^+ ion when it dissociates in water; the other 3 H atoms stay covalently bonded to the acetate ion:

$$\mathrm{H{-}\overset{\overset{\large H}{|}}{\underset{\underset{\large H}{|}}{C}}{-}\overset{\overset{\large O}{\|}}{C}{-}OH} \longrightarrow \mathrm{H{-}\overset{\overset{\large H}{|}}{\underset{\underset{\large H}{|}}{C}}{-}\overset{\overset{\large O}{\|}}{C}{-}O^-} + H^+ \qquad (4.9.4)$$

Some of the most widely produced industrial chemicals are acids. Sulfuric acid ranks first among all chemicals produced in the U.S., with production of almost 40 million metric tons per year. Its greatest single use is in the production of phosphate fertilizers, and it has applications in many other areas including petroleum refining, alcohol synthesis, iron and steel pickling (corrosion removal), and storage battery manufacture. Nitric acid ranks about 10th among U.S. chemicals, with annual production of 7–8 million metric tons, and hydrochloric acid is about 25th at about 3 million metric tons (annual production and rank vary from year to year).

The naming of acids is addressed in more detail in Chapter 6. Briefly, acids that contain only H and another atom are "hydro-ic" acids, such as hydrochloric acid, HCl. For acids that contain two different amounts of oxygen in the anion part, the one with more oxygen is an "-ic" acid and the one with less is an "-ous" acid. This is illustrated by nitric acid, HNO_3, and nitrous acid, HNO_2. An even greater amount of oxygen is denoted by a "per-" prefix and less by the "hypo-ous" name. The guidelines discussed above are illustrated for acids formed by chlorine ranging from 0 to 4 oxygen atoms per acid molecule as follows: HCl, hydrochloric acid; HClO, hypochlorous acid; $HClO_2$, chlorous acid; $HClO_3$, chloric acid; $HClO_4$, perchloric acid.

Bases

A **base** is a substance that contains hydroxide ion OH^-, or produces it when dissolved in water. Most of the best-known inorganic bases have a formula unit composed of a metal cation and 1 or more hydroxide ions. Typical of these are sodium hydroxide, NaOH, and calcium hydroxide, $Ca(OH)_2$, which dissolve in water to yield hydroxide ion and their respective metal ions. Other bases, such as ammonia, NH_3, do not contain hydroxide ion, but react with water,

$$NH_3 + H_2O \longrightarrow NH_4^+ + OH^- \quad (4.9.5)$$

to produce hydroxide ion (this reaction proceeds only to a limited extent; most of the NH_3 is in solution as the NH_3 molecule).

Bases are named for the cation in them plus "hydroxide." Therefore, KOH is potassium hydroxide.

Salts

A **salt** is an ionic compound consisting of a cation other than H^+ and an anion other than OH^-. A salt is produced by a chemical reaction between an acid and a base. The other product of such a reaction is always water. The following are typical salt-producing reactions:

$$\underset{\text{base}}{KOH} + \underset{\text{acid}}{HCl} \longrightarrow \underset{\text{a salt, potassium chloride}}{KCl} + \underset{\text{water}}{H_2O} \quad (4.9.6)$$

$$\underset{\text{base}}{Ca(OH)_2} + \underset{\text{acid}}{H_2SO_4} \longrightarrow \underset{\text{a salt, calcium sulfate}}{CaSO_4} + \underset{\text{water}}{2H_2O} \quad (4.9.7)$$

Except for those compounds in which the cation is H^+ (acids) or the anion is OH^- (bases), the compounds that consist of a cation and an anion are salts. Therefore, the rules of nomenclature discussed for ionic compounds in Section 4.8 are those of salts. The salt product of Reaction 4.9.6, above, consists of K^+ cation and Cl^- anion, so the salt is called potassium chloride. The salt product of Reaction 4.9.7, above, is made up of Ca^{2+} cation and SO_4^{2-} anion and is called calcium sulfate. The reaction product of LiOH base with H_2SO_4 acid is composed of Li^+ ions and SO_4^{2-} ion. It takes 2 singly charged Li^+ ions to compensate for the 2− charge of the SO_4^{2-} anion, so the formula of the salt is Li_2SO_4. It is called simply lithium sulfate. It is not necessary to call it dilithium sulfate, because the charges on the ions denote the relative numbers of ions in the formula.

CHAPTER SUMMARY

The chapter summary below is presented in a programmed format to review the main points covered in this chapter. It is used most effectively by filling in the blanks, referring back to the chapter as necessary. The correct answers are given at the end of the summary.

Chemical bonds are normally formed by the transfer or sharing of [1]__________________, which are those in the [2]______________________________. An especially stable group of electrons attained by many atoms in chemical compounds is an [3]__________________. An ion consists of [4]____. A cation has [5]______________________________ and an anion has [6]__________________. An ionic compound is one that contains [7]____________________ and is held together by [8]____________________. Both F^- and Mg^{2+} have the electron configuration [9]______________ identical to that of the neutral atom [10]________________. In visualizing a neutral metal atom reacting with a neutral nonmetal atom to produce an ionic compound, the three major energy factors are [11]______________________________. The energy required to separate all of the ions in a crystalline ionic compound and remove them a sufficient distance from each other so that there is no interaction between them is called the [12]____________________. The ion formed from Ca is [13]____________, the ion formed from Cl is [14]____________, and the formula of the compound formed from these ions is [15]______________. A covalent bond may be described as [16]____. In the figure,

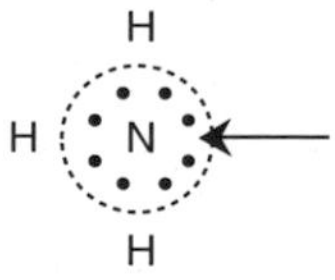

each pair of dots between N and H represents [17]______________________________ the arrow points to [18]____________________________ and the circle outlines [19]______________________________. A central atom refers to [20]______________________________. A triple bond consists of [21]________________ and is represented in a structural chemical formula as [22]____________. With increasing bond order (single < double < triple), bond length [23]_______________ and bond strength [24]_______________. Electronegativity refers to [25]__________________________. A polar covalent bond is one in which [26]______________________. A coordinate covalent bond is [27]____. Three major exceptions to the octet rule are [28]____. Resonance structures are those for which [29]____. Chemical formulas consist of [30]____ that tell the following about a compound: [31] ____. The percentage elemental composition of a chemical compound is calculated by [32] ____. The empirical formula of a chemical compound is calculated by computing the masses of each consituent element in [33] ________________, dividing each of the resulting values by [34]______________________________, dividing each value by [35]____, and rounding to [36]____________________. The prefixes for numbers 1–10 used to denote relative numbers of atoms of each kind of atom in a chemical formula are [37]____. The compound N_2O_5 is called [38]________________. The ionic compound $AlCl_3$ is called [39]______________________, which does not contain a "tri-" because [40]____. Ammonium ion, NH_4^+, is classified as a [41] ______________________, meaning that it consists of [42]_________________________________. An acid is [43]____. A base is [44]____. A salt is [45] ____ and is produced by [46]____. The other product of such a reaction is always [47]_______________.

Answers to Chapter Summary

1. valence electrons
2. outermost shell of the atom
3. octet of outer electrons

4. an atom or group of atoms having an unequal number of electrons and protons and, therefore, a net electrical charge
5. a positive charge
6. a negative charge
7. cations and anions
8. ionic bonds
9. $1s^2 2s^2 2p^6$
10. neon
11. ionization energy, electron affinity, and lattice energy
12. lattice energy
13. Ca^{2+}
14. Cl^-
15. $CaCl_2$
16. one that joins 2 atoms through the sharing of 1 or more pairs of electrons between them
17. a pair of electrons shared in a covalent bond
18. an unshared pair of electrons
19. a stable octet of electrons around the N atom
20. an atom to which several other atoms are bonded
21. 3 pair (total of 6) electrons shared in a covalent bond
22. $\equiv$ or $\vdots\vdots$
23. decreases
24. increases
25. the ability of a bonded atom to attract electrons to itself
26. the electrons involved are not shared equally
27. one in which only 1 of the 2 atoms contributes the two electrons in the bond
28. molecules with an uneven number of valence electrons, molecules in which an atom capable of forming an octet has fewer than 8 outer

electrons, and molecules in which an atom has more than 8 outer electrons

29. it is possible to draw two or more equivalent arrangements of electrons
30. atomic symbols, subscripts, and sometimes parentheses and charges
31. elements in it, relative numbers of each kind of atom, charge, if an ion
32. dividing the mass of a mole of a compound into the mass of each of the constituent elements in a mole
33. 100 g of the element
34. the atomic mass of the element
35. the smallest value
36. the smallest whole number for each element
37. 1 = mono, 2 = di, 3 = tri, 4 = tetra, 5 = penta, 6 = hexa, 7 = hepta, 8 = octa, 9 = nano, 10 = deca
38. dinitrogen pentoxide
39. aluminum chloride
40. the charges on ions are used to deduce chemical formulas
41. polyatomic ion
42. 2 or more atoms per ion
43. a substance that dissolves in water to produce hydrogen ion, H^+ (*aq*)
44. a substance that contains hydroxide ion OH^-, or produces it when dissolved in water
45. an ionic compound consisting of a cation other than H^+ and an anion other than OH^-.
46. a chemical reaction between an acid and a base
47. water

QUESTIONS AND PROBLEMS

1. Chlorofluorocarbons (Freons) are composed of molecules in which Cl and F atoms are bonded to 1 or 2 C atoms. These compounds do not break down well in the atmosphere until they drift high into the

stratosphere, where very short-wavelength ultraviolet electromagnetic radiation from the Sun is present, leading to the production of free Cl atoms that react to deplete the stratospheric ozone layer. Recalling what has been covered so far about the energy of electromagnetic radiation as a function of wavelength, what does this say about the strength of C—Cl and C—F bonds?

2. Illustrate the octet rule with examples of (a) a cation, (b) an anion, (c) a diatomic elemental gas, and (d) a covalently bound chemical compound.

3. When elements with atomic numbers 6 through 9 are covalently or ionically bound, or when Na, Mg, or Al have formed ions, which single element do their outer electron configurations most closely resemble? Explain your answer in terms of filled orbitals.

4. Ionic bonds exist because of (a) ______________________ between (b) ______________________.

5. In the crystal structure of NaCl, what are the number and type of ions that are nearest neighbors to each Na^+ ion?

6. What major aspect of ions in crystals is not shown in Figure 4.5?

7. What are five energy factors that should be considered in the formation of ionic NaCl from solid Na and gaseous Cl_2?

8. What are two major factors that increase the lattice energy of ions in an ionic compound?

9. Is energy released or is it absorbed when gaseous ions come together to form an ionic crystal?

10. What is incomplete about the statement that "the energy change from lattice energy for NaCl is 785 kJ of energy released"?

11. How do the sizes of anions and cations compare with their parent atoms?

12. How do the sizes of monatomic (one-atom) cations and anions compare in the same period?

13. Define covalent bond.

14. Explain the energy minimum in the diagram illustrating the H—H covalent bond in Figure 4.9.

15. What may be said about the likelihood of H atoms being involved in double covalent bonds?

16. What is represented by a dash, —, in a chemical formula?

17. What is represented by the two dots in the formula of phosphine, PH_3:

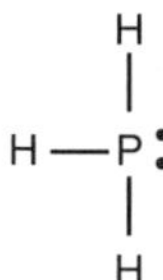

18. What are multiple bonds? Which three elements are most likely to form multiple bonds?

19. What can be said about the nature of covalent bonds between (a) 2 atoms with almost identical electronegativity values and (b) 2 atoms with substantially different electronegativity values?

20. What symbols are used to show bond polarity?

21. What is the "ultimate" in polar bonds?

22. A molecule of NH_3 will combine with one of BF_3. Describe the kind of bond formed in the resulting compound.

23. What is required for an atom to form a compound with more than 8 electrons in the central atom's outer shell?

24. What are resonance structures?

25. How many total valence electrons are in the nitrate ion, NO_3^-? What are the resonance structures of this ion?

26. Summarize the steps involved in calculating the percentage composition of a compound from its formula.

27. Phosgene, $COCl_2$, is a poisonous gas that was used for warfare in World War I. What is its percentage composition?

28. The molecular formula of acetylsalicylic acid (aspirin) is $C_9H_8O_4$. What is its percentage composition?

29. Hydrates are compounds in which each formula unit is associated with a definite number of water molecules. A typical hydrate is copper(II) sulfate pentahydrate, $CuSO_4 \cdot 5H_2O$. The water of hydration can be driven off by heating, leaving the anhydrous compound. Answer the following pertaining to $CuSO_4 \cdot 5H_2O$: (a) mass of 1 mole of the compound; (b) mass of H_2O in 1 mole of $CuSO_4 \cdot 5H_2O$; (c) percentage of H_2O in $CuSO_4 \cdot 5H_2O$.

30. What are the percentages of oxygen in (a) perchloric acid, (b) chloric acid, (c) chlorous acid, and (d) hypochlorous acid (these acids were discussed in this chapter).

31. What is the simplest (empirical) formula of sodium oxalate, $Na_2C_2O_4$?

32. What is the formula of dichlorine heptoxide? What is its percentage composition?

33. Chlorine dioxide, ClO_2, is used as a substitute for chlorine gas in the disinfection of drinking water. What is the percentage composition of ClO_2?

34. Summarize in steps the calculation of empirical formula from the percentage composition of a compound.

35. If the empirical formula and the actual formula mass (molecular mass) of a compound are known, how is the true formula calculated?

36. A chlorofluorocarbon gas is 9.93% C, 31.43% F, and 58.64% Cl. What is its empirical formula?

37. A 100.0 g portion of the chlorofluorocarbon gas (preceding question) was found to occupy 25.3 L at 100°C and 1.000 atm pressure. Assuming that the gas behaved ideally and using the ideal gas equation, how many moles of the gas are in 100.0 of the gas? What is its molar mass? What is its molecular formula? (To answer this question, it may be necessary to refer back to a discussion of the gas laws in Section 2.6.)

38. A compound is 5.88% H and 94.12% O. What is its empirical formula?

39. The compound from the preceding problem has a molecular mass of 34.0. What is its molecular formula?

40. A pure liquid compound with an overpowering vinegar odor is 40.0% C, 6.67% H, and 53.3% O. Its molecular mass is 60.0. What is its empirical formula? What is its molecular formula?

41. A compound is 29.1% Na, 40.5% S, and 30.4% O. It has a formula mass of 158.0. Fill out the table below pertaining to the compound, give its true formula, and give its actual formula. It is an ionic compound. What is the anion?

Element	Grams of element in 100 g compound	Mol element in 100 g compound	Mol element with fewest moles
Na	(a) ________	(d) ________	(g) ________
S	(b) ________	(e) ________	(h) ________
O	(c) ________	(f) ________	(i) ________

42. An ionic compound is 41.7% Mg, 54.9% O, and 3.4% H. What is its empirical formula? Considering ions mentioned in this chapter, what is the actual formula?

43. Ethylenediamine is 40.0% C, 13.4% H, and 46.6% N; its formula mass is 60.1. What are its empirical and molecular formulas?

44. The empirical formula of butane is C_2H_5 and its molecular mass is 58.14. What is the molecular formula?

45. Suppose that you were asked to give the Lewis formula of formic acid, H_2CO_2, where the atomic number of H is 1, that of C is 6, and that of O is 8. What is the total number of valence (outer shell) electrons that would have to be placed correctly in the structure: A. 6; B. 18; C. 20; D. 7; E. 16?

46. Given the Lewis symbols of hypothetical elements X and Z:

$$X\cdot \quad :\dot{Z}\cdot$$

What is the compound most likely formed by these two elements?

A. Covalently bound compound, formula X_2Z_3

B. Covalently bound compound, formula XZ

C. Ionic compound, formula X_2Z

D. Ionic compound XZ_2

E. Ionic compound XZ

47. Of the following, the statement that is **untrue** regarding ionic NaCl and its formation from gaseous Na and Cl atoms is:

A. Energy called electron repulsion is consumed in putting an electron on a Cl atom to produce a Cl^- ion.

B. Energy called ionization energy is required to remove an electron from a Na atom to produce an Na^+ ion.

C. A relatively large amount of energy (lattice energy) is released when the Na^+ and Cl^- ions come together to form crystalline NaCl.

D. A particular Cl^- ion in the crystal of NaCl has 6 Na^+ ions as its nearest neighbors.

E. Every ion in the crystal of NaCl is closest to ions of opposite charge, resulting in forces of attraction that account for the stability of ionic bonds.

48. Consider the single, double, and triple bonds connecting the 2 carbon atoms in the 3 compounds C_2H_6, C_2H_4, and C_2H_2. Of the following pertaining to these bonds, which is the **untrue** statement?

A. The C=C bond is shorter than the C—C bond.

B. The C≡C bond is stronger than the C=C bond.

C. Because it must accomodate a total of 6 electrons rather than 2, the C≡C bond is longer than the C—C bond.

D. As bond multiplicity increases, the bond strength increases and the bond length decreases.

E. The bonds can act like springs in that the atoms connected by the bonds vibrate when exposed to the right wavelength of infrared radiation.

49. Remember that O and S atoms all have the same number of valence electrons. Considering the structures (Lewis formulas)

$$:\underset{\cdot\cdot}{\ddot{O}}-\ddot{S}=\underset{\cdot}{\dot{O}}\colon \qquad \colon\underset{\cdot}{\dot{O}}=\ddot{S}-\underset{\cdot\cdot}{\ddot{O}}:$$

which is the true statement?

A. Only one of these structures can be correct.

B. Both of these structures are incorrect.

C. The structure on the left shows the incorrect number of outer shell electrons for the O atom on the left.

D. The structures are equivalent resonance structures.

E. The text showed the sulfur atom on one end, not in the middle.

5. CHEMICAL REACTIONS, EQUATIONS, AND STOICHIOMETRY

5.1. THE SENTENCES OF CHEMISTRY

As noted earlier, chemistry is a language. Success in the study of chemistry depends upon how well chemical language is learned. This chapter presents the last of the most basic parts of the chemical language. When it has been learned, the reader will have the essential tools needed to speak and write chemistry and to apply it in environmental and other areas.

Recall that the discussion of chemical language began by learning about the *elements*, the *atoms* composing the elements, and the *symbols* used to designate these elements and their atoms. Atoms of the elements bond together in various combinations to produce *chemical compounds*. These are designated by *chemical formulas* consisting of symbols for the kinds of atoms in the compound and subscripts indicating the relative numbers of atoms of each kind in the compound. In chemical language, the symbols of the elements are the letters of the chemical alphabet and the formulas are the words of chemistry.

Chemical Reactions and Equations: The Sentences of the Chemical Language

The formation of chemical compounds, their decomposition, and their interactions with one another fall under the category of **chemical reactions**. Chemical reactions are involved in the annual production of millions of kilograms of industrial chemicals, bacterially mediated degradation of water pollutants, the chemical analysis of the kinds and quantities of components of a sample, and practically any other operation involving chemicals. To a very large extent, chemistry is the study of chemical reactions expressed on paper as chemical equations. A **chemical equation** is a sentence of chemistry, made up of words consisting of chemical formulas. A sentence should be put together according to rules understood by all those literate in the language. The rules of the chemical language are particularly rigorous. Although a grammatically sloppy sentence in a spoken language can still convey a meaningful message, a chemical equation with even a small error is misleading and often meaningless.

Quantitative Calculations from Chemical Equations

Chemistry is a quantitative science, and it is important to know how to do some of the basic chemical calculations early in a beginning chemistry course. Among the most important of these are the calculations of the quantities of substances consumed or produced in a chemical reaction. Such calculations are classified as **stoichiometry**. Heat is normally evolved or taken up in the course of a chemical reaction. The calculation of the quantity of heat involved in a reaction falls in the branch of chemistry called **thermochemistry**.

5.2. THE INFORMATION IN A CHEMICAL EQUATION

Chemical Reactions

A chemical reaction is a process involving the breaking and/or formation of chemical bonds and a change in the chemical composition of the materials participating in the reaction. A chemical reaction might involve the combination of two elements to form a compound. An example of this is the reaction of elemental hydrogen and oxygen to produce the compound water. Passage of an electrical current through water can cause the compound to break down and produce elemental hydrogen and oxygen. When wood burns, cellulose, a compound in the wood, reacts with elemental oxygen in air to produce two compounds: carbon dioxide and water. If the carbon dioxide produced is bubbled through a solution of the compound calcium hydroxide dissolved in water, it produces the compounds calcium carbonate (a form of limestone) and water. Energy is involved in chemical reactions; some reactions produce energy, others require it in order for them to occur.

Expressing a Chemical Reaction as a Chemical Equation

A **chemical equation** is a means of expressing what happens when a chemical reaction occurs. It tells what reacts, what is produced, and the relative quantities of each. The information provided can best be understood by examining a typical chemical equation. For example, consider the burning of propane, a gas extracted from petroleum that is widely used for heating, cooking, grain drying, and other applications in which a clean-burning fuel is needed in areas where piped natural gas is not available. When propane burns in a camp stove, it reacts with oxygen in the air. The chemical equation for this reaction and the information in it are the following:

$$C_3H_8 + 5O_2 \rightarrow 3CO_2 + 4H_2O \tag{5.2.1}$$

- Propane reacts with oxygen to give carbon dioxide and water.
- There are two reactants on the left-hand side of the equation: propane, chemical formula C_3H_8, and oxygen, chemical formula O_2.
- There are two products on the right side of the equation: carbon dioxide, chemical formula CO_2, and water, chemical formula H_2O.
- For the smallest possible unit of this reaction, 1 propane molecule reacts with 5 oxygen molecules to produce 3 carbon dioxide molecules, and 4 water molecules, as shown by the respective numbers preceding the chemical formulas (there is understood to be a 1 in front of the C_3H_8).
- There are 3 C atoms altogether on the left side of the equation, all contained in the C_3H_8 molecule, and 3 C atoms on the right side contained in 3 molecules of CO_2.
- There are 8 H atoms among the reactants, all in the C_3H_8 molecule, and 8 H atoms among the 4 molecules of H_2O in the products.
- There are 10 O atoms in the 5 O_2 molecules on the left side of the equation and 10 O atoms on the right side. The 10 O atoms in the products are present in 3 CO_2 molecules and 4 H_2O molecules.

Like mathematical equations, the left-hand side of a chemical equation must be equivalent to the right-hand side. Chemical equations are **balanced** in terms of atoms:

A correctly written chemical equation has equal numbers of each kind of atom on both sides of the equation.

As has just been seen, the chemical equation under discussion has 3 carbon atoms, 8 hydrogen atoms and 10 oxygen atoms on both the left- and right-hand sides. Therefore, the equation is balanced. Balancing a chemical equation is a very important operation that is discussed in the next section.

Symbols Used in Chemical Equations

Several symbols are used in chemical equations. The two sides of the equation may be separated by an arrow, $\rightarrow$ or a double arrow, $\rightleftarrows$. The double arrow denotes a reversible reaction, that is, one that can go in either direction. The physical state (see Section 2.5) of a reaction component is indicated by letters in parentheses immediately following the formula. Thus, (*s*) stands for a solid, (*l*) for a liquid, (*g*) for a gas, and (*aq*) for a substance dissolved in aqueous (water) solution. An arrow pointing up, $\uparrow$, immediately after the formula of a product indicates that the product is evolved as a gas, whereas ↓ shows that it is a precipitate (a solid forming from a reaction in solution and settling to the bottom of the container). These two symbols are not used extensively in this book, but they are encountered in some of the older chemical literature. The symbol Δ over the arrow dividing products from reactants shows that heat is applied to the reaction. As an example of the uses of some of the symbols just defined above, consider the following reaction:

$$CaCO_3(s) + H_2SO_4(aq) \xrightarrow{\Delta} CaSO_4(s) + CO_2(g) + H_2O(l) \quad (5.2.2)$$

This reaction shows that solid calcium carbonate reacts with a heated solution of sulfuric acid dissolved in water to form solid calcium sulfate, carbon dioxide gas, and liquid water.

5.3. BALANCING CHEMICAL EQUATIONS

As indicated in the preceding section, a correctly written chemical equation has equal numbers of atoms of each element on both sides of the equation. Balancing a chemical equation is accomplished by placing the correct number in front of each formula in the chemical equation. However, the following must be remembered:

> *Only the numbers in front of the chemical formulas may be changed to balance a chemical equation. The chemical formulas themselves (subscript numbers) may not be changed in balancing the equation.*

Balancing an equation is best accomplished by considering one element at a time, balancing it by changing the numbers preceding the formulas in which it is contained, then successively balancing other elements in the formulas contained in the equation.

Balancing the Equation for the Reaction of Hydrogen Sulfide with Sulfur Dioxide

As an example of how to balance a chemical equation, consider the reaction of hydrogen sulfide gas (H_2S) with sulfur dioxide (SO_2) to yield elemental sulfur (S) and water (H_2O). This reaction is the basis of the Claus process by which commercially valuable elemental sulfur is recovered from pollutant sulfur dioxide and from toxic hydrogen sulfide in "sour" natural gas. The steps used in balancing the equation are the following:

1. Write the correct formulas of the reactants and products on either side of the equation. *These must remain the same throughout the balancing process.*

$$\underbrace{\underset{\text{Sulfur dioxide}}{SO_2} \quad + \quad \underset{\text{Hydrogen Sulfide}}{H_2S}}_{\text{Reactants}} \quad \rightarrow \quad \underbrace{\underset{\text{Sulfur}}{S} \quad + \quad \underset{\text{Water}}{H_2O}}_{\text{Products}}$$

2. Choose an element to balance initially, preferably one that is contained in only one reactant and one product. In this case, oxygen may be chosen. The 2 oxygen atoms in the SO_2 molecule on the left may be balanced by placing a 2 in front of the H_2O product:

$$SO_2 + H_2S \rightarrow S + 2H_2O$$

3. Choose another element in one of the formulas involved in the preceding operation and balance it on both sides of the equation. In this case, the H in H_2O can be balanced by placing a 2 in front of H_2S:

$$SO_2 + 2H_2S \rightarrow S + 2H_2O$$

4. Proceed to the remaining element. So far, sulfur has not yet been considered. There are 3 sulfur atoms on the left, contained in 1 SO_2 molecule and $2H_2S$ molecules. We have already considered these molecules in preceding operations and should avoid changing the numbers of either one. However, sulfur can be balanced by placing a 3 in front of the S product:

$$SO_2 + 2H_2S \rightarrow 3S + 2H_2O$$

5. Add up the numbers of each kind of element on both sides of the equation to see if they balance. In this case, it is seen that there are 3 S atoms, 4 H atoms, and 2 O atoms on both sides, so that the equation is, in fact, balanced.

Some Other Examples of Balancing Equations

Two other examples of balancing equations are considered here. The first of these is for the combustion in a moist atmosphere of aluminum phosphide, AlP, to give aluminum oxide and phosphoric acid, H_3PO_4. The unbalanced equation for this reaction is

$$AlP + O_2 + H_2O \rightarrow Al_2O_3 + H_3PO_4$$

The steps in balancing this equation are the following, starting with Al:

$$2AlP + O_2 + H_2O \rightarrow Al_2O_3 + H_3PO_4$$

Balance P:

$$2AlP + O_2 + H_2O \rightarrow Al_2O_3 + 2H_3PO_4$$

Balance H:

$$2AlP + O_2 + 3H_2O \rightarrow Al_2O_3 + 2H_3PO_4$$

Balance O:

$$2AlP + 4O_2 + 3H_2O \rightarrow Al_2O_3 + 2H_3PO_4 \tag{5.3.1}$$

Check each element for balance:

Reactants	**Products**
2 Al in 2 AlP	2 Al in 1 Al_2O_3
11 O in 4 O_2 and 3 H_2O	11 O in 1 Al_2O_3 and 2 H_3PO_4
6 H in 3 H_2O	6 H in 2 H_3PO_4

The second example of balancing equations is illustrated for trimethylchlorosilane, a flammable liquid used to produce high-purity silicon for semiconductor applications. Transportation accidents have resulted in spillage of this chemical and fires that produce carbon dioxide and a fog of silicon dioxide and hydrogen chloride dissolved in water droplets. The unbalanced equation for this reaction is

$$(CH_3)_3SiCl + O_2 \rightarrow CO_2 + H_2O + SiO_2 + HCl$$

Si and Cl are balanced as the equation stands, so balance H:

$$(CH_3)_3SiCl + O_2 \rightarrow CO_2 + 4H_2O + SiO_2 + HCl$$

Balance C:

$$(CH_3)_3SiCl + O_2 \rightarrow 3CO_2 + 4H_2O + SiO_2 + HCl$$

Balance O:

$$(CH_3)_3SiCl + 6O_2 \rightarrow 3CO_2 + 4H_2O + SiO_2 + HCl \tag{5.3.2}$$

Checking the quantities of each of the elements in the products and reactants shows that the equation is balanced.

In some cases, the presence of a diatomic species such as O_2 necessitates doubling quantities of everything else. As an example, consider the combustion of methane (natural gas, CH_4) in an oxygen-deficient atmosphere such that toxic carbon monoxide is produced. The unbalanced equation is

$$CH_4 + O_2 \rightarrow CO + H_2O$$

Balancing C and H gives

$$CH_4 + O_2 \rightarrow CO + 2H_2O$$

This puts a total of 3 O atoms on the right, so the addition of a single oxygen atom to the reactants side to give a total of 3 Os on the left would balance oxygen. This can be done by taking $\frac{3}{2}\,O_2$ to give

$$CH_4 + 3/2\,O_2 \rightarrow CO + 2H_2O$$

Ordinarily, however, integer numbers should be used for the coefficients. The fraction can be eliminated by multiplying everything by 2 to give the balanced equation:

$$2CH_4 + 3O_2 \rightarrow 2CO + 4H_2O \qquad (5.3.3)$$

Exercise: Balance the following:

1. $Fe_2O_3 + CO \rightarrow Fe + CO_2$
2. $FeSO_4 + O_2 + H_2O \rightarrow Fe(OH)_3 + H_2SO_4$
3. $C_2H_2 + O_2 \rightarrow CO_2 + H_2O$
4. $Mg_3N_2 + H_2O \rightarrow Mg(OH)_2 + NH_3$
5. $NaAlH_4 + H_2O \rightarrow H_2 + NaOH + Al(OH)_3$
6. $Zn(C_2H_5)_2 + O_2 \rightarrow ZnO + CO_2 + H_2O$

Answers: (1) $Fe_2O_3 + 3CO \rightarrow 2Fe + 3CO_2$; (2) $4Fe(SO)_4 + O_2 + 10H_2O \rightarrow 4Fe(OH)_3 + 4H_2SO_4$; (3) $2C_2H_2 + 5O_2 \rightarrow 4CO_2 + 2H_2O$; (4) $Mg_3N_2 + 6H_2O \rightarrow 3\,Mg(OH)_2 + 2NH_3$; (5) $NaAlH_4 + 4H_2O \rightarrow 4H_2 + NaOH + Al(OH)_3$; (6) $Zn(C_2H_5)_2 + 7O_2 \rightarrow ZnO + 4CO_2 + 5H_2O$.

Summary of Steps in Balancing an Equation

Below is a summary of steps that can be followed to balance a chemical equation. Keep in mind that there is a limit to the usefulness of following a set of rules for this procedure. Ultimately, it is a matter of experience and good judgment. In general, the best sequence of steps to take is the following:

1. Express the equation in words representing the compounds, elements, and (where present) the ions participating in the reaction.
2. Write down the correct formulas of all the reactants and all the products.
3. Examine the unbalanced equation for groups of atoms, such as those in the SO_4^{2-} ion, that go through the reaction intact. Balancing is simplified by considering these atoms as a group.
4. Examine the unbalanced equation for diatomic molecules, such as O_2, whose presence may require doubling the numbers in front of the other reaction participants.

5. Choose an element, preferably one that is found in only one reactant and one product, and balance that element by placing the appropriate numbers in front of both the reactants and products involved.

6. Balance another element that appears in one of the species balanced in the preceding step.

7. Continue the balancing process, one element at a time, until all the elements have been balanced.

8. Check to make sure that the same numbers of atoms of each kind of element appear on both sides of the equation and that the charges from charged species (ions, whose appearance in chemical equations will be considered later) in the equation also balance on the left and right.

Exercise: The reaction of liquid hydrazine, N_2H_4, with liquid dinitrogen tetroxide, N_2O_4, to produce nitrogen gas and water is used in some rocket engines for propulsion. Balance the equation for this reaction by going through the following steps:

(a) What is the unbalanced equation for the reaction?

(b) What is the equation after balancing O?

(c) What is the equation after balancing H?

(d) What is the equation after balancing N?

(e) How many atoms of each element are on both sides of the equation after going through these steps?

Answers: (a) $N_2O_4 + N_2H_4 \rightarrow H_2O + N_2$; (b) $N_2O_4 + N_2H_4 \rightarrow 4H_2O + N_2$; (c) $N_2O_4 + 2N_2H_4 \rightarrow 4H_2O + N_2$; (d) $N_2O_4 + 2N_2H_4 \rightarrow 4H_2O + 3N_2$; (e) 6 N, 4 O, 8 H.

5.4. WILL A REACTION OCCUR?

It is possible to write chemical equations for reactions that do not occur, or which occur only to a limited extent. This can be illustrated with a couple of examples. Consider the laboratory problem faced by a technician performing studies of plant nutrient metal ions leached from soil by water. The technician was using atomic absorption analysis, a sensitive instrumental technique for the determination of metal ions in solution. While determining the concentration of zinc ion, Zn^{2+}, dissolved in the soil leachate, the technician ran out of standard zinc solution used to provide known concentrations of zinc to calibrate the instrument, so that its readings would give known values from the sample solutions. Each liter of the standard solution contained exactly 1 mg of zinc in the form of dissolved zinc chloride, $ZnCl_2$. The technician reasoned that such a solution could be prepared by weighing out 100 mg of pure zinc metal, dissolving it in a solution of hydrochloric acid (HCl), diluting the solution to a volume of 1000 mL, and in turn diluting 10mL of that solution to 1000 mL to give the desired solution containing 1mg of zinc per liter. After thinking a bit, the technician came up with the following equation to describe the chemical reaction:

$$Zn(s) + 2HCl(aq) \rightarrow H_2(g) + ZnCl_2(aq) \tag{5.4.1}$$

When the 100 mg piece of zinc metal was added to some hydrochloric acid in a flask, bubbles of hydrogen gas were evolved, the zinc dissolved as zinc chloride, and the standard solution containing the desired concentration of dissolved zinc was prepared according to the plan.

Later in the investigation, the technician ran out of standard copper solution containing 1 mg of copper per liter in the form of dissolved copper(II) chloride, $CuCl_2$. The same procedure that was used to prepare the standard zinc solution was tried, with a 100 mg piece of copper wire substituted for the zinc metal. However, nothing happened to the copper metal when it was placed in hydrochloric acid. No amount of heating, stirring, or waiting could persuade the copper wire to dissolve. The technician wrote the following chemical equation for the reaction, analogous to that of zinc:

$$Cu(s) + 2HCl(aq) \rightarrow H_2(g) + CuCl_2(aq) \tag{5.4.2}$$

But copper metal and hydrochloric acid simply do not react with each other (Figure 5.1). Even though a plausible chemical equation can be written for a reaction, it does not tell whether the chemical reaction will, in fact, occur. Consideration of whether particular reactions take place is considered in the realms of chemical thermodynamics and chemical equilibrium, covered in later chapters.

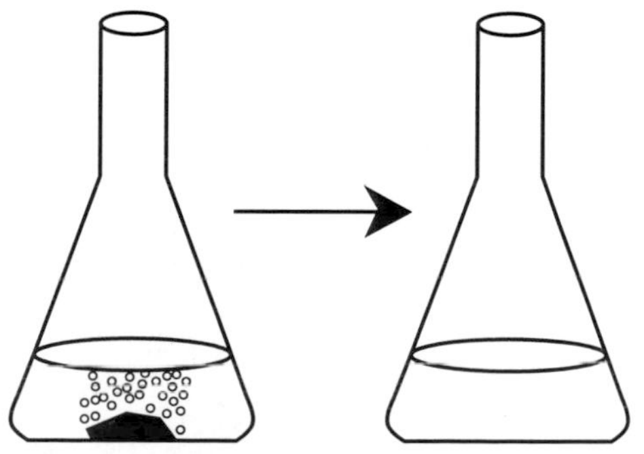

A piece of zinc metal in contact with hydrochloric acid reacts rapidly, giving off hydrogen gas and going into solution as $ZnCl_2$.

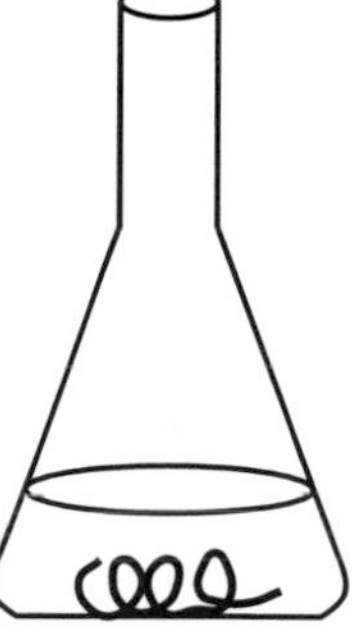

A piece of copper metal (wire) placed in hydrochloric acid does not react.

Figure 5.1 Illustrating the variable reactivities of substances: zinc metal reacts readily with hydrochloric acid, whereas copper metal does not react at all with this acid.

5.5. HOW FAST DOES A REACTION GO?

Consider a spoonful of table sugar, sucrose, exposed to air (Figure 5.2). Will sucrose, chemical formula $C_{12}H_{22}O_{11}$, react with oxygen in the air? The following equation can be written for the chemical reaction that might occur:

$$C_{12}H_{22}O_{11}(s) + 12O_2(g) \rightarrow 12CO_2(g) + 11H_2O(l) \quad (5.5.1)$$

One may have an intuitive feeling that this reaction should occur from having seen sugar burn in a fire, or from knowing that the human body "burns" sugar to obtain energy. Furthermore, it is true that, from the standpoint of energy, the atoms shown in the above equation are more stable when present as 12 molecules of CO_2 and 11 molecules of H_2O, rather than as one molecule of $C_{12}H_{22}O_{11}$ and 12 molecules of O_2. But anyone knows from experience that a spoonful of sugar can be exposed to dry air for a very long time without the occurrence of any visible change.

The answer to the question raised above lies in the **rate of reaction**. Sucrose does, indeed, tend to react with O_2 as shown in the chemical equation above. But, at room temperature, the reaction is just too slow to be significant. Of course, if the sugar were thrown into a roaring fire in a fireplace, it would burn rapidly. Special proteins known as enzymes inside of living cells can bring about the reaction of sugar and oxygen at body temperature of about 37°C, enabling the body to use the energy from the reaction. The enzymes themselves are not used up in the reaction, although they speed it up greatly; a substance that acts in such a manner is called a **catalyst**.

An important distinction must be made between reactions, such as that between copper and hydrochloric acid, that will not occur under any circumstances and others that "want to occur," but that are just too slow to be perceptible at moderate temperatures or in the absence of a catalyst. The latter type of reaction often does take place under the proper conditions or with a catalyst. Rates of reactions are quite important in chemistry.

5.6. CLASSIFICATION OF CHEMICAL REACTIONS

Chemical reactions may involve several kinds of processes. Reactions may consist of putting elements together to make compounds, or taking compounds apart to produce the component elements. Reactions may occur between compounds, between

Figure 5.2 Sugar exposed to air at room temperature does not react with the oxygen in the air at a detectable rate.

compounds and elements, or between ions. Many reactions involve the transfer of electrons, whereas others do not. Given these possibilities, plus others, it is helpful to categorize chemical reactions in several major classes. These are defined below.

A **combination reaction** is one in which two reactants bond together to form a single product. An example of such a reaction is provided by the burning of elemental phosphorus as one of the steps in the manufacture of phosphoric acid, a widely used industrial chemical and fertilizer ingredient. The reaction is

$$P_4 + 5O_2\,(g) \rightarrow P_4O_{10} \qquad (5.6.1)$$

in which elemental phosphorus, P_4, burns in oxygen to produce tetraphosphorus decoxide, P_4O_{10}. This is one example of the many combination reactions in which two elements combine together to form a compound. The general classification, however, may be applied to combinations of two compounds or of a compound and an element to form a compound. For example, the P_4O_{10} produced in the preceding reaction is combined with water to yield phosphoric acid:

$$6H_2O + P_4O_{10} \rightarrow 4H_3PO_4 \qquad (5.6.2)$$

A **decomposition reaction** is the opposite of a combination reaction. An example of a decomposition reaction in which a compound decomposes to form the elements in it is provided by the manufacture of carbon black. This material is a finely divided form of pure carbon, C, and is used as a filler in rubber tire manufacture and as an ingredient in the paste used to fill electrical dry cells. It is made by heating methane (natural gas) to temperatures in the range of 1260–1425°C in a special furnace, causing the following reaction to occur:

$$CH_4(g) \xrightarrow{\Delta} C(s) + 2H_2(g) \qquad (5.6.3)$$

The finely divided carbon black product is collected in a special device called a cyclone collector, shown in Figure 5.3, and the hydrogen gas by-product is recycled as a fuel to the furnace that heats the methane. As illustrated by the preceding reaction, a reaction in which a compound is broken down into its component elements is a decomposition reaction.

Decomposition reactions may also involve the breakdown of a compound to another compound and an element, or to two or more compounds. As an example of the latter, consider the reaction for the manufacture of calcium oxide, CaO, commonly called quicklime:

$$\underset{\substack{\text{Calcium}\\\text{carbonate}\\\text{(limestone)}}}{CaCO_3(s)} \xrightarrow{\Delta} \underset{\substack{\text{Calcium oxide}\\\text{(quicklime)}}}{CaO(s)} + CO_2(g) \qquad (5.6.4)$$

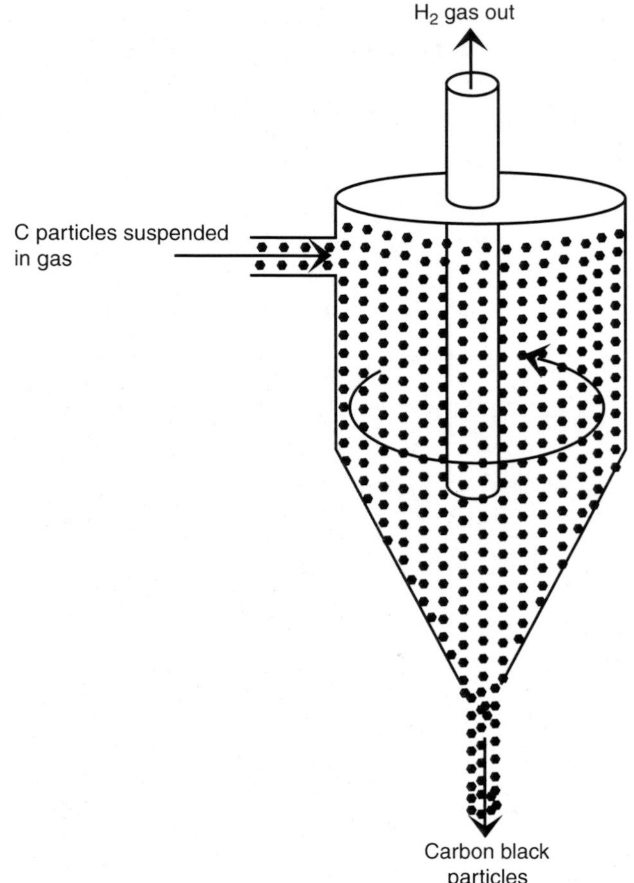

Figure 5.3 A cyclone collector is used to collect carbon black particles from a gas stream.

In this reaction, high temperatures are used to decompose limestone ($CaCO_3$) to CaO and carbon dioxide gas. This is an important reaction because quicklime ranks second only to sulfuric acid in annual chemical production. It is used in the manufacture of cement, for water treatment, and in many other applications.

Substitution or **replacement** reactions occur when a component of a chemical compound is replaced by something else. For example, zinc displaces hydrogen in Reaction 5.4.1. A **double replacement** or **metathesis** reaction occurs when there is a two-way exchange of ions between compounds. This happens, for example, when a solution of sulfuric acid reacts with a solution of barium hydroxide to yield solid barium sulfate and water:

$$H_2SO_4(aq) + Ba(OH)_2(aq) \rightarrow BaSO_4(s) + 2H_2O(l) \tag{5.6.5}$$

This reaction also falls into two other categories. Because it involves the combination of H^+ ions (from H_2SO_4) and OH^- ions (from $Ba(OH)_2$) to produce water, it is a neutralization reaction. This particular neutralization reaction is also a precipitation reaction because of the formation of a solid, $BaSO_4$. Such a solid formed by the reaction of two dissolved chemicals is called a precipitate.

The reaction of H^+ from any acid with OH^- to produce water is a neutralization reaction.

Evolution of a gas can also be used as a basis for classifying reactions. An example is provided by the treatment of industrial wastewater containing dissolved ammonium chloride, NH_4Cl. This compound is composed of the ammonium ion, NH_4^+, and the chloride ion, Cl^-. The commercially valuable byproduct ammonia gas can be recovered from such water by the addition of calcium hydroxide, resulting in the evolution of ammonia gas, NH_3, which can be recovered:

$$Ca(OH)_2(aq) + 2NH_4Cl(aq) \rightarrow 2NH_3(g) + H_2O(l) + CaCl_2(aq) \qquad (5.6.6)$$

Exercise: Classify each of the following reactions as combination, decomposition, substitution, metathesis, neutralization, precipitation, or evolution of a gas. In some cases, a reaction will fit into more than one category.

(a) $2Mg(s) + O_2(g) \rightarrow 2MgO(s)$

(b) $2KClO_3(s) \xrightarrow{\Delta} 2KCl(s) + 3O_2(g)$

(c) $SO_2(g) + H_2O(l) \rightarrow H_2SO_3(aq)$

(d) $CaCO_3(s) + 2HCl(aq) \rightarrow CaCl_2(aq) + H_2O(l) + CO_2(g)$

(e) $Fe(s) + CuCl_2(aq) \rightarrow Cu(s) + FeCl_2(aq)$

(f) $NaOH(aq) + HCl(aq) \rightarrow NaCl(aq) + H_2O(l)$

(g) $MgCl_2(aq) + 2NaOH(aq) \rightarrow Mg(OH)_2(s) + 2NaCl(aq)$

Answers: (a) combination; (b) decomposition, evolution of a gas; (c) combination; (d) metathesis, evolution of a gas; (e) substitution; (f) neutralization, metathesis; (g) precipitation, metathesis.

5.7. QUANTITATIVE INFORMATION FROM CHEMICAL REACTIONS

Review of Quantitative Chemical Terms

So far, chemical equations have been described largely in terms of individual atoms and molecules. Chemistry deals with much larger quantities, of course. On an industrial level, kilograms, tons, or even thousands of tons are commonly used. It is easy to scale up to such large quantities, because the relative quantities of materials involved remain the same, whether one is dealing with just a few atoms and molecules, or train-car loads of material. Before proceeding with the discussion of quantitative calculations with chemical equations, it will be helpful to consider some terms that have been defined previously:

Formula mass: The sum of the atomic masses of all the atoms in a formula unit of a compound. Although the average masses of atoms and molecules may be expressed in atomic mass units (amu or u), formula mass is generally viewed as being relative and without units.

Molar mass: When *X* is the formula mass, the molar mass is *X* grams of an element or compound, that is, the mass in grams of 1 mole of the element or compound.

Mole: The fundamental unit for quantity of material. Each mole contains Avogadro's number (6.022×10^{23}) of formula units of the element or compound.

Formula	Formula Mass	Molar Mass	Number of Formula Units per Mole
H	1.01	1.01 g/mol	6.022×10^{23} H atoms/mol
N	14.01	14.01 g/mol	6.022×10^{23} N atoms/mol
N_2	$2 \times 14.01 = 28.02$	28.02 g/mol	6.022×10^{23} N_2 molecules/mol
NH_3	$14.01 + 3 \times 1.01 = 17.04$	17.04 g/mol	6.022×10^{23} NH_3 molecules/mol
CaO	$40.08 + 16.00 = 56.08$	56.08 g/mol	6.022×10^{23} formula units CaO/mol[a]

[a] Since CaO consists of Ca^{2+} and O^{2-} ions, there are not really individual CaO molecules, so it is more correct to refer to a formula unit of CaO consisting of 1 Ca^{2+} ion and 1 O^{2-} ion.

Calcination of Limestone

To illustrate some of the quantitative information that may be obtained from chemical equations, consider the calcination of limestone (calcium carbonate, $CaCO_3$) to make quicklime (CaO) for water treatment (Figure 5.4):

$$CaCO_3(s) \xrightarrow{\Delta} CaO(s) + CO_2(g) \tag{5.7.1}$$

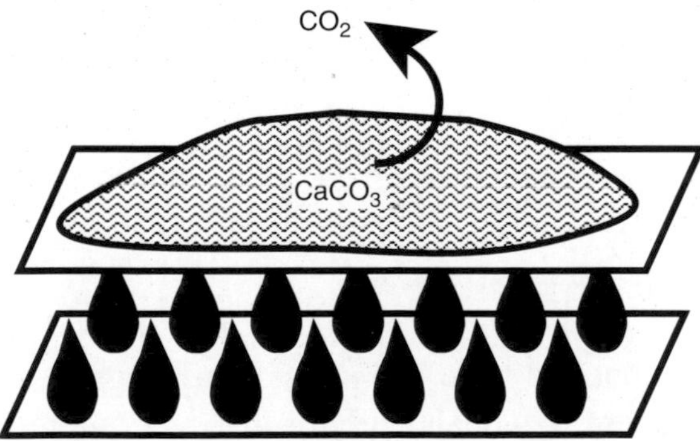

Figure 5.4 Heating calcium carbonate, $CaCO_3$, to a high temperature results in the production of quicklime, CaO, and the evolution of carbon dioxide. The reaction is a decomposition reaction, and the process is called calcination.

The quantitative information contained in this equation can be summarized as follows:

$$CaCO_3(s) \xrightarrow{\Delta} CaO(s) + CO_2(g)$$

At the formula unit (molecular) *level*		
1 formula unit	1 formula unit	1 molecule
1 Ca atom (atomic mass 40.1)	1 Ca atom	1 C atom
1 C atom (atomic mass 12.0)	1 O atom	2 O atoms
3 O atoms (atomic mass 16.0)		
100.1 u	56.1 u	44.0 u
At the mole level		
1 mole of $CaCO_3$	1 mole of CO	1 mole of CO_2
100.1 g of $CaCO_3$	56.1 g of CO	44.0 g of CO_2

From the quantities above, it can be seen that the equation can be viewed in terms as small as the smallest number of molecules and formula units. In this case, that involves simply 1 formula unit of $CaCO_3$ (*s*), 1 formula unit of CaO (*s*), and 1 molecule of CO_2. This would involve a total of 1 Ca atom, 1 C atom, and 3 O atoms. From such a small scale, it is possible to expand to moles by scaling up by 6.022×10^{23} (Avogadro's number), giving 100.1 g of $CaCO_3$, 56.1 g of CaO, and 44.0 g of CO_2. Actually, these quantitative relationships are applicable to any amount of matter, and they enable the calculation of the amounts of material reacting and produced in a chemical reaction. Next, it is shown how these kinds of calculations are performed.

5.8. WHAT IS STOICHIOMETRY AND WHY IS IT IMPORTANT?

Stoichiometry is the calculation of the quantities of reactants or products involved in a chemical reaction. The importance of stoichiometry can be appreciated by visualizing industrial operations that process hundreds or thousands of tons of chemicals per day. The economics of many chemical manufacturing processes are such that an unnecessary excess of only a percent or so of a reacting chemical can lead to waste that can make the operation unprofitable. Obtaining accurate values in chemical analysis, which may need to be known to within about a part per thousand, often involves highly exacting stoichiometric calculations.

Stoichiometric calculations are based upon **the law of conservation of mass**, which states the following:

The total mass of reactants in a chemical reaction equals the total mass of products.

Matter is neither created nor destroyed in chemical reactions.

The key to performing stoichiometric calculations correctly is the following:

> *The relative masses (or number of moles, number of atoms, or molecules) of the participants in a designated chemical reaction remain the same, regardless of the overall quantities of reaction participants.*

The Mole Ratio Method of Stoichiometric Calculations

In a chemical reaction, there is a definite ratio between the number of moles of a particular reactant or product and the number of moles of any other reactant or product. These ratios are readily seen by simply examining the coefficients in front of the reaction species in the chemical equation. Normally, a stoichiometric calculation is performed to relate the quantities of only two of the reaction participants. The objective may be to determine how much of one reactant will react with a given quantity of another reactant. Or, a particular quantity of a product may be desired, so that it is necessary to calculate the quantity of a specific reactant needed to give the amount of product. To perform stoichiometric calculations involving only two reaction participants, it is necessary only to know the relative number of moles of each and their molar masses. The most straightforward type of stoichiometric calculation is the **mole ratio method** defined as follows:

> *The **mole ratio method** is a means of performing stoichiometric calculations based upon the constant ratios of the numbers of moles of various reactants and products, regardless of the overall quantity of reaction taking place.*

The mole ratio method greatly simplifies stoichiometric calculations. It can even be used to relate relative quantities of reaction participants in a series of reactions. For example, if a particular quantity of reactant is involved in a reaction followed by one or more additional reactions, the amount of a product in the final reaction is readily calculated by the mole ratio method.

To illustrate the mole ratio method, consider a typical reaction, namely, that of hydrogen gas and carbon monoxide gas to produce methane:

$$\begin{array}{cccc} 3H_2 + & CO \rightarrow & CH_4 + & H_2O \\ 3\text{ mol} & 1\text{ mol} & 1\text{ mol} & 1\text{ mol} \\ 6.0\text{ g} & 28.0\text{ g} & 16.0\text{ g} & 18.0\text{ g} \end{array} \tag{5.8.1}$$

This reaction is called the **methanation reaction**, and is used in the petroleum and synthetic fuels industry for the manufacture of synthetic natural gas (CH_4, the least polluting of the fossil fuels). From examination of the chemical equation, it is easy to obtain the ratio of the number of moles of any reaction participant to the number of moles of any other reaction participant, as shown in Table 5.1.

Table 5.1. Mole Ratios Used in Calculations with the Methanation Reaction

Equality of Number of Moles	Mole Ratios	
$3 \text{ mol } H_2 = 1 \text{ mol CO}$	$\frac{3 \text{ mol } H_2}{1 \text{ mol CO}}$	$\frac{1 \text{ mol CO}}{3 \text{ mol } H_2}$
$3 \text{ mol } H_2 = 1 \text{ mol } CH_4$	$\frac{3 \text{ mol } H_2}{1 \text{ mol } CH_4}$	$\frac{1 \text{ mol } CH_4}{3 \text{ mol } H_2}$
$3 \text{ mol } H_2 = 1 \text{ mol } H_2O$	$\frac{3 \text{ mol } H_2}{1 \text{ mol } H_2O}$	$\frac{1 \text{ mol } H_2O}{3 \text{ mol } H_2}$
$1 \text{ mol CO} = 1 \text{ mol } CH_4$	$\frac{1 \text{ mol CO}}{1 \text{ mol } CH_4}$	$\frac{1 \text{ mol } CH_4}{1 \text{ mol CO}}$
$1 \text{ mol CO} = 1 \text{ mol } H_2O$	$\frac{1 \text{ mol CO}}{1 \text{ mol } H_2O}$	$\frac{1 \text{ mol } H_2O}{1 \text{ mol CO}}$
$1 \text{ mol } CH_4 = 1 \text{ mol } H_2O$	$\frac{1 \text{ mol } CH_4}{1 \text{ mol } H_2O}$	$\frac{1 \text{ mol } H_2O}{1 \text{ mol } CH_4}$

The ratios given in Table 5.1 enable calculation of the number of moles of any reaction participant, if the number of moles of any other participant is known. For example, if it is known that 1.00 mol of H_2 reacts, then the calculation of the number of moles of CH_4 produced is simply the following:

$$1.00 \text{ mol } H_2 \times \frac{1 \text{ mol } CH_4}{3 \text{ mol } H_2} = 0.333 \text{ mol } CH_4 \quad (5.8.2)$$

Calculation of the mass of a substance requires conversion between moles and mass. Suppose that one needs to know the mass of H_2 required to produce 4.00 g of CH_4. The first step is to convert the mass of CH_4 to moles, given that its molar mass is 16.0 g/mol:

$$4.00 \text{ g } CH_4 \times \frac{1 \text{ mol } CH_4}{16.0 \text{ g } CH_4} = 0.250 \text{ mol } CH_4 \quad (5.8.3)$$

The next step is to multiply by the mole ratio of H_2 to CH_4:

$$0.250 \text{ mol } CH_4 \times \frac{3 \text{ mol } H_2}{1 \text{ mol } CH_4} = 0.750 \text{ mol } H_2 \quad (5.8.4)$$

The last step is to multiply by the molar mass of H_2:

$$0.750 \text{ mol } H_2 \times \frac{2.00 \text{ g } H_2}{1 \text{ mol } H_2} = 1.50 \text{ g } H_2 \quad (5.8.5)$$

All of these steps can be performed at once, as shown below:

$$4.00 \text{ g } CH_4 \times \frac{1 \text{ mol } CH_4}{16.0 \text{ g } CH_4} \times \frac{3 \text{ mol } H_2}{1 \text{ mol } CH_4} \times \frac{2.00 \text{ g } H_2}{1 \text{ mol } H_2} = 1.50 \text{ g } H_2 \quad (5.8.6)$$

Several problems will be presented that illustrate the mole ratio method. First, however, it will be helpful to learn the following steps used in solving a stoichiometric problem by this method:

1. Write the balanced chemical equation for the reaction involved.
2. Identify the reactant or product whose quantity is known (known substance) and the one whose quantity is being calculated (desired substance).
3. Express the number of moles of the known substance, which usually must be calculated from its mass.
4. Multiply the number of moles of the known substance by the mole ratio of the desired substance to obtain the number of moles of the latter:

   Moles desired substance = moles of known substance × mole ratio of desired substance to known substance

5. Calculate the number of grams of the desired substance by multiplying its molar mass by the number of moles:

   Mass in grams of desired substance = molar mass of desired substance × moles of desired substance

To illustrate these steps, consider the preparation of ammonia, NH_3, from hydrogen gas and atmospheric nitrogen. The reaction in words is

Hydrogen plus nitrogen yields ammonia

Insertion of the correct formulas gives

$$H_2 + N_2 \rightarrow NH_3 \tag{5.8.7}$$

and the equation is balanced by placing the correct coefficients in front of each formula:

$$\underset{3\ \text{mol}}{3H_2} + \underset{1\ \text{mol}}{N_2} \rightarrow \underset{2\ \text{mol}}{2NH_3} \tag{5.8.8}$$

As an example, calculatc thc number of grams of H_2 required for the synthesis of 4.25 g of NH_3 using the following steps:

1. Calculate the number of moles of NH_3 (molar mass 17.0 g/mol).

$$\text{Mol NH}_3 = 4.25\ \text{g NH}_3 \times \frac{1\ \text{mol NH}_3}{17.0\ \text{g NH}_3} = 0.250\ \text{g NH}_3$$

2. Express the mole ratio of H_2 to NH_3 from examination of Equation 5.8.8:

$$\frac{3 \text{ mol } H_2}{2 \text{ mol } NH_3}$$

3. Calculate the number of moles of H_2:

$$\text{Mol } H_2 = 0.250 \text{ mol } NH_3 \times \frac{3 \text{ mol } H_2}{2 \text{ mol } NH_3} = 0.375 \text{ mol } H_2$$

4. Calculate the mass of H_2:

$$0.375 \text{ mol } H_2 \times \frac{2.00 \text{ g } H_2}{1 \text{ mol } H_2} = 0.750 \text{ g } H_2$$

Once the individual steps involved are understood, it is easy to combine them all into a single calculation as follows:

$$4.25 \text{ g } NH_3 \times \frac{1 \text{ mol } NH_3}{17.0 \text{ g } NH_3} \times \frac{3 \text{ mol } H_2}{2 \text{ mol } NH_3} \times \frac{2.00 \text{ g } H_2}{1 \text{ mol } H_2} = 0.750 \text{ g } H_2 \quad (5.8.9)$$

As another example of a stoichiometric calculation by the mole ratio method, consider the reaction of iron(III) sulfate, $Fe(SO_4)_3$, with calcium hydroxide, $Ca(OH)_2$. This reaction is used in water treatment processes for the preparation of gelatinous iron(III) hydroxide, $Fe(OH)_3$, which settles in the water, carrying solid particles with it. The iron(III) hydroxide acts to remove suspended matter (turbidity) from water. The $Ca(OH)_2$ (slaked lime) is added as a base (source of OH^- ion) to react with iron(III) sulfate. The reaction is

$$Fe_2(SO_4)_3\ (aq) + 3Ca(OH)_2\ (aq) \rightarrow 2Fe(OH)_3\ (s) + 3CaSO_4\ (s) \quad (5.8.10)$$

Suppose that a mass of 1000 g of iron(III) sulfate is to be used to treat a tankful of water. What mass of calcium hydroxide is required to react with the iron(III) sulfate? The steps required to solve this problem are as follows:

1. Formula mass $Fe_2(SO_4)_3$

$$= \underbrace{2 \times 55.8}_{\text{2 Fe atoms}} + \underbrace{3 \times 32.0}_{\text{3 S atoms}} + \underbrace{12 \times 16.0}_{\text{12 O atoms}} = 399.6$$

Formula mass $Ca(OH)_2$

$$= \underbrace{1 \times 40.1}_{\text{1 Ca atom}} + \underbrace{2 \times 16.0}_{\text{2 O atoms}} + \underbrace{2 \times 1.0}_{\text{2 H atoms}} = 74.1$$

2. $$\text{Mol } Fe_2(SO_4)_3 = 1000 \text{ g } Fe_2(SO_4)_3 \times \frac{1 \text{ mol } Fe_2(SO_4)_3}{399.6 \text{ g } Fe_2(SO_4)_3}$$

3. $$\text{Mole ratio} = \frac{3 \text{ mol } Ca(OH)_2}{1 \text{ mol } Fe_2(SO_4)_3}$$

$$\text{4. Mass of } Ca(OH)_2 = 1000\text{ g } Fe_2(SO_4)_3 \times \frac{1\text{ mol } Fe_2(SO_4)_3}{399.6\text{ g } Fe_2(SO_4)_3} \times \frac{3\text{ mol } Ca(OH)_2}{1\text{ mol } Fe_2(SO_4)_3} \times \frac{74.1\text{ g } Ca(OH)_2}{1\text{ mol } Ca(OH)_2} = 556\text{ g } Ca(OH)_2$$

Exercise: Gelatinous aluminum hydroxide for the treatment of water can be generated by the following reaction with sodium bicarbonate:

$$Al_2(SO_4)_3\ (aq) + 6NaHCO_3\ (aq) \rightarrow 2Al(OH)_3\ (s) + 3Na_2SO_4\ (aq) + 6CO_2\ (g)$$

Calculate the mass in kilograms of $NaHCO_3$ required to react with 80.0 g of $Al_2(SO_4)_3$ and form a precipitate of $Al(OH)_3$.

Answer: The molar mass of $NaHCO_3$ is 72.0 and that of $Al_2(SO_4)_3$ is 342.3. From the above reaction, it can be seen that 6 mol of $NaHCO_3$ are required per mol of $Al_2(SO_4)_3$. Therefore, the calculation is as follows:

$$\text{Mass of } NaHCO_3 = 80.0\text{ g } Al_2(SO_4)_3 \times \frac{1\text{ mol } Al_2(SO_4)_3}{342.3\text{ g } Al_2(SO_4)_3} \times \frac{6\text{ mol } NaHCO_3}{1\text{ mol } Al_2(SO_4)_3} \times \frac{72.0\text{ g } NaHCO_3}{1\text{ mol } NaHCO_3} = 101.0\text{ g } NaHCO_3$$

CHAPTER SUMMARY

The chapter summary below is presented in a programmed format to review the main points covered in this chapter. It is used most effectively by filling in the blanks, referring back to the chapter as necessary. The correct answers are given at the end of the summary.

The formation of chemical compounds, their decomposition, and their interactions with one another fall under the category of [1]________________________. A chemical equation may be viewed as a [2]____________________________ in chemical language. Calculations of the quantities of substances consumed or produced in a chemical reaction are classified as [3]__________________________. The calculation of the quantity of heat involved in a reaction falls in the branch of chemistry called [4]____. Substances on the left side of a chemical equation are called [5]________________________________, whereas those on the right are [6]____. A correctly written chemical equation has equal numbers of [7]____ on each side. When used in a chemical equation, the symbols $\rightarrow$, $\rightleftarrows$, (*s*), (*l*), (*g*), (*aq*), $\uparrow$, $\downarrow$ and Δ stand for, [8]____

____, respectively. Chemical formulas [9]________________________ in balancing chemical equations. After expressing an equation in words, the next step in balancing it is to [10]________________________. ________________________ Balancing an equation may be simplified by considering groups of atoms that [11]________________________ ________________. The presence of [12]________________________ may require doubling the numbers in front of the other reaction participants. The final step in balancing an equation is [13]________________________ ____ ____ ____.

Simply because it is possible to write a chemical equation for a reaction does not necessarily mean that [14]________________________. Rates of reactions [15]________________________, but may be increased by a [16]________________ ________. The reaction $C + O_2 \rightarrow CO_2$ is an example of a [17]________________ ________ reaction. The reaction $2H_2O \rightarrow 2H_2 + O_2$ is an example of a [18]________________________ reaction. The reaction $H_2SO_4 + Fe \rightarrow FeSO_4$ $(s) + H_2$ is a [19]________________________. A neutralization reaction is [20]____, and a precipitation reaction is described as [21]________________________ ________________________. Formula mass is [22]________________ ____ ________________________. Molar mass is [23]________________ ____. Each mole of an element or compound contains [24]________________________ ____. The molar mass of NH_3 is [25]________________, a mass that contains [26]________________ individual H atoms. For the smallest whole numbers of moles, the equation

$$CaCO_3\,(s) \xrightarrow{\Delta} CaO\,(s) + CO_2\,(g)$$

states that [27]________ g of $CaCO_3$ are heated to produce [28]________ g of CaO and [29]________ g of CO_2. Stoichiometry is [30]________________________ ________________________. Stoichiometric calculations are based upon the law of conservation of mass, which states that [31]________________________ ____ ____. The "key to doing stoichiometric calculations correctly" is the fact that [32]________________________ ____ ____.

In a chemical reaction, there is a [33]____________________ between the number of moles of a particular reactant or product and the number of moles of any other reactant or product. Using the mole ratio method for stoichiometric calculations, the steps involved after writing the balanced chemical equation for the reaction are [34]____

____

____

____.

Answers to Chapter Summary

1. chemical reactions
2. sentence
3. stoichiometry
4. thermochemistry
5. reactants
6. products
7. each kind of atom
8. division between reactants and products, reversible reaction, solid, liquid, gas, substance dissolved in water, evolution of a gas, formation of a precipitate, and application of heat
9. cannot be changed
10. write down the correct formulas of all the reactants and all the products
11. go through the reaction intact
12. diatomic molecules, such as O_2
13. to check to make sure that the same number of atoms of each kind of element appear on both sides of the equation and that the charges from ions also balance on the left and right
14. the reaction will occur
15. vary greatly
16. catalyst
17. combination
18. decomposition

19. displacement

20. the reaction of H^+ from any acid with OH^- to produce water

21. one in which a solid forms from a reaction in solution

22. the sum of the atomic masses of all the atoms in a formula unit of a compound

23. the mass in grams of 1 mole of an element or compound

24. Avogadro's number (6.022×10^{23}) of formula units of the element or compound

25. 17.0 g/mol

26. $3 \times 6.022 \times 10^{23}$

27. 100.1

28. 56.1

29. 44.0

30. the calculation of the quantities of reactants or products involved in a chemical reaction

31. the total mass of reactants in a chemical reaction equals the total mass of products; that is, matter is neither created nor destroyed in chemical reactions

32. the relative masses (or number of moles, number of atoms, or molecules) of the participants in a designated chemical reaction remain the same, regardless of the overall quantities of reaction participants

33. definite ratio

34. (1) identify the reactant or product whose quantity is known and the one whose quantity is being calculated; (2) express the number of moles of the known substance; (3) multiply the number of moles of known substance times the mole ratio of desired substance to known substance to obtain the number of moles of desired substance; (4) calculate the number of grams of desired substance by multiplying its molar mass times its number of moles

QUESTIONS AND PROBLEMS

1. Describe chemical reactions and chemical equations and distinguish between them.

2. Briefly summarize the information in the chemical equation

$$CS_2 + 3O_2 \rightarrow CO_2 + 2SO_2$$

3. Summarize the information in the chemical equation, $H_2O(l)$ + $Na(s) \rightarrow NaOH(aq) + H_2(g)$, including the meanings of the terms in italics.

4. What are the meanings of the following symbols in a chemical equation: $\rightarrow$, $\rightleftarrows$, $\uparrow$, (s), (l), (g), (aq), $\downarrow$, and Δ?

5. A typical word statement of a chemical reaction is, "Ammonia reacts with sulfuric acid to yield ammonium sulfate." The compounds or the ions involved in them have been given earlier in this book. Using correct chemical formulas, show the unbalanced and balanced chemical equations for this reaction.

6. What is wrong with balancing $Ca + O_2 \rightarrow CaO$ as $Ca + O_2 \rightarrow CaO_2$?

7. Iron(II) sulfate dissolved in acid mine water, a pollutant from coal mines, reacts with oxygen from air according to the unbalanced equation $FeSO_4 + H_2SO_4 + O_2 \rightarrow Fe_2(SO_4)_3 + H_2O$. Examination of this equation reveals two things about groups of atoms that simplify the balancing of the equation. What are these two factors, and how can awareness of them help to balance the equation?

8. Balance the chemical equation from Question 7.

9. Balance each of the following: (a) $C_2H_6 + O_2 \rightarrow CO_2 + H_2O$; (b) $KClO_3 \rightarrow KCl + O_2$; (c) $Ag_2SO_4 + BaI_2 \rightarrow AgI + BaSO_4$; (d) $KClO_4 + C_6H_{12}O_6 \rightarrow KCl + CO_2 + H_2O$; (e) $Fe + O_2 \rightarrow Fe_2O_3 + H_2O$; (f) $P + Cl_2 \rightarrow PCl_3$.

10. Having studied Section 5.4, and knowing something about silver metal and its uses, suggest what would happen if a small item of silver jewelry were placed in a solution of hydrochloric acid.

11. From the information given about Reactions 5.4.1 and 5.4.2, suggest a reaction that might occur if a piece of zinc metal were placed in a solution of $CuCl_2$. Explain. If a chemical reaction does occur, write an equation describing the reaction.

12. Steel wool heated in a flame and thrust into a bottle of oxygen gas burns vigorously, implying that a reaction such as $4Fe + 3O_2 \rightarrow 2Fe_2O_3$ does in fact occur. Why does a steel girder, such as one used for bridge construction, not burn when exposed to air?

13. What does a catalyst do?

14. Classify each of the following reactions according to the categories that are given in Section 5.6:

    (a) $NaCl(l, \textit{melted liquid}) \xrightarrow[\text{current}]{\text{Electrical}} 2Na(l) + Cl_2(g)$

    (b) $Ca(OH)_2(aq) + CO_2(g) \rightarrow CaCO_3(s) + H_2O(l)$

    (c) $Zn(s) + CuCl_2(aq) \rightarrow Cu(s) + ZnCl_2(aq)$

    (d) $2NaOH(aq) + H_3PO_4(aq) \rightarrow Na_2HPO_4(aq) + 2H_2O(l)$

    (e) $BaCl_2(aq) + Ag_2SO_4(aq) \rightarrow BaSO_4(s) + 2AgCl(s)$

    (f) $(NH_4)_2SO_4(s) + Ca(OH)_2(s) \xrightarrow{\Delta} 2NH_3(g) + 2H_2O(g) + CaSO_4(s)$

15. Define (a) molecular mass (formula mass when the formula unit is a molecule), (b) mole, and (c) molar mass.

16. What are the three main entities that can compose a formula unit of a substance?

17. A total of 7.52 g of $AlCl_3$ contains (a)________ moles and (b) ________ formula units of the compound.

18. A total of 336 g of methane, CH_4, contains (a)____________ moles and (b) ______________ molecules of the compound.

19. Ammonium nitrate, NH_4NO_3, can be made by a reaction between NH_3 and HNO_3. A "recipe" for the manufacture of ammonium nitrate specifies mixing 270 kg of NH_3 with 1000 kg of HNO_3. What is the significance of the relative quantities of these two ingredients?

20. What are some reasons that stoichiometric ratios of reactants are used for many industrial processes? In what cases are stoichiometric ratios not used?

21. What is the relationship constituting the basis of stoichiometry between the masses of the reactants and products in a chemical reaction?

22. In addition to the law of conservation of mass, another important stoichiometric relationship involves the proportions in mass of the reaction participants. What is this relationship?

23. Briefly define and explain the mole ratio method of stoichiometric calculations.

24. Carbon disulfide, CS_2, burns rapidly in air because of the reaction with oxygen. What is the mole ratio of O_2 to CS_2 in this reaction?

25. What is the first step required to solve a problem by the mole ratio method?

26. What is the mass (g) of NO_2 produced by the reaction of 10.44 g of O_2 with NO, yielding NO_2?

27. Plants utilize light energy in the photosynthesis process to synthesize glucose, $C_6H_{12}O_6$, from CO_2 and H_2O by way of the reaction $6CO_2 + 6H_2O \rightarrow C_6H_{12}O_6 + 6O_2$. How many grams of CO_2 are consumed in the production of 90 g of glucose?

28. Bacteria in water utilize organic material as a food source, consuming oxygen in a process called respiration. If glucose sugar is the energy source, the reaction is $C_6H_{12}O_6 + 6O_2 \rightarrow 6CO_2 + 6H_2O$ (chemically, the reverse of the photosynthesis reaction in the preceding problem). At 25°C, the maximum amount of oxygen in 1.00 L of water due to dissolved air is 8.32 mg. What is the mass of glucose in mg that will cause all the oxygen in 1.50 L of water to be used up by bacterial respiration?

29. In a blast furnace, the overall reaction by which carbon in coke is used to produce iron metal from iron ore is $2Fe_2O_3 + 3C \rightarrow 4Fe + 3CO_2$. How many tons of C are required to produce 100 tons of iron metal?

30. The gravimetric chemical analysis of NaCl may be carried out by precipitating dissolved chloride from solution by the following reaction with silver nitrate solution: $AgNO_3(aq) + NaCl \rightarrow AgCl(s) + NaNO_3(aq)$. If this reaction produced 1.225g of AgCl, what was the mass of NaCl?

31. For the reaction of 100.0 g of NaOH with Cl_2, $2NaOH + Cl_2 \rightarrow NaClO + NaCl + H_2O$, give the masses of each of the products.

32. Hydrochloric acid (HCl gas dissolved in water) reacts with calcium carbonate in a piece of limestone as follows: $CaCO_3 + 2HCl \rightarrow CaCl_2 + CO_2 + H_2O$. If 14.6 g of CO_2 are produced in this reaction, what is the total mass of reactants and the total mass of the products?

33. Given the reaction in the preceding problem, how many moles of HCl are required to react with 0.618 moles of $CaCO_3$?

34. Given the reaction in Problem 32, how many moles of $CaCO_3$ are required to produce 100.0 g of $CaCl_2$?

35. Silicon tetrachloride, $SiCl_4$, is used to make organosilicon compounds (silicones) and produces an excellent smokescreen for military operations. In the latter application, the $SiCl_4$ reacts with atmospheric moisture (water), $SiCl_4(g) + 2H_2O(g) \rightarrow SiO_2(s) + 4HCl(g)$ to form particles of silicon dioxide and hydrogen chloride gas. The HCl extracts additional moisture from the atmosphere to produce droplets of hydrochloric acid,

which, along with small particles of SiO_2, constitute the "smoke" in the smokescreen. Air in a smokescreen was sampled by drawing 100 m^3 of the air through water to collect HCl and to cause any unreacted $SiCl_4$ to react according to the above reaction. After the sampling was completed, the water was found to contain 1.85 g of HCl. What was the original concentration of $SiCl_4$ in the atmosphere (before any of the above reactions occurred) in units of milligrams of $SiCl_4$ per cubic meter?

36. Match each reaction below with the type of reaction that it represents:

A. Combination (addition)

B. Decomposition

C. Metathetical (double displacement)

D. Neutralization

(1) $2NaOH(aq) + H_2SO_4(aq) \rightarrow Na_2SO_4(aq) + 2H_2O(l)$

(2) $Pb(NO_3)_2(aq) + H_2SO_4(aq) \rightarrow PbSO_4(s) + 2HNO_3(aq)$

(3) $CuSO_4 \cdot 5H_2O(s) \xrightarrow{\Delta} CuSO_4(s) + 5H_2O(g)$

(4) $4Al(s) + 3O_2(g) \rightarrow 2Al_2O_3(s)$

37. Given atomic masses of Al 27 and O 16.0, the mass of O_2 consumed and the mass of Al_2O_3 produced when 29.5 g of Al undergo the reaction $4Al + 3O_2 \rightarrow 2\,Al_2O_3$ are

A. 26.2 g O_2 and 55.7 g Al_2O_3

B. 23.4 g O_2 and 49.7 g Al_2O_3

C. 28.7 g O_2 and 64.7 g Al_2O_3

D. 30.6 g O_2 and 65.1 g Al_2O_3

E. 20.3 g O_2 and 49.8 g Al_2O_3

38. Given atomic masses H 1.0, S, 32.0, O 16.0, the mass of SO_2 that must react to yield 46.8 g of S by the reaction $2H_2S + SO_2 \rightarrow 3S + 2H_2O$ is

A. 35.1 g SO_2

B. 19.1 g SO_2

C. 31.2 g SO_2

D. 40.6 g SO_2

E. 28.3 g SO_2

39. Given atomic masses H 1.0, S, 32.0, O 16.0, the number of moles of H^+ ion that can be obtained from 24.0 g of H_2SO_4 is

 A. 0.669 mol H^+

 B. 0.490 mol H^+

 C. 1.22 mol H^+

 D. 0.555 mol H^+

 E. 0.380 mol H^+

40. Of the following, mark the statement that is **incorrect** regarding the equation below:

$$CaCO_3(s) + H_2SO_4(aq) \xrightarrow{\Delta} CaSO_4(s) + CO_2(g) + H_2O(l)$$

 A. $CaSO_4$ is dissolved in water.

 B. $CaCO_3$ is a reactant.

 C. CO_2 is a gas product.

 D. The reaction mixture was heated.

 E. H_2SO_4 was dissolved in water.

41. Given atomic masses of H 1.0, Fe 55.8, and O 16.0, the number of *moles* of CO_2 produced when 79.8 g of Fe_2O_3 undergo the reaction $Fe_2O_3 + 3CO \rightarrow 2Fe + 3CO_2$ are

 A. 2.56 mol of CO_2

 B. 1.36 mol of CO_2

 C. 1.50 mol of CO_2

 D. 1.14 mol of CO_2

 E. 1.62 mol of CO_2

42. Consider the combustion of tetraethyllead, $Pb(C_2H_5)_4$, formerly used as a gasoline octane booster, which burns in the presence of O_2 to give PbO, H_2O, and CO_2. What is the balanced chemical equation for this reaction? Assume that 1.00 g of tetraethyllead was burned. What mole ratio would be used to calculate the mass of H_2O produced? What mass of H_2O was produced?

6. ACIDS, BASES, AND SALTS

6.1. THE IMPORTANCE OF ACIDS, BASES, AND SALTS

Almost all inorganic compounds and many organic compounds can be classified as acids, bases, or salts. Some of these types of compounds were mentioned in earlier chapters, and they are discussed in greater detail in this chapter.

Acids, bases, and salts are vitally involved with life processes, agriculture, industry, and the environment (Figure 6.1). The most widely produced chemical is

Figure 6.1. Acid rain resulting from the introduction of sulfuric, nitric, and hydrochloric acids into the atmosphere by the burning of fossil fuels damages buildings, statues, crops, and electrical equipment in some areas of the world, including parts of the northeastern U.S.

an acid, sulfuric acid. The second-ranking chemical, lime, is a base. Another base, ammonia, ranks fourth in annual chemical production. Among salts, sodium chloride is widely produced as an industrial chemical; potassium chloride is a source of essential potassium fertilizer; and sodium carbonate is used in huge quantities for glass and paper manufacture and for water treatment.

The salt content and the acid–base balance of blood must stay within very narrow limits to keep a person healthy, or even alive. Soil with too much acid or excessive base will not support good crop growth. Too much salt in irrigation water may prevent crops from growing. This is a major agricultural problem in arid regions of the world such as the Middle East and California's Imperial Valley. The high salt content of irrigation water discharged to the Rio Grande River has been a source of dispute between the U.S. and Mexico that has been resolved to a degree by installation of a large desalination (salt removal) plant by the U.S.

From the above discussion, it can be seen that acids, bases, and salts are important to human health and welfare. This chapter discusses their preparation, properties, and naming.

6.2. THE NATURE OF ACIDS, BASES, AND SALTS

Hydrogen Ion and Hydroxide Ion

Recall that an ion is an atom or group of atoms having an electrical charge. In discussing acids and bases, two very important ions are involved. One of these is the **hydrogen ion, H^+**. It is always produced by acids. The other is the **hydroxide ion, OH^-**. It is always produced by bases. These two ions react together to produce water:

$$H^+ + OH^- \rightarrow H_2O \tag{6.2.1}$$

This is called a **neutralization reaction.** It is one of the most important of all chemical reactions.

Acids

An **acid** is a substance that produces hydrogen ions. For example, HCl in water is entirely in the form of H^+ ions and Cl^- ions. These two ions in water form hydrochloric acid. Acetic acid, which is present in vinegar, also produces hydrogen ions in water:

$$\mathrm{H{-}\overset{\displaystyle H}{\underset{\displaystyle H}{C}}{-}\overset{\displaystyle O}{\overset{\|}{C}}{-}OH} \rightleftharpoons \mathrm{H{-}\overset{\displaystyle H}{\underset{\displaystyle H}{C}}{-}\overset{\displaystyle O}{\overset{\|}{C}}{-}O^-} + H^+ \tag{6.2.2}$$

Acetic acid demonstrates two important characteristics of acids. First, many acids contain H that is not released by the acid molecule to form H^+. Of the 4 hydrogens in CH_3CO_2H, only the one bonded to oxygen is ionizable to form H^+. The second important point about acetic acid has to do with how much of it is ionized to form H^+ and acetate ion, $CH_3CO_2^-$. Most of the acetic acid remains as molecules of CH_3CO_2H in solution. In a 1 molar solution of acetic acid (containing 1mol of acetic acid per liter of solution) only about 0.5% of the acid is ionized to produce an acetate ion and a hydrogen ion. Of a thousand molecules of acetic acid, 995 remain as un-ionized CH_3CO_2H. Therefore, acetic acid is said to be a weak acid. This term will be discussed later in the chapter.

A hydrogen ion in water is strongly attracted to water molecules. Hydrogen ions react with water to form H_3O^+ or clusters with even more water molecules such as $H_5O_2^+$ or $H_7O_3^+$:

$$\begin{array}{c}\text{H}\\ \text{H}:\ddot{\underset{\cdot\cdot}{\text{O}}}:\text{H}\end{array} + \text{H} \rightarrow \begin{array}{c}\text{H}\\ \text{H}:\ddot{\underset{\cdot\cdot}{\text{O}}}:\text{H}^+\end{array} \quad (6.2.3)$$

Hydronium ion H_3O^+

The hydrogen ion in water is frequently shown as H_3O +. In this book, however, it is simply indicated as H^+.

Bases

A **base** is a substance that produces hydroxide ion and/or accepts H^+. Many bases consist of metal ions and hydroxide ions. For example, solid sodium hydroxide dissolves in water to yield a solution containing OH^- ions:

$$NaOH\ (s) \rightarrow Na^+\ (aq) + OH^-\ (aq) \quad (6.2.4)$$

When ammonia gas is bubbled into water, a few of the NH_3 molecules remove hydrogen ion from water and produce ammonium ion, NH_4^+, and hydroxide ion, as shown by the following reaction:

$$NH_3 + H_2O \rightarrow NH_4^+ + OH^- \quad (6.2.5)$$

Only about 0.5% of the ammonia in a 1 molar solution goes to NH_4^+ and OH^-. Therefore, as discussed later in the chapter, NH_3 is called a **weak base.**

Salts

Whenever an acid and a base are brought together, water is always a product. But a negative ion from the acid and a positive ion from the base are always left over, as shown in the following reaction:

$$\underbrace{H^+ + Cl^-}_{\text{hydrochloric acid}} + \underbrace{Na^+ + OH^-}_{\text{sodium hydroxide}} \rightarrow \underbrace{Na^+ + Cl^-}_{\text{sodium chloride}} + \underbrace{H_2O}_{\text{water}} \quad (6.2.6)$$

Sodium chloride dissolved in water is a solution of a **salt**. A salt is made up of a positively charged ion called a *cation* and a negatively charged ion called an *anion*. If the water were evaporated, the solid salt made up of cations and anions would remain as crystals. A salt is a chemical compound made up of a cation (other than H^+) and an anion (other than OH^-).

Amphoteric Substances

Some substances, called **amphoteric substances**, can act both as an acid and a base. The simplest example is water. Water can split apart to form a hydrogen ion and a hydroxide ion:

$$H_2O \rightarrow H^+ + OH^- \tag{6.2.7}$$

Since it produces a hydrogen ion, water is an acid. However, the fact that it produces a hydroxide ion also makes it a base. This reaction occurs only to a very small extent. In pure water, only one out of 10 million molecules of water is in the form of H^+ and OH^-. Except for this very low concentration of these two ions that can exist together, H^+ and OH^- react strongly with each other to form water.

Another important substance that can be either an acid or base is glycine. Glycine is one of the amino acids that are essential components of the body's proteins. It can give off a hydrogen ion,

$$H_3N^+\text{–}CH_2\text{–}C(=O)\text{–}O^- \underset{}{\overset{H_2O}{\rightleftharpoons}} H_2N\text{–}CH_2\text{–}C(=O)\text{–}O^- + H^+ \tag{6.2.8}$$

or it can react with water to release a hydroxide ion from the water:

$$H_3N^+\text{–}CH_2\text{–}C(=O)\text{–}O^- + H_2O \rightleftharpoons H_3N^+\text{–}CH_2\text{–}C(=O)\text{–}OH + OH^- \tag{6.2.9}$$

Metal Ions as Acids

Some metal ions are acids. As an example, consider iron(III) ion, Fe^{3+}. This ion used to be commonly called ferric ion. When iron(III) chloride, $FeCl_3$, is dissolved in water it produces chloride ions and triply charged iron(III) ions:

$$FeCl_3 + 6H_2O \rightarrow Fe(H_2O)_6^{3+} + 3Cl^- \tag{6.2.10}$$

Each iron(III) ion is bonded to 6 water molecules. An iron(III) ion surrounded by water is called a **hydrated ion.** This hydrated iron(III) ion can lose hydrogen ions and form a slimy brown precipitate of iron(III) hydroxide, $Fe(OH)_3$:

$$Fe(H_2O)_6^{3+} \rightarrow Fe(OH)_3 + 3H_2O + 3H^+ \qquad (6.2.11)$$

It is this reaction that is partly responsible for the acid in iron-rich acid mine water. It is also used to purify drinking water. The gelatinous $Fe(OH)_3$ settles out, carrying the impurities to the bottom of the container, and the water clears up.

Salts That Act as Bases

Some salts that do not contain hydroxide ion produce this ion in solution. The most widely used of these is sodium carbonate, Na_2CO_3, which is commonly known as soda ash. Millions of pounds of soda ash are produced each year for the removal of hardness from boiler water, for the treatment of waste acid, and for many other industrial pro cesses. Sodium carbonate reacts in water to produce hydroxide ion:

$$\underbrace{2Na^+ + CO_3^{2-}}_{\text{sodium carbonate}} + H_2O \rightarrow \underbrace{Na^+ + HCO_3^-}_{\text{sodium bicarbonate}} + \underbrace{Na^+ + OH^-}_{\text{sodium hydroxide}} \qquad (6.2.12)$$

If H^+, such as from hydrochloric acid, is already present in the water, sodium carbonate reacts with it as follows:

$$\underbrace{2Na^+ + CO_3^{2-}}_{\text{sodium carbonate}} + \underbrace{H^+ + Cl^-}_{\text{hydrochloric acid}} \rightarrow \underbrace{Na^+ + Cl^-}_{\text{sodium chloride}} + \underbrace{Na^+ + HCO_3^-}_{\text{sodium bicarbonate}} \qquad (6.2.13)$$

Salts That Act as Acids

Some salts act as acids. Salts that act as acids react with hydroxide ions. Ammonium chloride, NH_4Cl, is such a salt. This salt is also called "sal ammoniac." As a "flux" added to solder used to solder copper plumbing or automobile radiators, ammonium chloride dissolves coatings of corrosion on the metal surfaces so that the solder can stick. In the presence of a base, NH_4Cl reacts with the hydroxide ion to produce ammonia gas and water:

$$\underbrace{NH_4^+ + Cl^-}_{\text{ammonium chloride}} + \underbrace{Na^+ + OH^-}_{\text{sodium hydroxide}} \rightarrow NH_3 + H_2O + \underbrace{Na^+ + Cl^-}_{\text{sodium chloride}} \qquad (6.2.14)$$

6.3. CONDUCTANCE OF ELECTRICITY BY ACIDS, BASES, AND SALTS IN SOLUTION

When acids, bases, or salts are dissolved in water, charged ions are formed. When HCl gas is dissolved in water, all of it goes to H^+ and Cl^- ions:

$$HCl(g) \xrightarrow{\text{Water}} H^+(aq) + Cl^-(aq) \tag{6.3.1}$$

Acetic acid in water also forms a few ions, but most of it stays as CH_3CO_2H:

$$CH_3CO_2H \xrightarrow{\text{Water}} H^+ + CH_3CO_2^- \tag{6.3.2}$$

Sodium hydroxide in water is all in the form of Na^+ and OH^- ions. The salt, NaCl, is all present as Na^+ and Cl^- ions in water.

One of the most important properties of ions is that they conduct electricity in water. Water containing ions from an acid, base, or salt will conduct electricity much like a metal wire. Consider what would happen if very pure distilled water were made part of an electrical circuit as shown on the left in Figure 6.2. The light bulb would not glow at all. This is because pure water does not conduct electricity. However, if a solution of salt water, such as oil well brine, is substituted for the distilled water as shown on the right in Figure 6.2, the bulb will glow brightly. Salty water conducts electricity because of the ions that it contains. Even tap water has

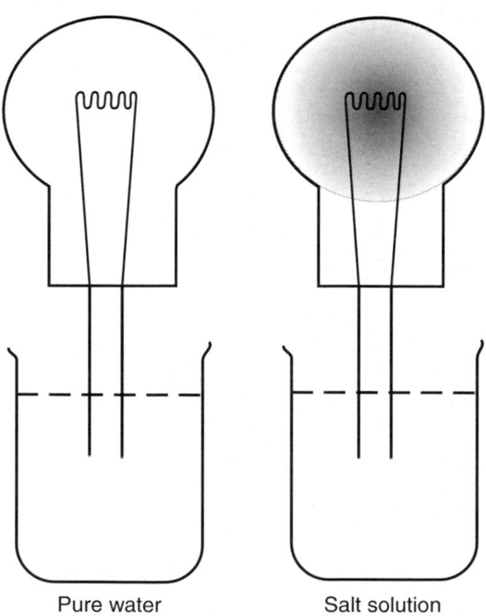

Figure 6.2. Pure water does not conduct electricity, whereas water containing dissolved salt conducts electricity very well.

some ions dissolved in it, which is why one may experience a painful, even fatal, electric shock by touching an electrical fixture while bathing.

Electrolytes

Materials that conduct electricity in water are called **electrolytes**. These materials form ions in water. The charged ions allow the electrical current to flow through the water. Materials, such as sugar, that do not form ions in water are called nonelectrolytes. Solutions of nonelectrolytes in water do not conduct electricity. A solution of brine conducts electricity very well because it contains dissolved NaCl. All of the NaCl in the water is in the form of Na^+ and Cl^-. The NaCl is completely ionized, and it is a strong electrolyte. An ammonia water solution (used for washing windows) does not conduct electricity very well. That is because only a small fraction of the NH_3 molecules react to form the ions that let electricity pass through the water:

$$NH_3 + H_2O \rightarrow NH_4^+ + OH^- \quad (6.3.3)$$

Ammonia is a weak electrolyte. (Recall that it is also a weak base.) Nitric acid, HNO_3 is a strong electrolyte because it is completely ionized to H^+ and NO_3^- ions.

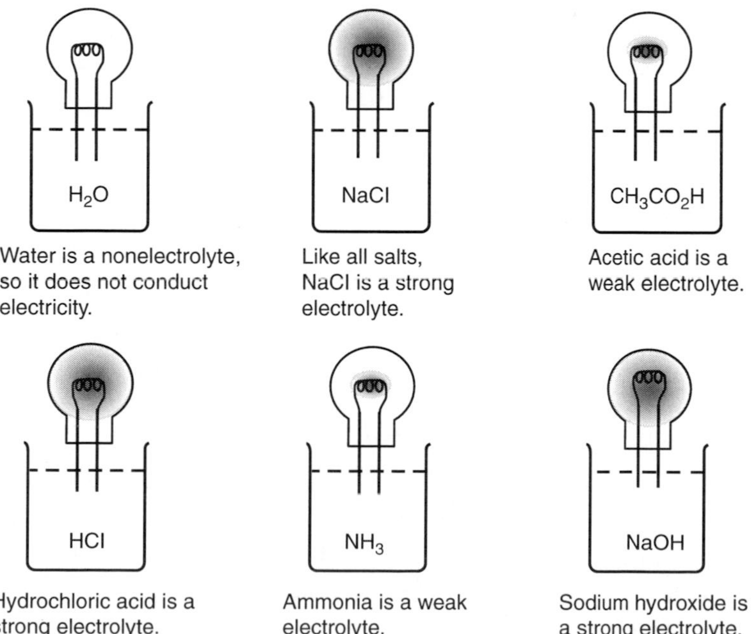

Figure 6.3. The electrical conductivity of a solution can be determined by placing the solution in an electrical circuit and observing how well it conducts electricity. **Strong electrolytes** conduct electricity well; **weak electrolytes** conduct it poorly. This principle is used in water analysis to determine the total salt concentrations in water.

Acetic acid is a weak electrolyte, as well as a weak acid. The base sodium hydroxide is a strong electrolyte. All salts are strong electrolytes, because they are always completely ionized in water. Acids and bases can be weak or strong electrolytes.

In the laboratory, the strength of an electrolyte can be measured by how well it conducts electricity in solution, as shown in Figure 6.3. The ability of a solution to conduct electrical current is called its **conductivity**.

When electricity is passed through solutions of acids, bases, or salts, chemical reactions occur. One such reaction is the breakdown of water to hydrogen and oxygen. Electricity passing through a solution is widely used to separate and purify various substances.

6.4. DISSOCIATION OF ACIDS AND BASES IN WATER

It has already been seen that acids and bases come apart in water to form ions. When acetic acid splits up in water, it forms hydrogen ions and acetate ions:

$$CH_3CO_2H \rightarrow CH_3CO_2^- + H^+ \qquad (6.4.1)$$

The process of forming ions is called **ionization**. Another term is also commonly employed. When the acetic acid molecule comes apart, it is said to **dissociate**. The process is called **dissociation**.

There is a great difference in how much various acids and bases dissociate. Some, such as HCl or NaOH, are completely dissociated in water. Because of this, hydrochloric acid is called a **strong acid**. Sodium hydroxide is a **strong base**. Some acids such as acetic acid are only partly dissociated in water. They are called weak acids. Ammonia, NH_3, reacts only slightly in water to form an ammonium ion (NH_4^+) and a hydroxide ion (OH^-). It is a weak base.

The extent of dissociation is a very important property of an acid or base. The 3% or so acetic acid solution used to make up oil and vinegar salad dressing lends a pleasant taste to the lettuce and tomatoes. There is not much of the H^+ ion in the acetic acid. If 3% HCl had been used instead, nobody could eat the salad. All of the H in HCl is in the form of H^+, and a 3% solution of hydrochloric acid is very sour indeed. Similarly, a several percent solution of NH_3 in water makes a good window-washing agent, helping to dissolve grease and grime on the window surface. If a similar concentration of sodium hydroxide were used to clean windows, they would soon become permanently fogged because the OH^- in the strong base eventually reacts with glass and etches it. However, sodium hydroxide solutions are used to clean ovens, where a very strong base is required to break down the charred, baked-on grease.

Table 6.1 shows some acids and the degree to which they are dissociated. It allows comparison of the strengths of these acids.

The percentage of acid molecules that are dissociated depends upon the concentration of the acid. The lower the concentration, the higher the percentage of

Table 6.1. Dissociation of Acids

Acid Formula	Acid Name	Common Uses	Percent Dissociated in 1 M Solution[a]	Strength
H_2SO_4	Sulfuric	Industrial chemical	100	Strong
HNO_3	Nitric	Industrial chemical	100	Strong
H_3PO_4	Phosphoric moderately	Fertilizer, food additive	8	Weak
$H_3C_6H_5O_7$	Citric	Fruit drinks	3	Weak
CH_3CO_2H	Acetic	Foods, industry	0.4	Weak
HClO	Hypochlorous	Disinfectant	0.02	Weak
HCN	Hydrocyanic	Very poisonous, industrial chemical, electroplating waste	0.002	Very weak
H_3BO_3	Boric acid	Antiseptic, ceramics	0.002	Very weak

[a] A 1 M solution contains 1 mole of acid per liter of solution.

dissociated molecules. This may be understood by looking again at the reaction for the dissociation of acetic acid:

$$CH_3CO_2H \rightarrow CH_3CO_2^- + H^+ \tag{6.4.2}$$

At high concentrations, there will be more crowding together of H^+ and $CH_3CO_2^-$ ions. This forces them back together to form CH_3CO_2H again. At low concentrations, there are fewer H^+ and $CH_3CO_2^-$ ions. They are freer to roam around the solution alone, and there is less pressure for them to form CH_3CO_2H. It is somewhat like the seating that occurs on a bus. If there are few passengers, they will spread out and not sit next to each other; that is, they will be dissociated. If there are many passengers, they will, of course, have to occupy adjacent seats.

An idea of the effect of concentration upon the dissociation of a weak acid can be obtained from the percentage of acid molecules that have dissociated to ions at several different concentrations. This is shown for acetic acid in Table 6.2.

Table 6.2 shows that, in a 1 M solution (1 mol/L), less than 1% of acetic acid is dissociated. In a one-thousandth M solution (1×10^{-3} mol/L), 12 out of 100 molecules of acetic acid are in the form of H^+ and $CH_3CO_2^-$. In a one-millionth M solution (1×10^{-6} mol/L), only 5 out of 100 acetic acid molecules are present as CH_3CO_2H.

It is important to know the difference between the strength of an acid or base in solution and the concentration of the solution. A strong acid is one that is all in the form of H^+ ions and anions. It may be very concentrated or very dilute. A weak acid does not give off much H^+ to water solution. It may also range in concentration from a very dilute solution to a very concentrated one. Similar arguments apply to bases.

Table 6.2. Percent Dissociation of Acetic Acid at Various Concentrations

Total Acetic Acid Concentration	Percent Dissociated to H^+ and $CH_3CO_2^-$
1 mol/L	0.4
0.1 mol/L	1.3
0.01 mol/L	4.1
1×10^{-3} mol/L	12
1×10^{-4} mol/L	34
1×10^{-5} mol/L	71
1×10^{-6} mol/L	95
1×10^{-7} mol/L	99

6.5. THE HYDROGEN ION CONCENTRATION AND BUFFERS

It is important to make the distinction between the concentration of H^+ and the concentration of an acid. To show this difference, compare 1 M solutions of acetic acid and hydrochloric acid. The concentration of H^+ in a 1 M solution of CH_3CO_2H is only 0.0042 mol/L. The concentration of H^+ in a 1 M solution of HCl is 1 mol/L. A liter of a 1 M solution of HCl contains 240 times as many H^+ ions as a liter of a 1 M solution of acetic acid.

Consider, however, the amount of NaOH that will react with 1.00 L of 1.00 M acetic acid. The reaction is

$$\underset{\text{acetic acid}}{CH_3CO_2H} + \underset{\text{sodium hydroxide}}{Na^+ + OH^-} \rightarrow \underset{\text{sodium acetate}}{Na^+ + CH_3CO_2^-} + H_2O \quad (6.5.1)$$

Exactly 1.00 mol of NaOH reacts with the 1.00 mol of acetic acid contained in 1.00 L of a 1.00 M solution of this acid. Exactly the same amount of NaOH reacts with the HCl in 1.00 L of 1.00 M HCl:

$$\underset{\text{hydrochloric acid}}{H^+ + Cl^-} + \underset{\text{sodium hydroxide}}{Na^+ + OH^-} \rightarrow \underset{\text{sodium chloride}}{Na^+ + Cl^-} + H_2O \quad (6.5.2)$$

Therefore, even though acetic acid is a weaker acid than hydrochloric acid, equal volumes of each, with the same molar concentration, will react with the same number of moles of base.

In many systems, the concentration of H^+ is very important. For a person to remain healthy, the H^+ concentration in their blood must stay within a very narrow range. If the H^+ concentration is too high in a boiler system, then the pipes may corrode through in a short time. If the H^+ concentration becomes too high or too low in a lake, then plant and animal life cannot thrive in it.

Buffers

Fortunately, there are mixtures of chemicals that keep the H^+ concentration of a solution relatively constant. Reasonable quantities of acid or base added to such solutions do not cause large changes in H^+ concentration. Solutions that resist changes in H^+ concentration are called **buffers.**

To understand how a buffer works, consider a typical buffer system. A solution containing both acetic acid and sodium acetate is a good buffer. The acetic acid in the solution is present as undissociated CH_3CO_2H. The H^+, which is in solution, is there because a very small amount of the CH_3CO_2H has dissociated to H^+ and $CH_3CO_2^-$ ions. The sodium acetate is present as Na^+ ion and $CH_3CO_2^-$ ion. If some base, such as NaOH, is added, some of the acetic acid reacts:

$$\underset{}{CH_3CO_2H} + \underbrace{Na^+ + OH^-}_{\text{sodium hydroxide}} \rightarrow \underbrace{Na^+ + CH_3CO_2^-}_{\text{sodium acetate}} + H_2O \tag{6.5.3}$$

This reaction changes some of the acetic acid to sodium acetate, but it does not change the hydrogen ion concentration much. If a small amount of hydrochloric acid is added to the buffer mixture of acetic acid and sodium acetate, some of the sodium acetate is changed to acetic acid.

$$\underbrace{Na^+ + CH_3CO_2^-}_{\text{sodium acetate}} + \underbrace{H^+ + Cl^-}_{\text{hydrochloric acid}} \rightarrow CH_3CO_2H + Na^+ + Cl^- \tag{6.5.4}$$

The acetate ion acts like a sponge for H^+ and prevents the concentration of the added hydrogen ion from becoming too high.

Buffers can also be made from a mixture of a weak base and a salt of the base. A mixture of NH_3 and NH_4Cl is such a buffer. Mixtures of two salts can be buffers. A mixture of NaH_2PO_4 and Na_2HPO_4 is a buffer made from salts. It is one of the very common phosphate buffers, such as those that occur in body fluids.

6.6. pH AND THE RELATIONSHIP BETWEEN HYDROGEN ION AND HYDROXIDE ION CONCENTRATIONS

Because of the fact that water itself produces both hydrogen ion and hydroxide ion,

$$H_2O \rightarrow H^+ + OH^- \tag{6.6.1}$$

there is always some H^+ and some OH^- in any water solution. Of course, in an acid solution, the concentration of OH^- must be very low. In a solution of base, the concentration of OH^- is very high and that of H^+ is very low. There is a definite relationship between the concentration of H^+ and the concentration of OH^-. It varies a little with temperature. At 25°C (about room temperature), the following relationship applies:

$$[H^+][OH^-] = 1.00 \times 10^{-14} = K_w \quad \text{(at 25°C)} \tag{6.6.2}$$

where square brackets [X] denote the concentration of substance X. If the value of either $[H^+]$ or $[OH^-]$ is known, the value of the other can be calculated by substituting into the K_w expression. For example, in a solution of 0.100 M HCl, in which $[H^+] = 0.100$ mol/L,

$$[OH^-] = \frac{K_w}{[H^+]} = \frac{1.00 \times 10^{-14}}{0.100} = 1.00 \times 10^{-13} \text{ mol/L} \tag{6.6.3}$$

Acids, such as hydrochloric acid and sulfuric acid, produce H^+ ion, whereas bases, such as sodium hydroxide and calcium hydroxide, produce OH^-. Molar concentrations of hydrogen ion, $[H^+]$, range over many orders of magnitude and are conveniently expressed by pH, defined as

$$\text{pH} = -\log[H^+] \tag{6.6.4}$$

In absolutely pure water, the value of $[H^+]$ is exactly 1×10^{-7} mol/L; therefore, the pH of pure water is 7.00, and the solution is **neutral** (neither acidic nor basic). **Acidic** solutions have pH values of less than 7 and **basic** solutions have pH values of greater than 7. Table 6.3 gives some example hydrogen ion concentrations and the corresponding pH values.

As can be seen in Table 6.3, when the H^+ ion concentration is 1 times 10 to a power (the superscript number, such as −2, −7, etc.), the pH is simply the negative

Table 6.3. Values of $[H^+]$ and Corresponding pH Values

$[H^+]$ (mol/L)	pH
1.00	0.00
0.100	1.00
1.00×10^{-3}	3.00
2.25×10^{-6} ($10^{-5.65}$)	5.65
1.00×10^{-7}	7.00
1.00×10^{-9}	9.00
5.17×10^{-9} ($10^{-8.29}$)	8.29
1.00×10^{-13}	13.00
1.00×10^{-14}	14.00

value of that power. Thus, when $[H^+] = 1 \times 10^{-3}$, the pH is 3; when $[H^+] = 1 \times 10^{-4}$, the pH is 4. This is because the log of 1×10^{-3} is -3 and that of 1×10^{-4} is -4; therefore, the negative logs are 3 and 4, respectively, because the sign is reversed. What about the pH of a solution with a hydrogen ion concentration between 1×10^{-4} and 1×10^{-3}, such as 3.16×10^{-4}? Since $[H^+]$ is between 1×10^{-4} and 1×10^{-3} mol/L, the pH is obviously going to be between 3 and 4. The pH is calculated very easily on an electronic calculator by entering 3.16×10^{-4} on the keyboard and pressing the "log" button. The log of the number is -3.50, and so the pH is 3.50.

Acid–Base Equilibria

Many of the phenomena in aquatic chemistry and geochemistry involve **solution equilibrium**. In a general sense, solution equilibrium deals with the extent to which **reversible** acid–base, solubilization (precipitation), complexation, or oxidation–reduction reactions proceed in a forward or backward direction. This is expressed for a generalized equilibrium reaction

$$aA + bB \rightarrow cC + dD \tag{6.6.5}$$

There are several major kinds of equilibria in aqueous solution. The one under consideration here is acid–base equilibrium, as exemplified by the ionization of acetic acid, HAc,

$$HAc \rightleftharpoons H^+ + Ac^- \tag{6.6.6}$$

where

$$Ac^- \text{ represents } \quad \begin{array}{ccccccc} & & H & & O & & \\ & & | & & \| & & \\ H & - & C & - & C & - & O^- \\ & & | & & & & \\ & & H & & & & \end{array}$$

for which the acid dissociation constant is

$$\frac{[H^+][Ac^-]}{[HAc]} = K = 1.75 \times 10^{-5} \text{ (at 25°C)} \tag{6.6.7}$$

As an example of an acid–base equilibrium problem, consider water in equilibrium with atmospheric carbon dioxide. The value of $[CO_2\ (aq)]$ in water at 25°C in equilibrium with air that is 390 parts per million CO_2 (close to the concentration of this gas in the atmosphere) is 1.277×10^{-5} mol/L. The carbon dioxide dissociates partially in water to produce equal concentrations of H^+ and HCO_3^-:

$$CO_2 + H_2O \rightleftharpoons HCO_3^- + H^+ \tag{6.6.8}$$

so that

$$[H^+] = [HCO_3^-] \tag{6.6.9}$$

The concentrations of H^+ and HCO_3^- are calculated from K_{a1}, where

$$K_{a1} = \frac{[H^+][HCO_3^-]}{[CO_2]} = \frac{[H^+]^2}{1.277 \times 10^{-5}} = 4.45 \times 10^{-7} \tag{6.6.10}$$

Since $[H^+] = [HCO_3^-]$, this relationship simplifies to

$$[H^+] = [HCO_3^-] = (1.277 \times 10^{-5} \times 4.45 \times 10^{-7})^{1/2} = 2.38 \times 10^{-6} \tag{6.6.11}$$

and so

$$pH = 5.62$$

This calculation explains why pure water that has equilibrated with the unpolluted atmosphere is slightly acidic, with a pH somewhat less than 7.

6.7. PREPARATION OF ACIDS

Acids can be prepared in several ways. In discussing their preparation, it is important to keep in mind that acids usually contain nonmetals. All acids either contain ionizable hydrogen or produce it when dissolved in water. Furthermore, the hydrogen has to be ionizable; it must have the ability to form H^+ ion. Finally, more often than not, acids contain oxygen.

A simple way to make an acid is to react hydrogen with a nonmetal that forms a compound with hydrogen that will produce H^+ ion in water. Hydrochloric acid can be made by reacting hydrogen and chlorine,

$$H_2 + Cl_2 \rightarrow 2HCl \tag{6.7.1}$$

and adding the hydrogen chloride product to water. Other acids that consist of hydrogen combined with a nonmetal are HF, HBr, HI, and H_2S. Hydrocyanic acid, HCN, is an "honorary member" of this family of acids, even though it contains three elements.

Sometimes, a nonmetal reacts directly with water to produce acids. The best example of this is the reaction of chlorine with water to produce hydrochloric acid and hypochlorous acid:

$$Cl_2 + H_2O \rightarrow HCl + HClO \tag{6.7.2}$$

Many very important acids are produced when nonmetal oxides react with water. One of the common examples is the reaction of sulfur trioxide with water to produce sulfuric acid:

$$SO_3 + H_2O \rightarrow H_2SO_4 \tag{6.7.3}$$

Table 6.4. Important Acids Produced when Nonmetal Oxides React with Water

Oxide Reacted	Acid Formula	Acid Name	Use and Significance of Acid
SO_3 constituent	H_2SO_4	Sulfuric	Major industrial chemical, of acid rain
SO_2	H_2SO_3	Sulfurous	Paper making, scrubbed from stack gas containing SO_2
N_2O_5	HNO_3	Nitric	Synthesis of chemicals, constituent of acid rain
N_2O_3	HNO_2	Nitrous	Unstable, toxic to ingest, few uses
P_4O_{10}	H_3PO_4	Phosphoric	Fertilizer, chemical synthesis

Other examples are shown in Table 6.4.

Volatile acids—those that evaporate easily—can be made from salts and nonvolatile acids. The most common nonvolatile acid so used is sulfuric acid, H_2SO_4. When solid NaCl is heated in contact with concentrated sulfuric acid, hydrogen chloride gas is given off:

$$2NaCl(s) + H_2SO_4(l) \rightarrow 2HCl(g) + Na_2SO_4(s) \qquad (6.7.4)$$

This gas can be collected in water to make hydrochloric acid. Similarly, when calcium sulfite is heated with sulfuric acid, sulfur dioxide is given off as a gas:

$$CaSO_3(s) + H_2SO_4(l) \rightarrow CaSO_4(s) + SO_2(g) + H_2O \qquad (6.7.5)$$

It can be collected in water to produce sulfurous acid, H_2SO_3.

Organic acids, such as acetic acid, CH_3CO_2H, have the group

$$\text{—}\overset{\overset{\displaystyle O}{\|}}{C}\text{—OH} \quad \text{Carboxylic acid group, }–CO_2H$$

attached to a hydrocarbon group. These **carboxylic acids** are discussed further in Chapter 10.

6.8. PREPARATION OF BASES

Bases can be prepared in several ways. Many bases contain metals, and some metals react directly with water to produce a solution of base. Lithium, sodium, and potassium react very vigorously with water to produce their hydroxides:

$$2K + 2H_2O \rightarrow \underbrace{2K^+ + 2OH^-}_{\substack{\text{potassium hydroxide} \\ \text{(strong base)}}} + H_2(g) \qquad (6.8.1)$$

Many metal oxides form bases when they are dissolved in water. When waste liquor (a concentrated solution of salts and materials extracted from wood) from the sulfite paper-making process is burned to produce energy and reclaim magnesium hydroxide, the magnesium in the ash is recovered as MgO. This is added to water

$$MgO + H_2O \rightarrow Mg(OH)_2 \tag{6.8.2}$$

to produce the magnesium hydroxide used with other chemicals to break down the wood and produce paper fibers. Other important bases and the metal oxides from which they are prepared are given in Table 6.5.

Many important bases cannot be isolated as the hydroxides, but produce OH^- ion in water. A very good example is ammonia, NH_3. Ammonium hydroxide, NH_4OH, cannot be obtained in a pure form. Even when ammonia is dissolved in water, very little NH_4OH is present in the solution. However, ammonia does react with water to give an ammonium ion and a hydroxide ion:

$$NH_3 + H_2O \rightarrow NH_4^+ + OH^- \tag{6.8.3}$$

Since only a small percentage of the ammonia molecules react this way, ammonia is a weak base.

Many salts that do not themselves contain hydroxide ion act as bases by reacting with water to produce OH^-. Sodium carbonate, Na_2CO_3, is the most widely used of these salts. When sodium carbonate is placed in water, the carbonate ion reacts with water to form a hydroxide ion and a bicarbonate ion, HCO_3^-:

$$CO_3^{2+} + H_2O \rightarrow HCO_3^- + OH^- \tag{6.8.4}$$

Commercial-grade sodium carbonate, soda ash, is used very widely for neutralizing acid in water treatment and other applications. It is used in phosphate-free detergents. It is a much easier base to handle and use than sodium hydroxide. Whereas sodium hydroxide rapidly absorbs enough water from the atmosphere to dissolve

Table 6.5. Important Bases Produced when Metal Oxides React with Water

Oxide Reacted	Base Formula	Base Name	Use and Significance of Base
Li_2O	LiOH	Lithium hydroxide	Constituent of some lubricating greases
Na_2O	NaOH	Sodium hydroxide	Soap making, many industrial uses, removal of H_2S from petroleum
K_2O	KOH	Potassium hydroxide	Alkaline battery manufacture
MgO	$Mg(OH)_2$	Magnesium hydroxide	Paper making, medicinal uses
CaO	$Ca(OH)_2$	Calcium hydroxide	Water purification, soil treatment to neutralize excessive acid

itself to make little puddles of highly concentrated NaOH solution that are very harmful to the skin, sodium carbonate does not absorb water nearly so readily. It is not as dangerous to the skin.

Trisodium phosphate, Na_3PO_4, is a stronger base than sodium carbonate. The phosphate ion reacts with water to yield a high concentration of hydroxide ions:

$$PO_4^{3-} + H_2O \rightarrow HPO_4^{2-} + OH^- \quad (6.8.5)$$

This kind of reaction with water is called a **hydrolysis reaction**.

Many organic compounds are bases. Most of these contain nitrogen. One of these is trimethylamine, $(CH_3)_3N$. This compound is one of several that give dead fish their foul smell. It reacts with water to produce hydroxide ion:

$$(CH_3)_3N + H_2O \rightarrow (CH_3)_3NH^+ + OH^+ \quad (6.8.6)$$

Like most organic bases, it is a weak base.

6.9. PREPARATION OF SALTS

Many salts are important industrial chemicals. Others are used in food preparation or medicine. A huge quantity of Na_2CO_3 is used each year, largely to treat water and to neutralize acid. Over 1.5 million tons of Na_2SO_4 are used in applications such as inert filler in powdered detergents. Approximately 30 000 tons of sodium thiosulfate, $Na_2S_2O_3$, are used each year in developing photographic film and in other applications. Canadian mines produce more than 10 million tons of KCl each year for use as fertilizer. Lithium carbonate, Li_2CO_3, is used as a medicine to treat some kinds of bipolar affective (manic-depressive) illness. Many other examples of the importance of salts could be given.

Whenever possible, salts are obtained by simply mining them. Many kinds of salts can be obtained by evaporating water from a few salt-rich inland sea waters or from brines pumped from beneath the ground. However, most salts cannot be obtained so directly, and must be made by chemical processes. Some of these processes will be discussed here.

One way of making salts already discussed in this chapter is to react an acid and a base to produce a salt and water. Calcium propionate, which is used to preserve bread, is made by reacting calcium hydroxide and propionic acid, $HC_3H_5O_2$:

$$Ca(OH)_2 + 2HC_3H_5O_2 \rightarrow \underset{\text{calcium propionate}}{Ca(C_3H_5O_2)_2} + 2H_2O \quad (6.9.1)$$

Almost any salt can be made by the reaction of the appropriate acid and base. In some cases, a metal and a nonmetal will react directly to make a salt. If a strip of magnesium burns (explodes would be a better description) in an atmosphere of chlorine gas, the salt magnesium chloride is produced:

$$Mg + Cl_2 \rightarrow MgCl_2 \tag{6.9.2}$$

Metals react with acids to produce a salt and hydrogen gas. Calcium placed in sulfuric acid will yield calcium sulfate:

$$Ca + H_2SO_4 \rightarrow H_2(g) + CaSO_4(s) \tag{6.9.3}$$

Some metals react with strong bases to produce salts. Aluminum metal reacts with sodium hydroxide to yield sodium aluminate, Na_3AlO_3:

$$2Al + 6NaOH \rightarrow 2Na_3AlO_3 + 3H_2(g) \tag{6.9.4}$$

In cases where a metal forms an insoluble hydroxide, addition of a base to a salt of that metal can result in the formation of a new salt. If potassium hydroxide is added to a solution of magnesium sulfate, the insoluble magnesium hydroxide precipitates out of the solution, leaving potassium sulfate salt in solution:

$$2KOH + MgSO_4 \rightarrow Mg(OH)_2(s) + K_2SO_4(aq) \tag{6.9.5}$$

If the anion in a salt can form a volatile acid, a new salt can be formed by adding a nonvolatile acid, heating to drive off the volatile product, and collecting the volatile acid in water. If nonvolatile sulfuric acid is heated with NaCl, then HCl gas is given off and sodium sulfate remains behind:

$$H_2SO_4 + 2NaCl \rightarrow 2HCl(g) + Na_2SO_4 \tag{6.9.6}$$

Some metals will displace other metals from a salt. Advantage is taken of this for the removal of toxic heavy metals from water solutions of the metals' salts by reaction with a more active metal, a process called **cementation**. For example, metallic iron can be reacted with wastewater containing dissolved toxic cadmium sulfate to isolate solid cadmium metal and leave solid cadmium metal and a new salt, iron(II) sulfate:

$$Fe(s) + CdSO_4(aq) \rightarrow Cd(s) + FeSO_4(aq) \tag{6.9.7}$$

Finally, there are many special commercial processes for making specific salts. One such example is the widely used Solvay process for making sodium bicarbonate and sodium carbonate. In this process, a sodium chloride solution is saturated with ammonia gas, then saturated with carbon dioxide, and finally cooled. The reaction that occurs is

$$NaCl + NH_3 + CO_2 + H_2O \rightarrow NaHCO_3(s) + NH_4Cl \tag{6.9.8}$$

and sodium bicarbonate (baking soda) precipitates from the cooled solution. When the sodium bicarbonate is heated, it is converted into sodium carbonate:

$$2NaHCO_3 + \text{heat} \rightarrow Na_2CO_3 + H_2O(g) + CO_2(g) \qquad (6.9.9)$$

6.10. ACID SALTS AND BASIC SALTS

Acid Salts

Some compounds are crosses between acids and salts. Other salts are really crosses between bases and salts. The acid salts contain hydrogen ion. This hydrogen ion can react with bases. One example of this is sodium hydrogen sulfate, $NaHSO_4$, which reacts with sodium hydroxide to give sodium sulfate and water:

$$NaHSO_4 + NaOH \rightarrow Na_2SO_4 + H_2O \qquad (6.10.1)$$

Some other examples of acid salts are shown in Table 6.6.

Basic Salts

Some salts contain hydroxide ions. These are known as **basic salts**. One of the most important of these is calcium hydroxyapatite, $Ca_5OH(PO_4)_3$. Commonly known as *hydroxyapatite*, this salt occurs in the mineral rock phosphate, which is the source of essential phosphate fertilizer. The heavy metals, in particular, have a tendency to form basic salts. Many rock-forming minerals are basic salts.

6.11. WATER OF HYDRATION

Water is frequently bound to other chemical compounds. Water bound to a salt in a definite proportion is called **water of hydration**. An important example is sodium carbonate decahydrate: $Na_2CO_3 \cdot 10H_2O$. This salt is used in detergents, as

Table 6.6. Some Important Acid Salts

Acid Salt Formula	Acid Salt Name	Typical Use
$NaHCO_3$	Sodium hydrogen carbonate	Food preparation (baking soda)
NaH_2PO_4	Sodium dihydrogen phosphate	To prepare buffers
Na_2HPO_4	Disodium hydrogen phosphate	To prepare buffers
$KHC_4H_4O_6$	Potassium hydrogen tartrate	Dry acid in baking powder[a]

[a] Cream of tartar baking powder consists of a mixture of potassium hydrogen tartrate and sodium hydrogen carbonate. When this mixture contacts water in a batch of dough, the reaction

$$KHC_4H_4O_6 + NaHCO_3 \rightarrow KNaC_4H_4O_6 + H_2O + CO_2(g)$$

occurs, and the small bubbles of carbon dioxide that it produces cause the dough to rise.

a household cleaner, and for water softening. When dissolved in water, it yields a basic solution by reacting with the water to produce hydroxide ion, OH^-. Such a solution tends to dissolve grease, so it can be used as a household cleaner. Sodium carbonate decahydrate is used in detergents, because they work best in basic solutions. The carbonate ion from this salt reacts with calcium ion (which causes water hardness):

$$Ca^{2+} + CO_3^{2-} \rightarrow CaCO_3(s) \tag{6.11.1}$$

This removes hardness from the water by producing solid calcium carbonate, $CaCO_3$. Because of its ability to remove water hardness, sodium carbonate decahydrate is a good water softener, another reason that it is used in detergents. Sodium carbonate without water, Na_2CO_3, is said to be **anhydrous**. It should not be used in consumer products such as powdered detergents, because of its strong attraction for water. If anhydrous Na_2CO_3 were present in a detergent that was accidentally ingested, it would draw water from the tissue in the mouth and throat, greatly increasing the harm done.

6.12. NAMES OF ACIDS, BASES, AND SALTS

Acids

Acids consisting of hydrogen and one other element are named for the element with a *hydro-* prefix and an ending of *-ic*. These acids include HF, hydrofluoric; HCl, hydrochloric acid; HBr, hydrobromic acid; and H_2S, hydrosulfuric acid. Another acid named in this manner is HCN, hydrocyanic acid.

The names of many common acids end with the suffix *-ic*, as is the case with acetic acid, nitric acid, and sulfuric acid. In some cases where the anion of the acid contains oxygen, there are related acids with different numbers of oxygen atoms in the anion. When this occurs, the acid with one less oxygen than the "-ic" acid has a name ending with *-ous*. For example, H_2SO_4 is sulfuric acid and H_2SO_3 is sulfurous acid. Similarly, HNO_3 is nitric acid and HNO_2 is nitrous acid. One more oxygen in the anion than the "-ic" acid is indicated by a *per-* prefix and an *-ic* suffix on the acid name. One less oxygen in the anion than in the "-ous" acid gives the acid a name with a *hypo-* prefix and an *-ous* suffix. These rules are shown for the oxyacids (acids containing oxygen in the anion) of chlorine in Table 6.7.

Bases

Bases that contain hydroxide ion are named very simply by the rules of nomenclature for ionic compounds. The name consists of the name of the metal followed by "hydroxide." For example, LiOH is lithium hydroxide, KOH is potassium hydroxide, and $Mg(OH)_2$ is magnesium hydroxide.

Table 6.7. Names of the Oxyacids of Chlorine

Acid Formula	Acid Name[a]	Anion Name[a,b]
$HClO_4$	*Per*chlor*ic* acid	*Per*chlor*ate*
$HClO_3$	Chlor*ic* acid	Chlor*ate*
$HClO_2$	Chlor*ous* acid	Chlor*ite*
$HClO$	*Hypo*chlor*ous* acid	*Hypo*chlor*ite*

[a] Italicized letters are used in the names only to emphasize the prefixes and suffixes.
[b] Names of anions, such as ClO_4^-, formed by removal of an H^+ ion from the acid.

Salts

Salts are named according to the name of the cation followed by the name of the anion. The names of the more important ions are listed in Table 6.8. One important cation, ammonium ion, NH_4^+, and a number of anions are polyatomic ions, meaning that they consist of 2 or more atoms per ion. Table 6.8 includes some polyatomic ions. Note that most polyatomic ions contain oxygen as one of the elements.

As illustrated in Table 6.8, the names of anions are based upon the names of the acids from which they are formed by removal of H^+ ions. A "*hydro -ic*" acid yields an "*-ide*" anion and, therefore, an "*-ide*" salt; for example, hydrochloric acid reacts with a base to give a chloride salt. An "*-ic*" acid yields an "*-ate*" salt; for example, calcium sulf*ate* is the salt that results from a reaction of sulfur*ic* acid with calcium hydroxide. The anion contained in an "*-ous*" acid is designated by the suffix "*-ite*" in a salt; sulfur*ous* acid, H_2SO_3 reacts with NaOH to give the salt Na_2SO_3 sodium sulfite. A "*per -ic*" acid, such as perchloric acid, reacts with a base, such as NaOH, to give a "*per -ate*" salt, for example, sodium perchlorate, $NaClO_4$. A "*hypo -ous*" acid, such as hypochlorous acid, reacts with a base, KOH, for example, to give a "*hypo -ite*" salt, such as potassium hypochlorite, KClO. Some additional examples are illustrated by the following reactions:

$$\underset{\text{perchloric acid}}{2HClO_4} + Zn(OH)_2 \rightarrow \underset{\text{zinc perchlorate}}{Zn(ClO_4)_2} + 2H_2O \quad (6.12.1)$$

$$\underset{\text{nitric acid}}{2HNO_3} + Cu(OH)_2 \rightarrow \underset{\text{copper(II) nitrate}}{Cu(NO_3)_2} + 2H_2O \quad (6.12.2)$$

$$\underset{\text{nitrous acid}}{HNO_2} + KOH \rightarrow \underset{\text{potassium nitrite}}{KNO_2} + H_2O \quad (6.12.3)$$

$$\underset{\text{hypochlorous acid}}{2HClO} + Ca(OH)_2 \rightarrow \underset{\text{calcium hypochlorite}}{Ca(ClO)_2} + 2H_2O \quad (6.12.4)$$

Table 6.8. Some Important Ions

+1 Charge	+2 Charge	+3 Charge	−1 Charge	−2 Charge	−3 Charge
H^+ hydrogen	Mg^{2+} magnesium	Al^{3+} aluminum	H^- hydride	O^{2-} oxide	N^{3-} nitride
Li^+ lithium	Ca^{2+} calcium	Fe^{3+} iron(III)[a]	F^- fluoride	S^{2-} sulfide	PO_4^{3-} phosphate
Na^+ sodium	Ba^{2+} barium	Cr^{3+} chromium(III)[a]	Cl^- chloride	SO_4^{2-} sulfate	
K^+ potassium	Fe^{2+} iron(II)[b]		Br^- bromide	SO_3^{2-} sulfite	
NH_4^+ ammonium	Zn^{2+} zinc		I^- iodide	CO_3^{2-} carbonate	
Cu^+ copper(I)[b]	Cu^{2+} copper(II)[b]		$C_2H_3O_2^-$ acetate	CrO_4^{2-} chromate	
Ag^+ silver	Cr^{2+} chromium(II)[b]		OH^- hydroxide	$Cr_2O_7^{2-}$ dichromate	
	Pb^{2+} lead(II)[b]		CN^- cyanide	O_2^{2-} peroxide	
	Hg^{2+} mercury(II)[a]		NO_3^- nitrate	HPO_4^{2-} monohydrogen phosphate	
	Sn^{2+} tin(II)[b]		$H_2PO_4^-$ dihydrogen phosphate		
			HCO_3^- hydrogen carbonate[c]		
			HSO_4^- hydrogen sulfate[c]		
			MnO_4^- permanganate		

[a] These metals can also exist as ions with a lower charge, and may also be designated by their Latin names with an *-ic* ending as follows: Cu^{2+}, cupric; Hg^{2+}, mercuric; Fe^{3+}, ferric; Cr^{3+}, chromic.

[b] These metals can also exist as ions with a higher charge, and may also be designated by their Latin names with an *-ous* ending as follows: Cu^+, cuprous; Fe^{2+}, ferrous; Cr^{2+}, chromous; Pb^{2+}, plumbous; Mn^{2+}, manganous; Sn^{2+}, stannous.

[c] The ions HCO^{3-} and HSO^{4-} are also known as bicarbonate and bisulfate, respectively.

Using the information given in Table 6.8, it is possible to work out the formulas and give the names of a very large number of ionic compounds. To do so, simply observe the following steps:

1. Choose the cation and the anion of the compound. The name of the compound is simply the name of the cation followed by the name of the anion. For example, when Fe^{3+} is the cation and SO_4^{2+} is the anion, the name of the ionic compound is iron(III) sulfate.
2. Choose subscripts to place after the cation and anion in the chemical formula of the compound such that multiplying the subscript of the cation times the charge of the cation gives a number equal in magnitude and opposite in sign from that of the product of the anion's subscript times the anion's charge. In the example of iron(III) sulfate, a subscript 2 for Fe^{3+} gives $2 \times (3+) = 6+$, and a subscript 3 for SO_4^{2+} gives $2 \times (3-) = 6-$, thereby meeting the condition for a neutral compound.
3. Write the compound formula. If the subscript after any polyatomic ion is greater than 1, put the formula of the ion in parentheses to show that the subscript applies to all the atoms in the ion. Omit the charges on the ions, because they make the compound formula too cluttered. In the example under consideration, the formula of iron(III) sulfate is $Fe_2(SO_4)_3$.

Exercise: Match each cation in the left column below with each anion in the right column and give the formulas and names of each of the resulting ionic compounds.

(A) Na^+	1. Br^-
(B) Ca^{2+}	2. CO_3^{2-}
(C) Al^{2+}	3. PO_4^{3-}

Answers: (A-1) NaBr, sodium bromide; (A-2) Na_2CO_3, sodium carbonate; (A-3) Na_3PO_4, sodium phosphate; (B-1) $CaBr_2$, calcium bromide; (B-2) $CaCO_3$, calcium carbonate; (B-3) $Ca_3\ (PO_4)_2$, calcium phosphate; (C-1) $AlBr_3$, aluminum bromide; (C-2) $Al_2(CO_3)_3$, aluminum carbonate; (C-3) $AlPO_4$, aluminum phosphate.

The names and formulas of ionic compounds that contain hydrogen in the name and formula of the anion are handled just like any other ionic compound. Thus, $NaHCO_3$ is sodium hydrogen carbonate, $Ca(H_2PO_4)_2$ is calcium dihydrogen phosphate, and K_2HPO_4 is sodium monohydrogen phosphate. The acetate ion, $C_2H_3O^{2-}$, also contains hydrogen, but, as noted previously, its hydrogen is covalently bonded to a C atom, as shown by the structural formula,

```
   H  O
   |  ||
H- C- C- O⁻
   |
   H
```

Acetate anion with covalently bound, nonionizable hydrogens

and cannot form H^+ ions, whereas the anions listed with hydrogen in their names can produce H^+ ion when dissolved in water.

CHAPTER SUMMARY

The chapter summary below is presented in a programmed format to review the main points covered in this chapter. It is used most effectively by filling in the blanks, referring back to the chapter as necessary. The correct answers are given at the end of the summary.

[1]______ ion is produced by acids and [2]______ by bases. A neutralization reaction is [3]______________ . Hydrogen ion, H^+, in water is bonded to [4]____________ and is often represented as [5]________________. A **base** is a substance that accepts H^+ and produces hydroxide ion. Although NH_3 does not contain hydroxide ions, it undergoes the reaction [6]__________________________to produce OH^- in water. The two products produced whenever an acid and a base react together are [7]______ ________________. A salt is made up of [8]______________________ _____. An amphoteric substance is one that [9]______________________. In water, a metal ion is bonded to [10]____________________ in a form known as a [11]____________________. In terms of acid–base behavior, some metal ions act as [12]____________. In water solution, sodium carbonate acts as a [13]______________ and undergoes the reaction [14]________________________________. Salts that act as acids react with [15]________________. Pure water conducts electricity [16]____________, a solution of acetic acid conducts electricity [17]__________, and a solution of HCl conducts electricity [18]______________. These differences are due to differences in concentrations of [19]__________ in the water. Materials that conduct electricity in water are called [20]________________. Materials that do not form ions in water are called [21]________________. The reaction $CH_3CO_2H \rightarrow CH_3CO_2^- + H^+$ may be classified as [22]__________ or [23]__________, and when the acetic acid molecule comes apart, it is said to [24]__________. A base that is completely dissociated in water is called a [25]________________ and an acid that is only slightly dissociated is called a [26]________________. At high concentrations, the percentage of dissociation of a weak acid is [27]__________ than at lower concentrations. Buffers are [28]____________________________________. buffer can be made from a mixture of a weak base and [29]________________. The reaction that results in the production of very low concentrations of ions in even pure water is [30]________________ and the relationship between the concentrations of these ions in water is [31]____________________. In absolutely pure water at 25°C the value of $[H^+]$ is exactly [32]____________, the pH is [33]______, and the solution is said to be [34]____________. Acidic solutions have pH values of [35]______________ and basic solutions have pH values of [36]______________. In a general sense, solution equilibrium deals with the extent to which reversible acid–base, solubilization (precipitation), complexation, or oxidation–reduction

reactions [37]______________________. As an example of acid–base equilibrium, the reaction for the ionization of acetic acid, HAc, is [38]______________, for which the acid dissociation constant is [39]______________________. Some ways to prepare acids are [40]____. Some ways to prepare bases are [41]____. The reaction of an ion with water, such as $PO_4^{3-} + H_2O \rightarrow HPO_4^{2-} + OH^-$, is an example of a [42]______________. The most obvious way to prepare a salt is by [43]______________________. Active metals react with acids to produce [44]______________________. Other than reacting with acids, some metals react with [45]______________. If the anion in a salt can form a volatile acid, a new salt can be formed by [46]______________________________. Some metals will displace other metals from a salt. If magnesium, a highly reactive metal, is added to a solution of copper sulfate, the reaction that occurs is [47]______________________. $NaHSO_4$, which has an ionizable hydrogen, is an example of [48]______________, whereas $Ca_5OH(PO_4)_3$ is an example of [49]______________. The water in $CuSO_4 \cdot 5H_2O$ is called [50]______________________. The names of $HClO_4$, $HClO_3$, $HClO_2$, and HClO are, respectively, [51]____.The names of $NaClO_4$, $NaClO_3$, $NaClO_2$, and NaClO are, respectively, [52]____. The name of a base containing a metal consists of [53]______________________________. The name of a salt is [54]______________________________________. The names of the ions Ca^{2+}, Fe^{3+}, H^+, SO_3^{2-}, and $C_2H_3O_2^-$ are, respectively, [55]______________________. In writing the formulas of ionic compounds, choose subscripts to place after the cation and anion in the chemical formula of the compound such that multiplying the subscript of the cation by the charge of the cation gives a number [56]______________________ and opposite in sign from that of the product of [57]____.

Answers to Chapter Summary

1. H^+
2. OH^-
3. $H^+ + OH^- \rightarrow H_2O$
4. water molecules

5. hydronium ion, H_3O^+
6. $NH_3 + H_2O \rightarrow NH_4^+ + OH^-$
7. water and a salt
8. a cation (other than H^+) and an anion (other than OH^-)
9. can act as either an acid or a base
10. water molecules
11. hydrated ion
12. acids
13. base
14. $2Na^+ + CO_3^{2-} + H_2O \rightarrow Na^+ + HCO_3^- + Na^+ + OH^-$
15. hydroxide ions
16. not at all
17. poorly
18. very well
19. ions
20. electrolytes
21. nonelectrolytes
22. ionization
23. dissociation
24. dissociate
25. strong base
26. weak acid
27. lower
28. solutions that resist changes in H^+ concentration
29. a salt of the base
30. $H_2O \rightarrow H^+ + OH^-$
31. $[H^+][OH^-] = 1.00 \times 10^{-14} = K_w$
32. 1×10^{-7} mol/L

33. 7.00

34. neutral

35. less than 7

36. greater than 7

37. proceed in a forward or backward direction

38. $HAc \rightleftharpoons H^+ + Ac^-$

39. $\frac{[H^+][Ac^-]}{[HAc]} = K = 1.75 \times 10^{-5}$ (at 25°C)

40. reaction of hydrogen with a nonmetal, reaction of a nonmetal directly with water, reaction of a nonmetal oxide with water, production of volatile acids by reaction of salts of the acids with nonvolatile acids

41. reaction of active metals directly with water, reaction of metal oxides with water, reaction of a basic compound that does not itself contain hydroxide with water

42. hydrolysis

43. reaction of an acid with a base

44. a salt and hydrogen gas

45. strong bases

46. adding a nonvolatile acid

47. $Mg(s) + CuSO_4(aq) \rightarrow Cu(s) + MgSO_4(aq)$

48. an acid salt

49. a basic salt

50. water of hydration

51. perchloric acid, chloric acid, chlorous acid, and hypochlorous acid

52. sodium perchlorate, sodium chlorate, sodium chlorite, and sodium hypochlorite

53. the name of the metal followed by "hydroxide"

54. the name of the cation followed by the name of the anion

55. calcium, iron(III), hydride, sulfite, and acetate

56. equal in magnitude

57. the anion's subscript times the anion's charge

QUESTIONS AND PROBLEMS

1. Give the neutralization reaction for each of the acids in the left column reacting with each of the bases in the right column:

   (A) Hydrocyanic acid   1. Ammonia

   (B) Acetic acid   2. Sodium hydroxide

   (C) Phosphoric acid   3. Calcium hydroxide

2. Methylamine is an organic amine, formula $H_3C\text{-}NH_2$, that acts as a weak base in water. By analogy with ammonia, suggest how it might act as a base.

3. In a 1 molar solution of acetic acid (containing 1 mol of acetic acid per liter of solution) only about 0.5% of the acid is ionized to produce an acetate ion and a hydrogen ion.

4. What does H_3O^+ represent in water?

5. When exactly 1 mole of NaOH reacts with exactly 1 mole of H_2SO_4, the product is an acid salt. Show the production of the acid salt with a chemical reaction.

6. An amphoteric substance can be viewed as one that may either accept or produce an ion of H^+. Using that definition, explain how H_2O is amphoteric.

7. Explain how Fe^{3+} ion dissolved in water can be viewed as an acidic hydrated ion.

8. Cyanide ion, CN^-, has a strong attraction for H^+. Show how this explains why NaCN acts as a base.

9. Separate solutions containing 1 mole per liter of NH_3 and 1 mole per liter of acetic acid conduct electricity poorly, whereas when such solutions are mixed, the resulting solution conducts well. Explain.

10. What characteristic of solutions of electrolytes enables them to conduct electricity well?

11. A solution containing 6 moles of NH_3 dissolved in a liter of solution would be relatively highly concentrated. Explain why it would not be correct, however, to describe such a solution as a "strong base" solution.

12. The dissociation of acetic acid can be represented by

$$CH_3CO_2H \rightarrow CH_3CO_2^- + H^+$$

Explain why this reaction can be characterized as an ionization of acetic acid. Explain on the basis of the "crowding" concept why the percentage of acetic acid molecules dissociated is less in relatively concentrated solutions of the acid.

13. A solution containing 0.1 mole of HCl per liter of solution has a low pH of 1, whereas a solution containing 0.1 mole of acetic acid per liter of solution has a significantly higher pH. Explain.

14. Explain why a solution containing both NH_3 and NH_4Cl acts as a buffer. In so doing, consider reactions of NH_3, NH_4^+ ion, H^+ ion, OH^- ion, and H_2O.

15. NaH_2PO_4 and Na_2HPO_4 dissolved in water produce $H_2PO_4^-$ and HPO_4^{2-} ions, respectively. Show by reactions of these ions with H^+ and OH ions why a solution consisting of a mixture of both NaH_2PO_4 and Na_2PO_4 dissolved in water acts as a buffer.

16. What is the expression and value for K_w? What is the reaction upon which this expression is based?

17. On the basis of pH, distinguish among acidic, basic, and neutral solutions.

18. Give the pH values corresponding to each of the following values of $[H^+]$:

    (a) 1.00×10^{-4} mol/L, (b) 1.00×10^{-8} mol/L, (c) 5.63×10^{-9} mol/L, (d) 3.67×10^{-6} mol/L.

19. Why does solution equilibrium deal only with reversible reactions?

20. Write an equilibrium constant expression for the reaction

$$CO_3^{2-} + H_2O \rightleftharpoons HCO_3^- + OH^-$$

21. Calculate $[H^+]$ in a solution of carbon dioxide in which $[CO_2(aq)]$ is 3.25×10^{-4} moles/liter.

22. Cl_2 and F_2 are both halogens. Suggest acids that might be formed from the reaction of F_2 with H_2 and with H_2O.

23. Suggest the acid that might be formed by reacting S with H_2.

24. Suggest the acid or base that might be formed by the reaction of each of the following oxides with water: (a) N_2O_3, (b) CO_2, (c) SO_3, (d) Na_2O, (e) CaO, (f) Cl_2O.

25. Knowing that H_2SO_4 is a non-volatile acid, suggest the acid that might be formed by the reaction of H_2SO_4 with NaCl.

26. Acetic acid is a carboxylic acid. Formic acid is the lowest carboxylic acid, and it contains only 1 C atom per molecule. What is its formula?

27. A base can be prepared by the reaction of calcium metal with hot water. Give the reaction and the name of the base product.

28. A base can be prepared by the reaction of sodium oxide with water. Give the reaction and the name of the base product.

29. Give the reactions by which the following act as bases in water: (a) NH_3, (b) Na_2CO_3, (c) Na_3PO_4, and dimethylamine, $(CH_3)_2NH$.

30. Choosing from the reagents H_2SO_4, HCl, $Mg(OH)_2$, and LiOH give reactions that illustrate “the most straightforward” means of preparing salts.

31. Choosing from the reagents NaOH, HCl, H_2SO_4, NaCl, CaO, Mg, F_2 and Al give reactions that illustrate the preparation of salts by (a) reaction of a metal and a nonmetal that will react directly to make a salt, (b) reaction of a metal with acid, (c) reaction of a metal with strong base, (d) reaction of a salt with a nonvolatile acid, (e) cementation.

32. Describe what is meant by an acid salt.

33. Describe what is meant by a basic salt.

34. Explain how sodium carbonate decahydrate illustrates water of hydration. Why is it less hazardous to skin than is anhydrous sodium carbonate? Illustrate with a chemical reaction why it is also a basic salt.

35. Give the names of each of the following acids: (a) HBr, (b) HCN, (c) HClO, (d) $HClO_2$, (e) $HClO_3$, (f) $HClO_4$, (g) HNO_2.

36. Give the names of (a) LiOH, (b) $Ca(OH)_2$, and (c) $Al(OH)_3$.

37. Give the names of (a) $MgSO_3$, (b) NaClO, (c) $Ca(ClO_4)_2$, (d) KNO_3, (e) $Ca(NO_2)_2$

38. Match each cation in the left column below with each anion in the right column and give the formulas and names of each of the resulting ionic compounds.

(A) Li^+	1. CN^-
(B) Ca^{2+}	2. SO_3^{2-}
(C) Fe^{3+}	3. NO_3^-

39. What color is litmus in (A) acid and (B) base?

40. Give the formulas of each of the following:

(A) Magnesium acetate

(B) Calcium monohydrogen phosphate

(C) Aluminum sulfate

(D) Calcium hypochlorite

41. Using Lewis (electron-dot) structures, show the reaction between hydronium ion and hydroxide ion.

42. A common error in speaking the chemical language is to confuse acidic (pronounced uh-sid-ik) with acetic (pronounced uh-seat-ik). What is the correct meaning of each of these terms? What is the difference between ammoni*a* and ammoni*um*?

43. The following is a list that contains the names of three cations and three anions: hypochlorite, hydrogen, sodium, sulfate, calcium, nitrate. List the three cations. List the three anions. Give the formulas of nine compounds that can be made by various combinations of these.

44. Write a chemical reaction in which $NaHCO_3$ acts as an acid. Write another in which it acts as a base, remembering that if H_2CO_3 is produced in solution it largely goes to carbon dioxide gas and water.

45. Write the Lewis structure of the hydronium ion, H_3O^+.

46. Explain by chemical reactions how a mixture of $NaHCO_3$ and Na_2CO_3 in water would act as a buffer.

47. A solid known to be either NaCl or Na_2SO_4 was moistened with concentrated H_2SO_4 and heated, giving off a gas that turned moist blue litmus paper red. What was the solid?

48. Formulas of some chemical compounds are given in the left column. Match the formula of each compound with its correct name in the right column.

(A) CaO	1. Potassium sulfide
(B) SiO_2	2. Dinitrogen pentoxide
(C) K_2S	3. Nitrogen dioxide
(D) $AlCl_3$	4. Silicon dioxide
(E) NO_2	5. Potassium bromide
(F) N_2O_5	6. Sodium iodide
(G) NaI	7. Calcium oxide
(H) KBr	8. Magnesium fluoride
(I) MgF_2	9. Calcium fluoride
(J) CaF_2	10. Aluminum chloride

49. Match the names in the right column with the formulas in the left column.

(A) Na_2CO_3	1. Calcium sulfate
(B) $CaSO_3$	2. Potassium perchlorate
(C) $Al(OH)_3$	3. Sodium carbonate
(D) $CaSO_4$	4. Calcium phosphate
(E) $NaNO_2$	5. Aluminum hydroxide
(F) $Ca_3(PO_4)_2$	6. Calcium sulfite
(G) $NaNO_3$	7. Calcium hypochlorite
(H) $Ca(ClO)_2$	8. Sodium nitrate
(I) $KClO_4$	9. Sodium nitrite

50. Match the names in the right column with the formulas in the left column.

(A) K_2HPO_4	1. Sodium hydrogen sulfate
(B) $KHCO_3$	2. Sodium hydrogen oxalate
(C) $NaHSO_4$	3. Dipotassium hydrogen phosphate
(D) KH_2PO_4	4. Sodium hydrogen phthalate
(E) $NaHC_8H_4O_4$	5. Potassium hydrogen carbonate
(F) $NaHC_2O_4$	6. Potassium dihydrogen phosphate

51. Fill in each of the following blanks with the number corresponding to the meaning of each of the prefixes. The first one is done for you as an example.

tetra __4__ mono _____ octa _____ deca _____ di _____

penta _____ hepta _____ tri _____ nano _____ sexa _____

52. Give the correct name to each of the following compounds. The first one is done for you as an example.

N_2O_5 Dinitrogen pentoxide	N_2O_4 ____________________
NO_2 ____________________	N_2O_3 ____________________
NO ____________________	N_2O ____________________

53. Iron in a compound can also be designated as ferrous or ferric. Similarly copper (Cu) may be called cuprous or cupric. Tin (Sn) may be called stannous or stannic. Name each of the following compounds with two acceptable names. The first one is done for you as an example.

$FeCl_2$ ferrous chloride ______ or iron(II) chloride ______

$FeCl_3$ ____________________ or ____________________

CuCl ____________________ or ____________________

$CuCl_2$ ____________________ or ____________________

$SnCl_2$ ____________________ or ____________________

$SnCl_4$ ____________________ or ____________________

54. Dry $CaSO_4$ absorbs enough water to yield a product with a specific number or waters of hydration. Exactly 136 g. of $CaSO_4$ exposed to humid air gained enough water to weigh exactly 172 g. What is the formula of the product with the waters of hydration?

55. In each of the following chemical reactions fill in the formula of the missing compound. The rest of the chemical equation is balanced.

$2N_2 + 3O_2 \rightarrow$ ____________________ (nitrogen trioxide)

$KOH + SO_2 \rightarrow$ ____________________ (potassium hydrogen sulfite)

$KOH + H_3PO_4 \rightarrow$ ____________________ (potassium dihydrogen phosphate)

56. Give one or two examples of compounds in which each of the following prefixes or suffixes is used in the compound name.

-ic acetic acid hydrochloric acid

-ide ____________________ ____________________

-ous ____________________ ____________________

hypo- ____________________ ____________________

per- ____________________ ____________________

-ite ____________________ ____________________

-ate ____________________ ____________________

57. In this chapter, it was mentioned that a certain group was characteristic of organic acids. The Lewis structure of acetic acid was also given and it was

shown how many of its Hs can form H^+ ion. The Lewis formula of the formate ion produced by the ionization of formic acid is,

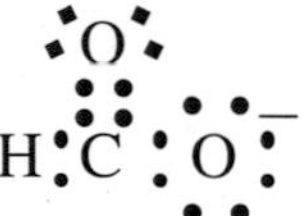

From this information, explain how many ionizable hydrogens formic acid has, and why it has that number.

58. How do you explain the fact that a 0.01 M solution of HCl contains more H^+ ions than does as 0.1 M (ten-fold higher concentration) of acetic acid?

59. How is it explained that a solution of $FeCl_3$ is acidic?

60. Explain how a mixture of NH_3 and NH_4Cl dissolved in solution can act as a buffer.

61. Hydrochloric acid, HCl, is often found in rainwater in ocean coastal areas in which the atmosphere is polluted by sulfuric acid. How might this occur?

62. Classify each of the following as strong electrolytes, weak electrolytes, or nonelectrolytes: (A) Solution of NaCl, (B) vinegar, (C) pure water, (D) solution of sugar, (E) solution of HCl, (F) solution of ammonia, (G) solution of sodium hydroxide.

63. Match the value of $[H^+]$ in the left column with the pH in the right column.

(A) 1.00×10^{-8} M   1. 8.37

(B) 1.00×10^{-9} M   2. 8.00

(C) 4.28×10^{-9} M   3. 7.70

(D) 2.00×10^{-8} M   4. 9.00

64. Exactly 1.00 mole of an acid was dissolved in 1.00 liter of water solution. The pH of the resulting solution was 2.00. Referring to Tables 6.1 and 6.3, classify the acid in regard to its strength.

7. SOLUTIONS AND SOLVENTS

7.1. WHAT ARE SOLUTIONS? WHY ARE THEY IMPORTANT?

To help understand solutions, consider a simple experiment. Run some water from a faucet into a glass and examine it. Depending on the condition of the local water supply, it will probably appear clear. If tasted, it will probably not have any particularly strong flavors. Now, add a teaspoon of sugar and stir the contents of the glass; you will notice that the sugar will begin to disappear, and the water around the sugar will start to appear cloudy or streaked. With continued stirring, the sugar will seem to disappear completely, and the water will look as clear as it did when it came out of the tap. However, it will have a sweet taste; some of its properties have therefore been changed by the addition of sugar.

This experiment illustrates several important characteristics of chemicals and how they are used. Sugar **dissolves** in water; when this happens, a **solution** is formed. The sugar molecules were originally contained in hard, rigid sugar crystals. In water, the molecules break away from the crystals and spread throughout the liquid. The sugar is still there, but it is dissolved in the water, which is called the **solvent**. The dissolved sugar is called the **solute**; it can no longer be seen, but it can be tasted. The water appears unchanged, but it is different. It has a different taste. Some of its other properties have changed, too; for example, it now boils at a higher temperature and freezes at a lower temperature than pure water. In some cases, there is readily visible evidence that a solute is present in a solvent. The solution may have a strong color, as is the case for intensely purple solutions of potassium permanganate, $KMnO_4$. It may have a strong odor, such as that of ammonia, NH_3, dissolved in water.

As illustrated in Figure 7.1, tap water itself is a solution containing many things. It has some dissolved oxygen and carbon dioxide in it. It almost certainly has some dissolved calcium compounds, making it "hard." There may be a small amount of iron present, which causes the water to stain clothing. It contains some chlorine, added to kill bacteria. In unfavorable cases, it may even contain some toxic lead and cadmium dissolved from plumbing and the solder used to connect copper pipes. Thus, it is clear that many of the liquids that people come in contact with are actually solutions.

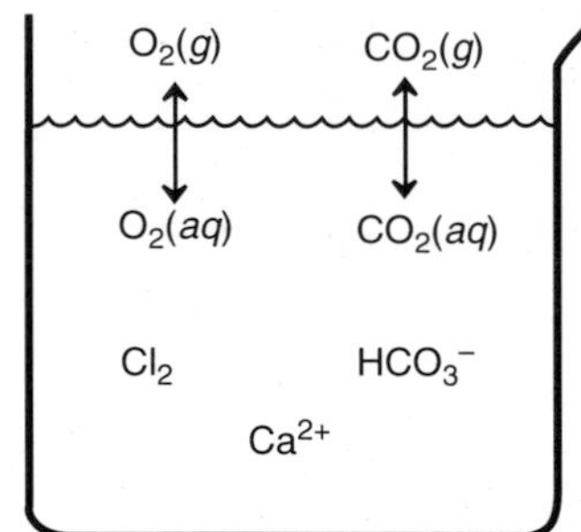

Figure 7.1. Tap water is actually a solution that contains, in small quantities, many chemical species besides H_2O.

There are many examples of important and useful solutions. Sugar, for instance, must be dissolved before it can be used by the body for food. Candy, which is sugar with added flavoring, dissolves in the mouth to form a solution of sugar with saliva. The coolant in an automobile's cooling system is a solution of antifreeze in water. This solution freezes at a much lower temperature than pure water, thus preventing the liquid from freezing and cracking the engine block. A solution containing the simple sugar glucose and some other substances may be injected directly into the veins of an ill or injured person who cannot take food through the mouth. Chemists use many different kinds of solutions that undergo chemical reactions with other kinds of chemicals. By measuring how much of solution is required to complete a reaction in a procedure called *titration*, the chemist can tell how much of a particular kind of chemical is in a solution.

Reactions in Solution

One of the most important properties of solutions is their ability to allow chemical species to come into close contact so that they can react. For example, if perfectly dry crystals of calcium chloride, $CaCl_2$, were mixed with dry crystals of sodium fluoride, NaF, a chemical reaction would not occur. However, if each is dissolved in separate solutions which are then mixed, a precipitation reaction occurs in which calcium chloride and sodium fluoride in aqueous solution (*aq*) react to produce calcium fluoride solid (*s*) and a solution of sodium chloride:

$$CaCl_2(aq) + 2NaF(aq) \rightarrow CaF_2(s) + 2NaCl(aq) \qquad (7.1.1)$$

This reaction occurs because in solution the Ca^{2+} ions (from dissolved $CaCl_2$) and the F^- ions (from dissolved NaF) move around and easily come together to form CaF_2. The calcium fluoride product does not stay in solution, but forms a precipitate; it is **insoluble**.

In other cases, solutions enable chemical reactions to occur that result in materials being dissolved. Some of these reactions are important in geology. Consider limestone, which is made of calcium carbonate, $CaCO_3$. Limestone does not react with dry CO_2 gas, nor is it soluble in pure water. However, when water containing dissolved CO_2 contacts limestone, a chemical reaction occurs:

$$CaCO_3(s) + CO_2(g) + H_2O \rightarrow Ca^{2+}(aq) + 2HCO_3^-(aq) \quad (7.1.2)$$

The calcium ion and the bicarbonate ion, HCO_3^-, remain dissolved in water; the solution of CO_2 in water dissolves the limestone, leaving a cave or hole in the limestone formation. In some regions, such as parts of southern Missouri, this has occurred to such an extent that the whole area is underlain by limestone caves and potholes; these are called *karst* regions.

Solutions in Living Systems

For living things, the most important function of solutions is to carry molecules and ions to and from cells. Body fluids consist of complex solutions. Digestion is largely a process of breaking down complex, insoluble food molecules to simple, soluble molecules that may be carried by the blood to the body cells, which need them for energy and production of more cellular material. On the return trip, the blood carries waste products, such as carbon dioxide, which are eliminated from the body.

Solutions in the Environment

Solutions are of utmost importance in the environment (Figure 7.2). Solutions transport environmental chemical species in the aquatic environment and are crucial

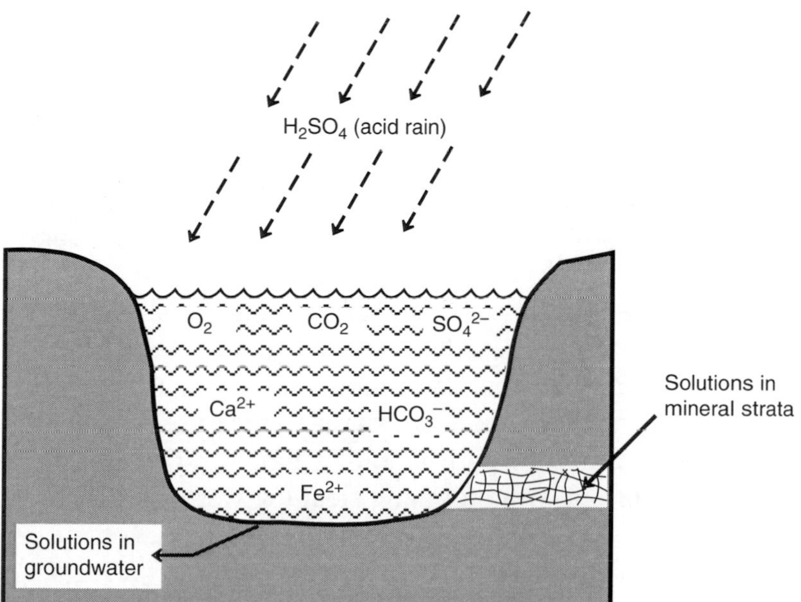

Figure 7.2. Many environmental chemical phenomena involve solutions.

participants in geochemical processes. Dissolution in rainwater is the most common process by which atmospheric pollutants are removed from air. Acid rain is a solution of strong mineral acids in water. Many important environmental chemical processes occur in solution and at the interface of solutions with solids and gases.

Pollutant pesticides and hazardous waste chemicals are transported in solution as surface water or groundwater. Many hazardous waste chemicals are dissolved in solution; often, the large amount of water in which they are dissolved makes their treatment relatively more difficult and expensive.

Industrial Uses of Solutions

Solutions are used throughout industry. Many chemical reactions that are part of the manufacture of important industrial chemicals occur in solutions, and the chemical processes that take place in solution—**solution chemistry**—are very important in the chemical industry. *Natural brines* are solutions that contain a lot of dissolved materials and that occur underground and in some lakes (saline lakes). Some natural brines are important sources of valuable chemicals. Commercially important chemicals that are recovered from brines include borax, a compound containing boron and oxygen, which is used as an antiseptic, in making ceramics, and in some cleaning formulations; bromine salts; and potassium salts, including those used in potassium fertilizer. Solutions of detergents are used for cleaning; some dyes are applied as solutions. Ammonia fertilizer may be added to the soil as a solution of NH_3 in water.

Many organic materials do not dissolve significantly in water. Such substances are usually soluble in organic solvents, such as benzene and carbon tetrachloride. When used to dissolve organic substances, organic liquids are commonly called solvents. Some aspects of solvents are discussed in the following section.

7.2. SOLVENTS

Water is the solvent for most of the solutions discussed in this chapter. However, as noted above, many other liquids are also used as solvents. Other than water, most solvents are organic (carbon-containing) liquids. Some of the more important organic solvents are shown in Table 7.1.

There are many uses for solvents. One of the most important of these is their role as media in which chemical reactions may occur. In the chemical industry, solvents are employed for purification, separation, and physical processing. Solvents are also used for cleaners; one important example is the use of organic solvents to dissolve grease and oil from metal parts after they have been fabricated. (In the past, although less so now, solvents used for parts cleaning were allowed to evaporate to the atmosphere, a major source of atmospheric organic pollutants.) The chemicals that make up synthetic fibers, such as rayon, are dissolved in solvents, and are then forced under very high pressure through small holes in a special die to make individual filaments of the fiber. One of the most important uses for solvents is in

Table 7.1. Important Organic Solvents

Solvent	Solvent Use (May Have Many Other Nonsolvent Uses)	Approximate Annual U.S. Production (Millions of kg)
Acetone	Solvent for spinning cellulose acetate fibers and for spreading paints and other protective coatings	1000
Benzene	Dissolves grease and other organic compounds	6500
Perchloroethylene	Best solvent for dry cleaning; also used for degreasing metals and extraction of fats	125
Stoddard solvent	Mixture of alkanes and aromatic hydrocarbons containing 7–12 carbon atoms per molecule; used as a solvent for organic materials	17
Toluene	Dissolves grease and other organic materials; substitute for benzene, but not so toxic	2700
Trichloroethylene	Vapor degreasing of metal parts; solvent for greases, oils, fats, waxes, and tars; fabric cleaner; waterless dying; ingredient of formulations of adhesives, lubricants, paints, varnishes, and paint strippers	90

coatings, which include paint, printing inks, lacquers, and antirust formulations. In order to apply these coatings, it is necessary to dissolve them in a solvent (**vehicle**) so that they may be spread around on the surface to be coated. The vehicle is a **volatile** liquid, one that evaporates quickly to form a vapor; when it evaporates, it leaves the coating behind as a thin layer.

Fire and toxicity are major hazards associated with the use of many solvents. Some organic solvents, such as benzene, are even more of a fire hazard than gasoline. Both benzene and carbon tetrachloride are toxic and can damage the body in cases of excess exposure. Benzene is suspected of causing leukemia, and worker exposure to this solvent is now carefully regulated. The toxicity hazard of solvents arises from absorption through the skin and inhalation through the lungs. One solvent, dimethylsulfoxide, is relatively harmless by itself, but has the property of carrying toxic solutes through the skin and into the body. Exposure to solvent vapor is limited by occupational health regulations, which include a threshold limiting value (TLV). This is the measure of solvent vapor concentration in the atmosphere considered safe for exposure to healthy humans over a normal 40-hour work week.

7.3. WATER—A UNIQUE SOLVENT

Water is a vitally important solvent in environmental chemistry, green chemistry, and industrial processes. Those properties of water that relate directly to its characteristics as a solvent are summarized in this chapter.

At room temperature H_2O is a colorless, tasteless, odorless liquid. It boils at 100°C (212°F) and freezes at 0°C (32°F). Water by itself is a very stable compound; it is very difficult to break up by heating. However, as explained in Section 8.6, when electrically conducting ions are present in water, a current may be passed through the water, causing it to break up into H_2 gas and O_2 gas.

Water is an excellent solvent for a variety of materials; these include many ionic compounds (acids, bases, and salts). Some gases dissolve well in water, particularly those that react with it chemically. Sugars and many other biologically important compounds are also soluble in water. However, greases and oils generally are not soluble in water but do dissolve in organic solvents.

Some of water's solvent properties can best be understood by considering the structure and bonding of the water molecule:

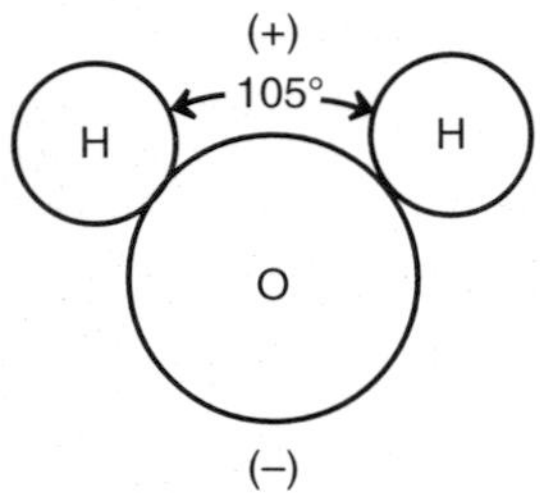

The water molecule is made up of two hydrogen atoms bonded to an oxygen atom. The three atoms are not in a straight line, but form an angle of 105°.

Because of water's bent structure and the fact that the oxygen atom attracts the negative electrons more strongly than do the hydrogen atoms, the water molecule behaves like a body having opposite electrical charges at either end or pole. Such a body is called a *dipole*. Owing to the fact that it has opposite charges at opposite ends, the water dipole may be attracted to either positively or negatively charged ions. Recall that NaCl dissolves in water to form positive Na^+ ions and negative Cl^- ions in solution. The positive sodium ions are surrounded by water molecules with their negative ends pointed at the ions, and the chloride ions are surrounded by water molecules with their positive ends pointing at the negative ions, as shown in Figure 7.3. This kind of attraction for ions is the reason why water dissolves many ionic compounds and salts that do not dissolve in other liquids. Some noteworthy examples

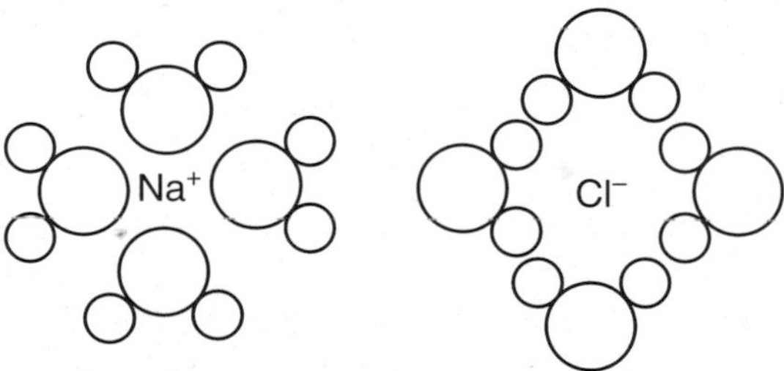

Figure 7.3. Polar water molecules surrounding an Na^+ ion (left) and a Cl^- ion (right).

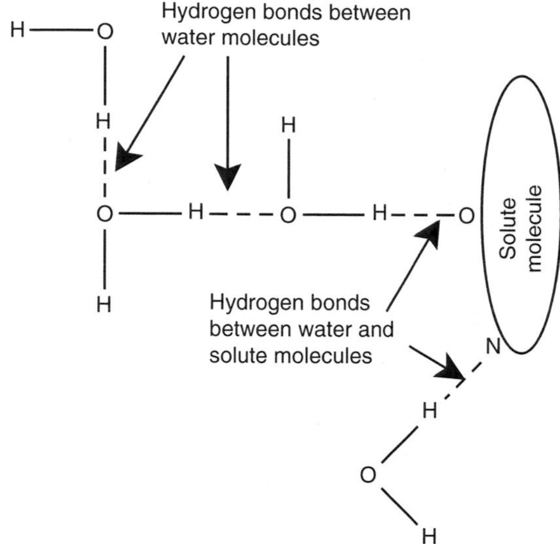

Figure 7.4. Hydrogen bonding between water molecules and between water molecules and a solute molecule in solution.

are sodium chloride in the ocean; waste salts in urine; calcium bicarbonate, which is very important in lakes and in geological processes; and widely used industrial acids (e.g., HNO_3, HCl, and H_2SO_4).

In addition to being a polar molecule, the water molecule has another important property that gives it many of its special characteristics: the ability to form **hydrogen bonds**. Hydrogen bonds are a special type of bond that can form between the hydrogen in one water molecule and the oxygen in another water molecule. This bonding takes place because the oxygen has a partly negative charge and the hydrogen a partly positive charge. Hydrogen bonds, shown in Figure 7.4 as dashed lines, hold the water molecules together in large groups.

Hydrogen bonds also help to hold some solute molecules or ions in solution. This happens when hydrogen bonds form between the water molecules and hydrogen or oxygen atoms on the solute molecule (see Figure 7.4). Hydrogen bonding is one of the main reasons that some proteins can be put in water solution or held suspended in water as extremely small particles called *colloidal particles* (see Section 7.9).

7.4. THE SOLUTION PROCESS AND SOLUBILITY

Very little happens to simple molecules, such as N_2 and O_2, when they dissolve in water. They mingle with the water molecules and occupy spaces that open up between water molecules to accommodate the N_2 and O_2 molecules. If the water is heated, some of the gases are driven out of solution. This may be observed as the small bubbles that appear in heated water just before it boils. A fish can extract some of the oxygen in water by "breathing" through its gills; just 6 or 7 parts of oxygen in a million parts of water is all that fish require. Water saturated with air at 25°C

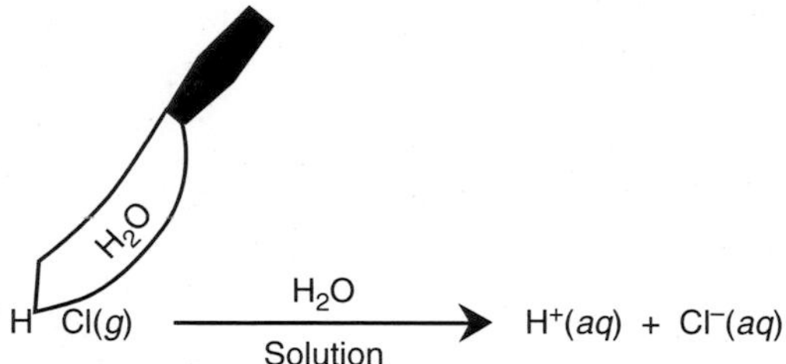

Figure 7.5. HCl dissolving in water. Water breaks apart a hydrogen chloride molecule to form a hydrogen ion, H^+, and a chloride ion, Cl^-.

contains about 8 parts per million oxygen. In bodies of water such as lakes, and in streams, a small amount of an oxygen-consuming substance can use up this tiny portion of oxygen dissolved in water and cause the fish to suffocate and die.

Although N_2 and O_2 dissolve in water in the simple form of their molecules, the situation is very different when hydrogen chloride gas, HCl, dissolves in water. The hydrogen chloride molecule consists of a hydrogen atom bonded to a chlorine atom with a covalent bond. (Recall that covalent bonds are formed by *sharing* electrons between atoms.) Water can absorb large amounts of hydrogen chloride: 100 g of water at 0°C will dissolve 82 g of this gas. When HCl dissolves in water (Figure 7.5), the solution is not simply hydrogen chloride molecules mixed with water molecules. The water has a strong effect upon the HCl molecule, breaking it into two parts, with the two electrons in the chemical bond staying with the chlorine. This results in the formation of a positively charged hydrogen ion, H^+, and a negatively charged chloride ion, Cl^-.

In water solution, the chloride ion is surrounded by the positive ends of the water molecules, which are attracted to the negatively charged Cl^- ion. This kind of attraction of water molecules for a negative ion has already been shown in Figure 7.3. The H^+ ion from the HCl molecule does not remain in water as an isolated ion; it attaches to an unshared electron pair on a water molecule. This water molecule, with its extra hydrogen ion and extra positive charge, becomes a different ion, with the formula H_3O^+; it is called a **hydronium ion** (Figure 7.6). Although a hydrogen ion in solution is indicated as H^+ for simplicity, it is really present as part of a hydronium ion or of larger ion aggregates (such as $H_5O_2^+$ and $H_7O_3^+$).

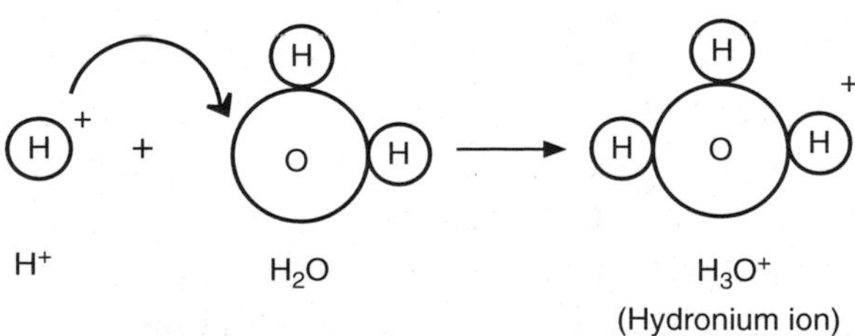

Figure 7.6. A hydrogen ion, H^+ bonds to a water molecule, H_2O, to produce a hydronium ion, H_3O^+. Bonding to additional water molecules may form larger aggregates, such as $H_5O_2^+$ and $H_7O_3^+$.

The solution of hydrogen chloride in water illustrates a case in which a neutral molecule dissolves and forms electrically charged ions in water. While this happens with other substances dissolved in water, the hydrogen ion resulting when substances such as HCl dissolve in water is particularly important because it results in the formation of a solution of acid. So, rather than calling this a solution of hydrogen chloride, it is called a **hydrochloric acid** solution.

7.5. SOLUTION CONCENTRATIONS

In describing a solution, it is necessary to do so *qualitatively*, that is, to specify what the solvent is and what the solutes are. For example, it has just been seen that HCl gas is the solute placed in water to form hydrochloric acid. It is also necessary to know what happens to the material when it dissolves. For instance, one needs to know that hydrogen chloride molecules dissolved in water form H^+ ions and Cl^- ions. In many cases, it is necessary to have *quantitative* information about a solution, that is, its **concentration**. The concentration of a solution is the *amount* of solute material dissolved in a particular amount of solution, or by a particular amount of solvent. Solution concentration may be expressed in a number of different ways. Solutions used in technical applications, such as for cleaning, are often made up of a specified number of grams of solute per 100 mL of solvent added. On the other hand, a person involved in crop spraying may mix the required solution by adding several pounds of pesticide to a specified number of barrels of water.

The concentrations of water pollutants frequently are given in units of milligrams per liter (mg/L). Most of the chemicals that commonly pollute water are harmful at such low levels that milligrams per liter of water is the most convenient way of expressing their concentrations. For example, water containing more than about one-third of a milligram of iron per liter of water can stain clothing and bathroom fixtures. To get an idea of how small this quantity is, consider that a liter of water (strictly speaking, at 4°C) weighs 1 million milligrams. Water containing one-third of a part per million of iron contains only 1 mg of iron in 3 million milligrams (3 liters) of water. Because 1 liter of water weighs 1 million milligrams, 1 mg of a solute dissolved in a liter is a **part per million**, abbreviated as **ppm**. The terms *part per million* and *milligrams per liter* are both frequently used in reference to levels of pollutants in water.

Some pollutants are so poisonous that their concentrations are given in micrograms per liter (μg/L). A particle weighing a microgram is so small that it cannot be seen with the naked eye. Since a microgram is a millionth of a gram, a liter of water weighs 1 billion micrograms. So, 1 microgram per liter is **1 part per billion (ppb)**. Sometimes, it is necessary to think in terms of concentrations that are this low; a good example of this involves the long-banned pesticide endrin. It is so toxic that, at a concentration of only two-thirds of a microgram per liter, it can kill half of the fingerlings (young fish) in a body of water over a four-day period.

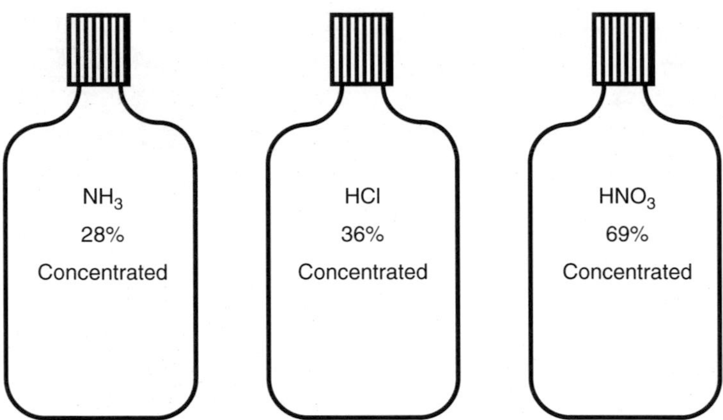

Figure 7.7. Percentages of solutes in commercial solutions of ammonia, hydrochloric acid, and nitric acid.

At the other end of the scale, it may be necessary to consider very high concentration; these are often given as *percent by mass*. As indicated in Figure 7.7, the concentrations of commercial acids and bases are often expressed in this way. For example, a solution of concentrated ammonia purchased for laboratory use is *28% by mass* NH_3; this means that out of 100 g of the ammonia solution, 28 g are NH_3 and 72 g are water. Commercial hydrochloric acid is about 36% by mass hydrogen chloride, HCl: of 100 g of concentrated hydrochloric acid, 36 g are HCl and 64 g are water.

Molar Concentration

The *mass* of a solute in solution often does not provide full information about its effect. For example, a solution of either NaCl or sodium iodide (NaI) will remove silver from a solution of silver nitrate ($AgNO_3$) when the two solutions are mixed. The two possible chemical reactions are

$$AgNO_3(aq) + NaCl(aq) \rightarrow NaNO_3(aq) + AgCl(s) \quad (7.5.1)$$

$$AgNO_3(aq) + NaI(aq) \rightarrow NaNO_3(aq) + AgI(s) \quad (7.5.2)$$

in which solid precipitates of AgCl and AgI, respectively, come out of solution. The molecular mass of NaCl is 58.5 and that of NaI is 149.9. By comparing the removal of silver from solution by solutions of NaCl and NaI, both containing the same mass of solid in a specific volume of solution, it can be seen that a given volume of NaCl solution removes more silver. In chemical reactions such as these, it is the number of ions or molecules that is important, not their masses in solution. It would be convenient to have a way of expressing concentrations in quantities that would be directly related to the number of molecules or ions in solution. Instead of expressing such huge numbers, however, the chemist normally works with *moles*. (Recall from Section 2.3 that a mole is the number of grams of a substance that is equal numerically to the mass in amu of the smallest unit of the substance. For example, since the molecular mass of NaCl is 58.5, a mole of NaCl weighs 58.5 g.) The **molar concentration** of a solution is *the number of moles of solute dissolved in a liter of solution.* Molar concentration is denoted by *M*, with units of M. A 1 M solution has 1 mole of solute dissolved in a liter of solution. To understand this concept better, consider the following:

- The molecular mass of HCl is 36.5; a 1 M solution of HCl has 1 mole of HCl (36.5 g) dissolved in a liter of solution.
- The molecular mass of NH_3 is 17; a 1 M solution of NH_3 has one mole of NH_3 (17 g) in a liter of solution.
- The molecular mass of glucose, $C_6H1_2O_6$, is 180; a 1 M solution of glucose has 1 mole of glucose (180 g) in a liter of solution.

Of course, solutions are not always exactly 1 M in concentration. However, it is easy to perform calculations involving molar concentration using the relationships expressed in the following equation:

$$M = \frac{\text{Moles solute}}{\text{Volume solution (L)}} = \frac{\text{Mass solute (g)}}{\text{Molar mass solute} \times \text{Volume solution (L)}} \quad (7.5.3)$$

Example: What is the molar concentration of a solution that contains 2.00 mol of HCl in 0.500 L of solution?

Answer:

$$M = \frac{2.00 \text{ mol}}{0.500 \text{ L}} = 4.00 \text{ mol/L}$$

Example: What is the mass of HCl, molar mass 36.5 g/mol, in 2.75 L of a 0.800 M solution?

Answer:

$$0.800 \text{ mol/L} = \frac{\text{Mass solute}}{36.5 \text{ g/mol} \times 2.75 \text{ L}}$$

$$\text{Mass solute} = 0.800 \text{ mol/L} \times 36.5 \text{ g/mol} \times 2.75 \text{ L} = 80.3 \text{ g}$$

Example: What is the molar concentration of a solution containing 87.6 g of HCl in a total volume of 3.81 L of solution?

$$M = \frac{87.6 \text{ g}}{36.5 \text{ g/mol} \times 3.81 \text{ L}} = 0.630 \text{ mol/L}$$

To summarize the steps involved in making up a certain quantity of a solution with a specified concentration, consider the task of making up 4.60 L of 0.750 M NaCl, which would involve the following steps:

Step 1. Calculate the mass of NaCl, molar mass 58.5 g/mol, required:
Mass solute = 0.750 mol/L × 58.5 g/mol × 4.60 L = 202 g
Step 2. Weigh out 202 g NaCl.
Step 3. Add NaCl to a container with a mark at 4.60 L.
Step 4. Add water, and mix to a final volume of 4.60 L.

Diluting Solutions

Often it is necessary to make a less concentrated solution from a more concentrated solution; this process is called **dilution**. This situation usually occurs in the laboratory because it is more convenient to store more-concentrated solutions in order to save shelf space. Also, laboratory acid solutions, such as those of hydrochloric acid, sulfuric acid, and phosphoric acid, are almost always purchased in highly concentrated form; this is more economical because there is more of the active ingredient per bottle. For standard solutions to use in chemical analysis, it is more accurate to weigh out a relatively large quantity of solute to make a relatively concentrated solution, and then dilute the solution quantitatively to prepare a more dilute standard solution.

For example, the concentration of commercial concentrated hydrochloric acid is 12 M. Suppose that a laboratory technician needs 1 L of 1 M hydrochloric acid. How much of the concentrated acid is required? The key to this problem is to realize that when a volume of the concentrated acid is diluted with water, the total *amount of solute acid* in the solution remains the same. The problem can then be solved by considering Equation 7.5.3:

$$M = \frac{\text{Moles solute}}{\text{Volume solution (L)}}$$

Using subscripts "1" and "2" to indicate values before and after dilution, respectively,

$$M_1 = \frac{(\text{Moles solute})_1}{\text{Volume solution } (\text{L}_1)} \quad (7.5.4)$$

$$M_2 = \frac{(\text{Moles solute})_2}{\text{Volume solution } (\text{L}_2)} \quad (7.5.5)$$

and setting moles of solute before and after dilution equal to each other give the following:

$$M_1 \times (\text{Volume solution})_1 = M_2 \times (\text{Volume solution})_2 \quad (7.5.6)$$

In the example cited above, the volume of HCl before dilution is to be calculated:

$$(\text{Volume solution})_1 = \frac{M_2 \times (\text{Volume solution})_2}{M_1} \tag{7.5.7}$$

$$(\text{Volume solution})_1 = \frac{1\text{ mol/L} \times 1\text{ L}}{12\text{ mol/L}} = 0.083\text{ L} \tag{7.5.8}$$

The result of this calculation shows that 0.083 L, or 83 mL, of 12 M HCl must be taken to make 1.00 L of 1.00 M solution.

The same general approach used in solving these dilution problems can be used when concentrations are expressed in units other than molar concentration. Concentrations of metals dissolved in water are frequently measured by *atomic absorption spectroscopy*, a technique that depends upon absorption of light by atoms. To measure the concentration of metal in an unknown solution, it is necessary to have a standard solution (see Section 7.6) of known concentration. As purchased, these standard solutions contain 1000 mg of the desired metal dissolved in 1 L of solution. Suppose that one had a standard solution of $CaCl_2$ containing 1000 mg of Ca^{2+} per liter of solution. How would this solution be diluted to make 5 L of a solution containing 20 mg of calcium per liter?

Step 1. Consider how many milligrams of calcium are in the desired 5 L of solution containing 20 mg of calcium per liter:

$$5\text{ L} \times 20\text{ mg calcium/L} = 100\text{ mg calcium} \tag{7.5.9}$$

Step 2. Find how much standard solution containing 1000 mg of calcium per liter contains 100 mg of calcium.

$$\frac{100\text{ mg calcium}}{1000\text{ mg calcium/L}} = 0.100\text{ L of standard solution} \tag{7.5.10}$$

Step 3. Dilute 0.100 L (100 mL) of the standard calcium solution to 5 L to obtain the desired solution containing 20 mg of calcium per liter.

Molar Concentration of H^+ Ion and pH

Concentrations are important in expressing the degree to which solutions are acidic or basic. Recall that **acids**, such as HCl and H_2SO_4, produce H^+ ion, whereas **bases**, such as sodium hydroxide and calcium hydroxide (NaOH and $Ca(OH)_2$, respectively), produce hydroxide ion, OH^-. Molar concentrations of hydrogen ion, $[H^+]$, range over many orders of magnitude and are conveniently expressed by pH defined as

$$\text{pH} = -\log[H^+] \tag{7.5.11}$$

In absolutely pure water the value of [H^+] is exactly 1×10^{-7} mol/L, the pH is 7.00, and the solution is **neutral** (neither acidic nor basic). **Acidic** solutions have pH values of less than 7 and **basic** solutions have pH values of greater than 7.

Solubility

If a chemist were to attempt to make a solution containing calcium ion by dissolving calcium carbonate, $CaCO_3$ in water, not much of anything would happen. The chemist would not observe any calcium carbonate dissolving, because its **solubility** in water is very low. Only about 5 mg (just a small white "speck") of calcium carbonate will dissolve in a liter of water.

This illustrates an important point that people working in laboratories should keep in mind. When preparing a solution, it is a good idea to look up the solubility of the compound being dissolved. Make sure that the solubility is high enough to give the concentration that is desired. Many futile hours have been spent shaking bottles, trying to get something to dissolve whose solubility is just too low to make the desired solution.

A solution that has dissolved as much of a solute as possible is said to be **saturated** with regard to that solute. The *concentration* of the substance in the *saturated* solution is the **solubility.** Solubilities of different substances in water vary enormously. At the low end of the scale, even glass will dissolve a little bit in water, although the concentration of glass in solution is very, very low. Even this very small solubility can cause difficulties in the analysis of silicon (one of the elements in glass). This is because the dissolved glass increases the silicon content of the water in the solution, which throws off the results of the analysis. Antifreeze (composed of an organic compound called *ethylene glycol*) is an example of a substance that is completely soluble in water. One could pour a cup of water into a barrel and keep adding antifreeze without ever getting a saturated solution of antifreeze in water, even if the barrel were filled to the point of overflowing.

Supersaturated Solutions

Consider what happens when a bottle of carbonated beverage at an elevated temperature is opened. As the cap comes off, there is a loud pop, and the contents pour forth in a geyser of foam. This is because the solution of carbon dioxide in the bottle is a **supersaturated** solution. The CO_2 is added to the cold beverage under pressure. When the contents of the bottle become warm and the pressure is released by removing the bottle cap, the solution suddenly contains more CO_2 than it can hold in an unpressurized solution. This is because as CO_2 becomes warmer, its solubility decreases.

Factors Affecting Solubility

Solubility is affected by several factors. The preceding example has shown two of the most important factors affecting *gas* solubility: temperature and pressure. The solubilities of gases such as CO_2 *decrease* with *increasing* temperature. A gas is much less soluble in water just about hot enough to boil than it is in water just about cold enough to freeze. However, once the solution freezes, the gas is not at all soluble in the ice formed. This can be illustrated vividly with soft drinks, which also contain CO_2 dissolved under pressure. A can of soda placed in the freezer to cool quickly and then forgotten may burst from the pressure of the CO_2 coming out of solution as the liquid freezes, causing havoc in its immediate surroundings.

The solubilities of most solids *increase* with *rising* temperature, although not in all cases. Sugar, for example, is much easier to dissolve in a cup of hot tea than in a glass of iced tea.

One of the most important things affecting solubility is the presence of other chemicals in the solution that react with the solute. For example, CO_2 is really not very soluble in water. But, if the water already contains some sodium hydroxide, NaOH, the CO_2 is very soluble because a chemical reaction occurs between it and NaOH to form highly soluble sodium bicarbonate, $NaHCO_3$:

$$NaOH + CO_2 \text{ (not very soluble)} \rightarrow NaHCO_3 \text{ (highly soluble)} \qquad (7.5.12)$$

7.6. STANDARD SOLUTIONS AND TITRATIONS

Standard solutions are those of known concentration that are widely used in chemical analysis. Basically, a standard solution is one against which a solution of unknown concentration can be compared to determine the concentration of the latter. One of the most common and straightforward means of comparison is by way of **titration**, which consists of measuring the amount of a standard solution that reacts with a sample using the apparatus illustrated in Figure 7.8.

As an example, consider a hospital incinerator with a water scrubbing system to clean exhaust gases. Suppose that it is observed that the scrub water contains hydrochloric acid from the burning of chlorine-containing plastics and that this acid is corroding iron drain pipes—a problem that can be eliminated by adding sodium hydroxide to neutralize the acid. The amount of base needed may be determined by *titration*, a procedure in which a measured volume of the hydrochloric acid-containing scrub solution is reacted with a standard solution of NaOH,

$$HCl\,(aq) + NaOH\,(aq) \rightarrow NaCl\,(aq) + H_2O \qquad (7.6.1)$$

until just enough base has been added to react with all the HCl in the measured volume of sample. The point at which this occurs is called the **end point** and is shown by an abrupt change in color of a dissolved dye called an **indicator**. Phenolphthalein, a dye that is red in base and colorless in acid, is commonly used as an indicator.

Suppose that 50.0 mL (0.0500 L) of waste incinerator scrubber water were titrated with 0.0100 M standard NaOH, requiring 40.0 mL (0.0400 L) of the titrant

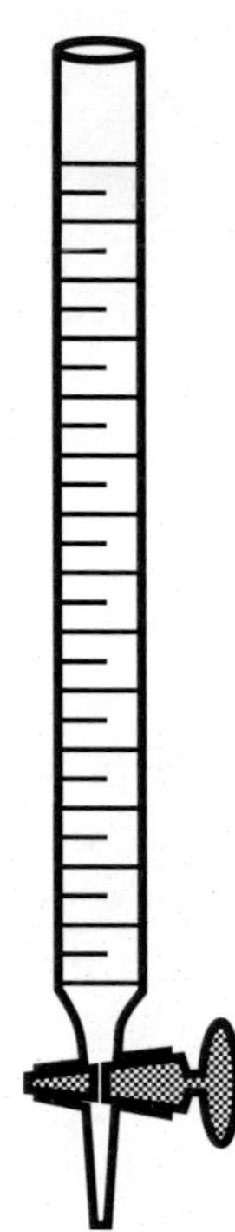

Figure 7.8. A buret, a long glass tube with marks on it and a stopcock on the end, is used to measure the volume of standard solution added to a sample during titration.

solution. The molar concentration of the HCl in the sample can be calculated from an equation derived from Equation 7.5.3:

$$M = \frac{\text{Moles solute}}{\text{Volume solution (L)}}$$

In this case, as shown by Reaction 7.6.1, the number of moles of HCl in the sample exactly equals the number of moles of NaOH at the end point, which leads to the relationship

$$M_{\mathrm{HCl}} \times \text{Volume HCl} = M_{\mathrm{NaOH}} \times \text{Volume NaOH} \tag{7.6.2}$$

where the subscript formulas denote HCl and NaOH solutions. This equation can be rearranged and values substituted into it to give the following:

$$M_{\mathrm{HCl}} = \frac{M_{\mathrm{NaOH}} \times \text{Volume NaOH}}{\text{Volume HC1}} \tag{7.6.3}$$

$$M_{\mathrm{HCl}} = \frac{0.0100\ \text{mol/L} \times 0.0400\ \text{L}}{0.0500\ \text{L}} = 0.00800\ \text{mol/L} \tag{7.6.4}$$

From this concentration, the quantity of base that must be added to neutralize the waste acid can be calculated.

7.7. PHYSICAL PROPERTIES OF SOLUTIONS

The presence of solutes in water can have profound effects upon the properties of the solvent. These effects include lowering the freezing point, elevating the boiling point, and osmosis. All such properties are called **colligative properties**; they depend upon the concentration of solute, rather than its particular identity. The effects of solutes and solution concentrations on colligative properties are addressed briefly here.

Freezing Point Depression

One of the most practical uses of solutions depends upon the effect that materials dissolved in water have upon the temperature at which the water, or the solution, freezes. Solutions freeze at lower temperatures than water does. This phenomenon is applied in the cooling systems of automobiles. Most automobile engines are water-cooled; that is, water circulates through the engine, picking up heat, and then goes to a radiator, where the excess heat is given off. If pure water is left in an engine at freezing temperatures, it will, of course, freeze. When water freezes, it expands, so that the ice that is formed has a larger volume than the original liquid. This increase in volume produces enormous forces that can crack the stoutest engine block as if it were an eggshell.

To prevent an engine block from damage by freezing and expansion of the coolant, antifreeze is mixed with water in the cooling system. The chemical name of antifreeze is ethylene glycol, and it has the chemical formula $C_2H_6O_2$. A solution containing 40% ethylene glycol and 60% water by mass freezes at −8°F, or −22.2°C. (Recall that pure water freezes at 32°F, or 0°C; it boils at 212°F, or 100°C.) A solution containing exactly half ethylene glycol and half water by mass freezes at −29°F.

Boiling Point Elevation

Solutions can also keep water from boiling. For instance, mixing antifreeze or other materials that do not boil easily with water raises the boiling temperature. Because of the higher boiling temperature, a solution of antifreeze in water makes a good "summer coolant." In fact, antifreeze is now marketed as "antifreeze–antiboil," and is virtually required in an engine to keep the cooling system from boiling in summer heat.

Osmosis

Human blood consists chiefly of red blood cells suspended in a fairly concentrated solution or **plasma**. The dissolved material making up the solution is mostly sodium chloride at around 0.15 molar concentration. The red blood cells also contain a solution much like the plasma in which they float. If one were to take some blood,

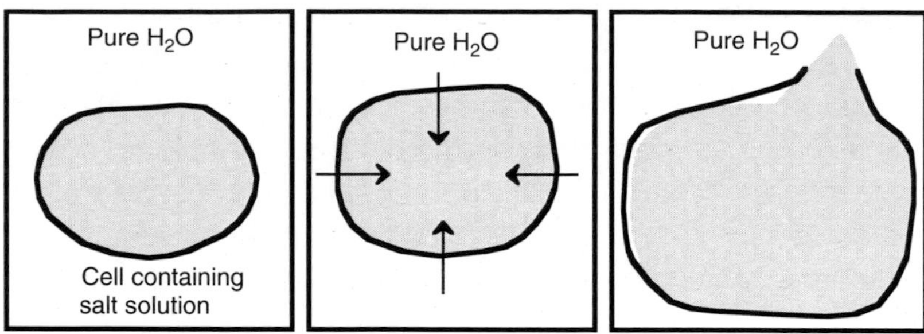

Figure 7.9. Osmosis causes pure water to be absorbed by a red blood cell, increasing the pressure in the cell and causing it to burst.

force the red cells to settle out by spinning test tubes full of blood samples in a centrifuge, and then place the red cells in pure water, a strange effect would be observed. When viewed under a microscope, the cells would be seen to swell and finally burst. This is an example of **osmosis**, the process whereby a substance passes through a membrane from an area of higher concentration of the substance to an area of lower concentration of the substance. Water can go through the membrane that holds the red blood cell together. When the cell is placed in pure water, the water tends to pass through the membrane to the more concentrated salt solution inside the cell. One way of looking at this is to say that the solution inside the cell has relatively less water, so that water from the outside has a tendency to enter the cell. Finally, so much water gets in that it breaks the cell apart (Figure 7.9). However, when a blood cell is placed in a very strong solution of sodium chloride, the opposite effect is observed: water passes from inside the cell to the outside, where there is less water, relative to sodium chloride, and the cell shrivels up (Figure 7.10).

Many biological membranes, such as the "wall" surrounding a red blood cell, allow material to pass through. Water in the ground tends to move through cell membranes in root cells and finally to plant leaves, where it evaporates to form water vapor in the atmosphere (a process called *transpiration*). Osmosis is a major part of the driving force behind these processes.

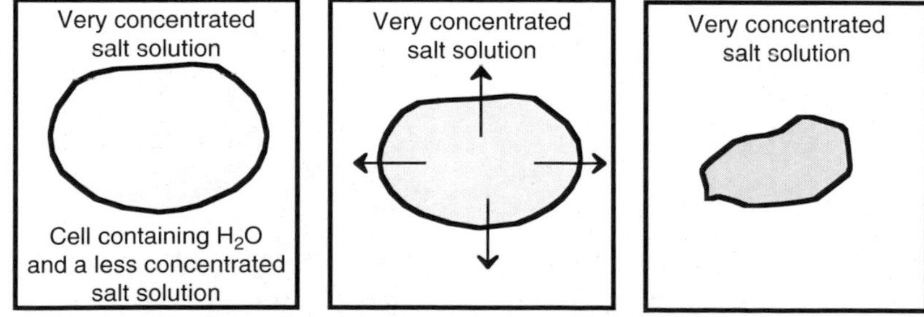

Figure 7.10. Osmosis causes water inside a red blood cell to pass out into a concentrated salt solution, causing the cell to shrivel.

The concentration of a solution relative to another solution determines which solution water will have a tendency to enter. *Water* tends to enter a solution of higher concentration (of some other substance), since it contains *less water*, relatively speaking. This can be particularly important in blood. Solutions injected into the veins for feeding, or to replace body fluids lost due to illness, must be prepared so that they will not cause water to move into or out of the cells by osmosis. Such a solution, which prevents osmosis from taking place, is called an **isotonic** solution. It is sometimes referred to as a *physiological* solution. Its use prevents damage to blood cells by either shrinking or swelling.

Osmosis may result in the buildup of a very high pressure called **osmotic pressure**. It is this pressure that can get so high in a red blood cell suspended in water that the cell bursts. Osmotic pressure can be several times atmospheric pressure.

Pure water tends to move by osmosis through a membrane to a more "contaminated" solution. Application of sufficient opposite pressure can "squeeze" water out of a contaminated solution, producing pure water. This process is called **reverse osmosis**, and is commonly used to purify small quantities of water. Some very large reverse osmosis installations are now being used to remove salt from seawater, a process called **desalination**.

7.8. SOLUTION EQUILIBRIA

Solution Equilibria

Many phenomena in aquatic chemistry and geochemistry involve **solution equilibrium**. In a general sense, solution equilibrium deals with the extent to which **reversible** acid–base, solubilization (precipitation), complexation, or oxidation–reduction reactions proceed in a forward or backward direction. This is expressed for a generalized equilibrium reaction,

$$aA + bB \rightleftharpoons cC + dD \tag{7.8.1}$$

by the following **equilibrium constant expression**:

$$\frac{[C]^c[D]^d}{[A]^a[B]^b} = K \tag{7.8.2}$$

where K is the **equilibrium constant**.

A reversible reaction may approach equilibrium from either direction. In the example above, if A were mixed with B, or C were mixed with D, the reaction would proceed in a forward or reverse direction such that the concentrations of species—[A], [B], [C], and [D]—substituted into the equilibrium expression gave a value equal to K.

As expressed by **Le Châtelier's principle**, a stress placed upon a system in equilibrium will shift the equilibrium to relieve the stress. For example, adding product “D” to a system in equilibrium will cause Reaction 7.8.1 to shift to the left, consuming “C” and producing “A” and “B,” until the equilibrium constant expression is again satisfied. This is called the **mass action effect**, and it is the driving force behind many environmental chemical phenomena.

There are several major kinds of equilibria in aqueous solution. One of these is **acid–base equilibrium** as exemplified by the ionization of acetic acid, HAc,

$$HAc \rightleftharpoons H^+ + Ac^- \tag{7.8.3}$$

for which the acid dissociation constant is K_a:

$$\frac{[H^+][Ac^-]}{[HAc]} = K_a = 1.75 \times 10^{-5} \tag{7.8.4}$$

Very similar expressions are obtained for the formation and dissociation of metal **complexes** or **complex ions**, formed by the reaction of a metal ion in solution with a **complexing agent** or **ligand**, both of which are capable of independent existence in solution. This can be shown by the reaction of iron(III) ion and thiocyanate ligand,

$$Fe^{3+} + SCN^- \rightleftharpoons FeSCN^{2+} \tag{7.8.5}$$

for which the **formation constant expression** is:

$$\frac{[FeSCN^{2+}]}{[Fe^{3+}][SCN^-]} = K_f = 1.07 \times 10^3 \text{ (at 25°C)} \tag{7.8.6}$$

The bright red color of the $FeSCN^{2+}$ complex formed could be used to test for the presence of iron(III) in acid mine water.

An example of an **oxidation–reduction reaction** (a reaction that involves the transfer of electrons between species) is

$$MnO_4^- + 5Fe^{2+} + 8H^+ \rightleftharpoons Mn + 5Fe^{3+} + 4H_2O \tag{7.8.7}$$

for which the equilibrium expression is

$$\frac{[Mn^{2+}][Fe^{3+}]^5}{[MnO_4^-][Fe^{2+}]^5[H^+]^8} = K = 3 \times 10^{62} \text{ (at 25°C)} \tag{7.8.8}$$

The value of K is calculated from the Nernst equation (see Chapter 8).

Distribution Between Phases

Many important environmental chemical phenomena involve distribution of species between phases. This most commonly involves the equilibria between species in solution and in a solid phase. **Solubility equilibria** deal with reactions such as

$$AgCl(s) \rightleftharpoons Ag^+ + Cl^- \tag{7.8.9}$$

in which one of the participants is a slightly soluble (virtually insoluble) salt and for which the equilibrium constant is K_{sp}:

$$[Ag^+][Cl^-] = K_{sp} = 1.82 \times 10^{-10} \text{ (at 25°C)} \tag{7.8.10}$$

a **solubility product**. Note that in the equilibrium constant expression, no value given for the solid AgCl. This is because the activity of a solid (in this case the tendency of Ag^+ and Cl^- ions to break away from solid AgCl) is constant at a specific temperature and is contained in the value of K_{sp}.

An important example of distribution between phases is that of a hazardous waste species partitioned between water and a body of immiscible organic liquid in a hazardous waste site. The equilibrium for such a reaction,

$$X(aq) \rightleftharpoons X(org) \tag{7.8.11}$$

is described by the **distribution law** expressed by a **distribution coefficient** or **partition coefficient** in the following form:

$$H_3N^+\text{-}CH_2\text{-}COO^- \underset{}{\overset{H_2O}{\rightleftharpoons}} H_2N\text{-}CH_2\text{-}COO + H^+ \tag{7.8.12}$$

Solubilities of Gases

The solubilities of gases in water are described by Henry's law, which states that *at constant temperature, the solubility of a gas in a liquid is proportional to the partial pressure of the gas in contact with the liquid.* For a gas "X," this law applies to equilibria of the type

$$X\,(g) \rightleftharpoons X\,(aq) \tag{7.8.13}$$

and does not account for additional reactions of the gas species in water, such as

$$NH_3 + H_2O \rightleftharpoons NH_4{}^+ + OH^- \tag{7.8.14}$$

$$SO_2 + HCO_3{}^- \text{ (from water alkalinity)} \rightleftharpoons CO_2 + HSO_3{}^- \tag{7.8.15}$$

which may result in much higher solubilities than predicted by Henry's law alone.

Mathematically, Henry's law is expressed as

$$[X(aq)] = KP_X \tag{7.8.16}$$

where $[X(aq)]$ is the aqueous concentration of the gas, P_X is the partial pressure of the gas, and K is the Henry's law constant applicable to a particular gas at a specified temperature. For gas concentrations in units of moles per liter and gas pressures in atmospheres, the units of K are mol L^{-1} atm^{-1}. Some values of K for dissolved gases that are significant in water are given in Table 7.2.

Table 7.2. Henry's Law Constants for Some Gases in Water at 25°C

Gas	K (mol L^{-1} atm^{-1})
O_2	1.28×10^{-3}
CO_2	3.38×10^{-2}
H_2	7.90×10^{-4}
CH_4	1.34×10^{-3}
N_2	6.48×10^{-4}
NO	2.0×10^{-4}

In calculating the solubility of a gas in water, a correction must be made for the partial pressure of water by subtracting it from the total pressure of the gas. At 25°C, the partial pressure of water is 0.0313 atm; values at other temperatures are readily obtained from standard handbooks. The concentration of oxygen in water saturated with air at 1.00 atm and 25°C can be found in an example of a simple gas solubility calculation. Considering that dry air is 20.95% by volume oxygen and factoring in the partial pressure of water gives the following:

$$P_{O_2} = (1.0000\,\text{atm} - 0.0313\,\text{atm}) \times 0.2095 = 0.2029\,\text{atm} \tag{7.8.17}$$

$$\begin{aligned}[O_2\,(aq)] &= KP_{O_2} = 1.28 \times 10^{-3}\,\text{mol L}^{-1}\,\text{atm}^{-1} \times 0.2029\,\text{atm} \\ &= 2.60 \times 10^{-4}\,\text{mol/L}\end{aligned} \tag{7.8.18}$$

Since the molecular mass of oxygen is 32, the concentration of dissolved oxygen in water in equilibrium with air under the conditions given above is 8.32 mg/L, or 8.32 parts per million (ppm).

The solubilities of gases decrease with increasing temperature. Account is taken of this factor with the **Clausius–Clapeyron** equation:

$$\log \frac{C_2}{C_1} = \frac{\Delta H}{2.303R} \left[\frac{1}{T_1} - \frac{1}{T_2}\right] \tag{7.8.19}$$

where C_1 and C_2 are the gas concentrations in water at absolute temperatures of T_1 and T_2, respectively, ΔH is the heat of solution, and R is the gas constant. The value of R is 1.987 cal deg^{-1} mol^{-1}, giving ΔH in units of cal/mol. Detailed calculations involving the Clausius–Clapeyron equation are beyond the scope of this chapter.

7.9. COLLOIDAL SUSPENSIONS

If some fine sand is poured into a bottle of water and shaken vigorously, the sand will float around in the water for a very brief time, and then rapidly settle to the bottom. Particles like sand, which float briefly in water and then rapidly settle out, are called **suspensions**. It has already been seen that molecules or ions that dissolve in water and mingle individually with the water molecules make up *solutions*. Between these two extremes are particles that are much smaller than can be seen with the naked eye, but much larger than individual molecules. Such particles are called **colloidal particles**. They play important roles in living systems, transformations of minerals, and a number of industrial processes. Colloidal particles in water form **colloidal suspensions**. Unlike sand stirred in a jar, or soil granules carried by a vigorously running stream, *colloidal particles do not settle out of colloidal suspension by gravity alone*. Colloidal particles range in diameter from about 0.001 micrometer (μm) to about 1 μm, have some characteristics of both species in solution and larger particles in suspension, and in general exhibit unique properties and behavior. An important characteristic of colloidal particles is their ability to scatter light. Such light scattering is called the **Tyndall effect** and is observed as a light blue hue at right angles to incident white light. This phenomenon results from colloidal particles being the same order of size as the wavelength of light.

Individual cells of bacteria are colloidal particles and can form colloidal suspensions in water. Many of the green algae in water are present as colloidal suspensions of individual cells. Milk, mayonnaise, and paint are all colloidal suspensions. Colloids are used in making rubber, glue, plastics, and grease. The formation of a colloidal suspension of butterfat in milk is the process by which milk is *homogenized*, so that the cream does not rise to the top.

Kinds of Colloidal Particles

Colloids can be classified as *hydrophilic colloids*, *hydrophobic colloids*, or *association colloids*. These three classes are briefly summarized below.

Hydrophilic colloids generally consist of macromolecules, such as proteins and synthetic polymers, that are characterized by strong interaction with water resulting in spontaneous formation of colloids when they are placed in water. In a sense, hydrophilic colloids are solutions of very large molecules or ions. Suspensions of hydrophilic colloids are less affected by the addition of salts to water than are suspensions of hydrophobic colloids.

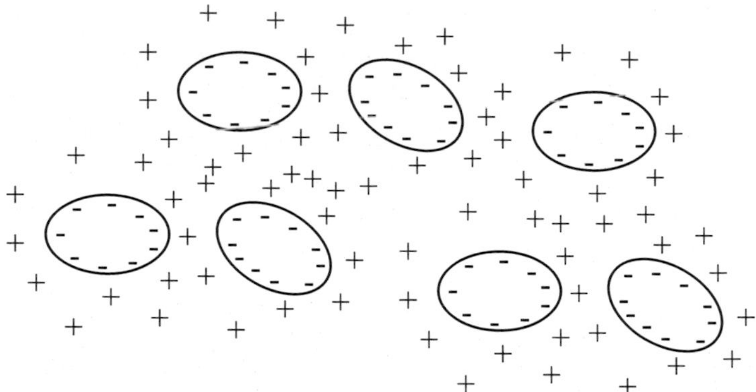

Figure 7.11. Representation of negatively charged hydrophobic colloidal particles surrounded in solution by positively charged counter-ions, forming an electrical double layer. (Colloidal particles suspended in water can have either a negative or a positive charge.)

Hydrophobic colloids interact to a lesser extent with water and are stable because of their positive or negative electrical charges, as shown in Figure 7.11. The charged surface of the colloidal particle and the **counter-ions** that surround it compose an **electrical double layer**, which causes the particles to repel each other. Hydrophobic colloids are usually caused to settle from suspension by the addition of salts. Examples of hydrophobic colloids are clay particles, petroleum droplets, and very small gold particles.

Association colloids consist of special aggregates of ions and molecules called **micelles**. To understand how these arise, consider sodium stearate, a typical soap with the following structural formula:

$$\begin{array}{c} \text{H H H H H H H H H H H H H H H H H O} \\ \text{H-C-C-C-C-C-C-C-C-C-C-C-C-C-C-C-C-C-C-O}^{-}\text{Na}^{+} \\ \text{H H H H H H H H H H H H H H H H H} \end{array}$$

Represented as $\sim\!\sim\!\sim\!\sim\!\sim\!\ominus$

The stearate ion has both a hydrophilic—CO_2^- head and a long organophilic $CH_3(CH_2)_{16}$— tail. As a result, stearate anions in water tend to form clusters consisting of as many as 100 anions clustered together, with their hydrocarbon "tails" on the inside of a spherical colloidal particle and their ionic "heads" on the surface in contact with water and with Na^+ counter-ions. This results in the formation of **micelles**, as illustrated in Figure 7.12.

Colloid Stability

The stability of colloids is a prime consideration in determining their behavior. It is involved in important aquatic chemical phenomena, including the formation of sediments, the dispersion and agglomeration of bacterial cells, and the dispersion and removal of pollutants (e.g., crude oil from an oil spill).

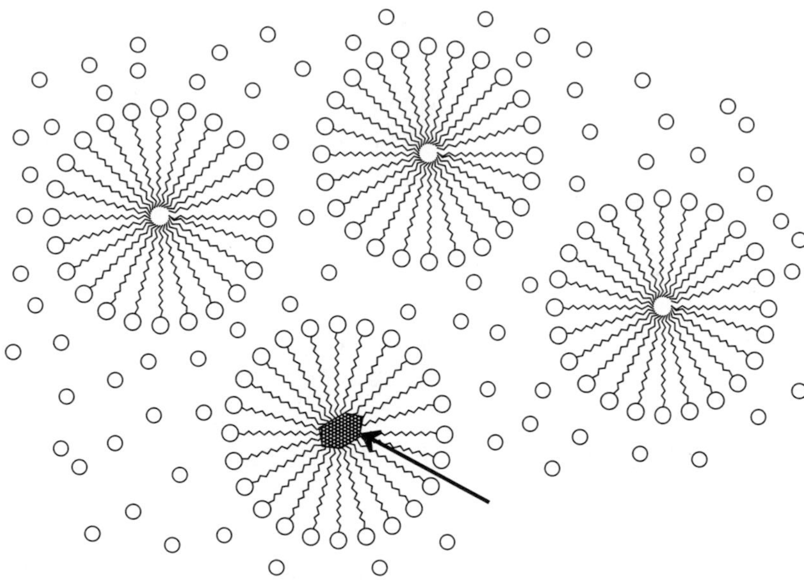

Figure 7.12. Representation of colloidal soap micelle particles. The stearate ions are shown as ~~~~~⊖.

As discussed above, the two main phenomena contributing to the stabilization of colloids are **hydration** and **surface charge**. The layer of water on the surface of hydrated colloidal particles prevents contact, which would result in the formation of larger units. A surface charge on colloidal particles may prevent aggregation, since like-charged particles repel each other. The surface charge is frequently pH-dependent; around pH7, most colloidal particles in natural waters are negatively charged. Negatively charged aquatic colloids include algal cells, bacterial cells, proteins, and colloidal petroleum droplets.

One of the three major ways in which a particle can acquire a surface charge is by **chemical reaction at the particle surface**. This phenomenon, which frequently involves hydrogen ions and is pH-dependent, is typical of hydroxides and oxides. As an illustration of pH-dependent charge on colloidal particle surfaces, consider the effects of pH on the surface charge of hydrated manganese oxide, $MnO_2(H_2O)$ (*s*). In a relatively acidic medium, the reaction

$$MnO_2(H_2O)(s) + H^+ \rightarrow MnO_2(H_3O)^+(s) \tag{7.9.1}$$

may occur on the surface, giving the particle a net positive charge. In a more basic medium, hydrogen ions may be lost from the hydrated oxide surface to yield negatively charged particles:

$$MnO_2(H_2O)(s) \rightarrow MnO_2(OH)^-(s) + H^+ \tag{7.9.2}$$

Ion absorption is a second way in which colloidal particles become charged. This phenomenon involves attachment of ions onto the colloidal particle surface by means other than conventional covalent bonding, including hydrogen bonding and London (van der Waals) interactions.

Ion replacement is a third way in which a colloidal particle can gain a net charge. For example, replacement of some of the Si(IV) with Al(III) in the basic SiO_2 chemical unit in the crystalline lattice of some clay minerals,

$$[SiO_2] + Al(III) \rightarrow [AlO_2^-] + Si(IV) \qquad (7.9.3)$$

yields sites with a net negative charge. Similarly, replacement of Al(III) by a divalent metal ion such as Mg(II) in the clay crystalline lattice produces a net negative charge.

Coagulation and Flocculation of Colloidal Particles

The **aggregation** of colloidal particles is very important in a number of processes, including industrial processes and aquatic chemical phenomena. The ways in which colloidal particles of minerals come together and settle out of water are crucial in the formation of mineral deposits. The bacteria in colloidal suspensions that devour sewage in the water at a sewage treatment plant must eventually be brought out of colloidal suspension to obtain purified water. Crude oil frequently comes out of the ground as a colloidal suspension in water; the colloidal particles of hydrocarbon must be separated from the water before the crude oil can be refined.

It is possible to force electrically charged colloidal particles to clump together and settle. This is done by adding charged ions, such as Na^+ and Cl^- ions from sodium chloride. These charged ions tend to neutralize the electrical charges on the colloidal particles and allow them to come together; this process is called **coagulation**. In some cases, colloidal particles are caused to settle by virtue of *bridging groups* called *flocculants* that join the particles together. This phenomenon is called **flocculation.**

CHAPTER SUMMARY

The chapter summary below is presented in a programmed format to review the main points covered in this chapter. It is used most effectively by filling in the blanks, referring back to the chapter as necessary. *The correct answers are given at the end of the summary.*

A solid that seems to disappear when stirred with water is said to [1]______________, forming a [2]__________. The substance that dissolves is called the [3]____________, and the liquid in which it dissolves is the [4]____________. A substance that does not dissolve appreciably in a solvent is said to be [5]_____________. Three typical organic solvents are [6]___________________________. Two characteristics of water molecules that make water an excellent solvent for many substances are [7]__________________ ________________. When O_2 dissolves in water, the oxygen molecules are in solution as [8]_______________. When NaCl dissolves in water, it is present as [9]_______________, and when HCl dissolves in water, it forms [10]_______________. The hydrogen ion in water is bonded to a water molecule to form a [11]_______________. The amount of solute in a particular amount of solution or solvent is the [12]_______________. Low concentrations of pollutants in water are often expressed as parts per [13]_________, abbreviated [14]_________, or, at even lower concentrations, as [15]_______________, abbreviated [16]_________. The molar concentration of a solution is the number of [17]_____________ of solute per [18]_______________. The molecular mass of NH_3 is [19]_____________, so the mass of a mole of ammonia is [20]__________. The molar concentration of a solution prepared by dissolving 8.5 g of NH_3 gas in 2 L of solution is [21]_____________ M. The relationship among the volume of a solution taken for dilution (V_1), its molar concentration (M_1), the volume of the diluted solution (V_2), and the concentration of the diluted solution (M_2) is [22]_______________________. The volume of 2 M HCl that must be diluted with water to make 5 L of 1.5 M HCl is [23]_______________ L. The equation that defines pH is [24]_______________. A pH less than 7 indicates a solution that is [25]__________. The maximum amount of solute that can dissolve in a particular volume of solution is called the solute's [26]__________. A solution containing the maximum amount of solute that it can normally hold is said to be [27]__________. A solution of known concentration used in chemical analyses is called a [28]_______________. An operation in which the quantity of such a solution required to react with a substance in another solution is measured is known as [29]_______________. The reaction is finished at the [30]___________, which is shown by a change in color of an [31]_____________. Effects, including lowering of freezing point, elevation of boiling point, and osmosis, that depend upon the concentration of solute, rather than its particular identity are called [32]___________. Solutes [33]__________ the freezing temperature of water. A solution of antifreeze in water freezes at a [34]___________ temperature and boils at a [35]_______________ temperature than pure water. Solution equilibrium deals with the extent to which [36]_______________ reactions proceed in a forward or backward direction. Four major classes of these reactions are [37]_______________. An expression such as

$$\frac{[C]^c[D]^d}{[A]^a[B]^b} = K$$

that describes an equilibrium reaction represented by $aA + bB \rightleftharpoons cC + dD$ is called [38]_______________________________. A statement of the law governing gas solubilities is [39]_______________________________, which is called [40]_______________________. For a gas, "X," a mathematical expression of this law is [41]____________________. Extremely small particles suspended uniformly in water form a [42]_____________ suspension. Colloidal particles are stabilized in water by electrical [43]_____________ or attraction to [44]________. Two foods that are colloidal suspensions are [45]____________________. Coagulation is a term given to the process by which colloidal particles [46]____________________. Water and solutes pass through certain membranes by a process called [47]____________. Because of this phenomenon, red blood cells placed in pure water [48]____________, and those placed in a solution containing much more salt than blood plasma contains will [49]____________.

Answers to Chapter Summary

1. dissolve
2. solution
3. solute
4. solvent
5. insoluble
6. benzene, perchloroethylene, and acetone
7. hydrogen bonding and the fact that water molecules are dipoles
8. O_2 molecules
9. Na^+ and Cl^- ions
10. H^+ and Cl^- ions
11. hydronium ion, H_3O^+
12. concentration
13. million
14. ppm
15. parts per billion
16. ppb
17. moles
18. liter of solution
19. 17
20. 17 g
21. 0.25
22. $M_1 \times V_1 = M_2 \times V_2$
23. 3.75

24. $pH = -\log[H^+]$

25. acidic

26. solubility

27. saturated

28. standard solution

29. titration

30. end point

31. indicator

32. colligative properties

33. lower

34. lower

35. higher

36. reversible

37. acid–base, solubilization (precipitation), complexation, oxidation–reduction

38. an equilibrium constant expression

39. at constant temperature the solubility of a gas in a liquid is proportional to the partial pressure of the gas in contact with the liquid

40. Henry's law

41. $[X\ (aq)] = KP_X$

42. colloidal

43. charge

44. water

45. milk and mayonnaise

46. aggregate or come together

47. osmosis

48. swell

49. shrink

QUESTIONS AND PROBLEMS

1. Of the following, the **untrue** statement is: (a) Sugar dissolves in water. (b) When this happens, the sugar becomes the solute. (c) A solution is formed. (d) In the solution, the sugar is present as aggregates of molecules in the form of very small microcrystals. (e) The water is the solvent.
2. Is tap water a solution? If so, what does it contain besides water?
3. What does a reaction such as Ca^{2+} *(aq)* + CO_3^{2-} *(aq)* → $CaCO_3$ *(s)* illustrate about the role of solutions in chemical reactions?
4. Summarize the industrial uses of solutions.
5. What is the significance of solutions in the environment?
6. Why are volatile solvents especially useful in industrial applications?
7. What are some of the health considerations involved with the use of industrial solvents?
8. Sketch the structure of the water molecule and explain why it is particularly significant in respect to water's solvent properties. Why is the water molecule called a dipole?
9. What are hydrogen bonds? Why are they significant in regard to water's solvent properties?
10. What is represented by the following species?

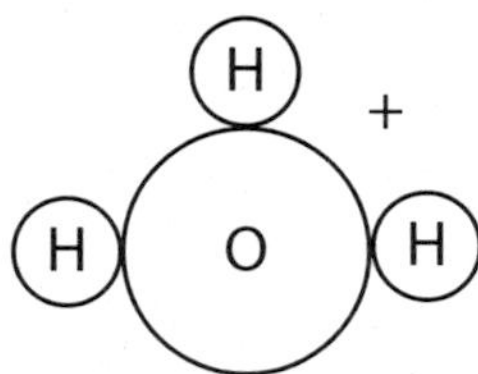

What is its formula, and what are the formulas of related species? How is it formed? What is it called?

11. Distinguish among the nature of the solution process and the species in solution for the dissolution of (a) O_2, (b) NaCl, and (c) HCl in water.
12. Why is a solution formed by dissolving HCl *(g)* in water not called a "hydrogen chloride" solution?
13. The molecular mass of ammonia, NH_3, is 17. How many grams of ammonia are in 1 L of 2 M solution?
14. How many grams of ammonia are in 3 L of 1 M solution?
15. How many grams of ammonia are in 3 L of 2 M solution?

16. If one wished to prepare 2 L of 1.2 M NaCl, how many liters of 2 M NaCl would be required?

17. How much of a standard solution of a sodium compound containing 1000 mg of Na^+ per liter should be taken to prepare 5 L of a standard sodium solution containing 10 mg of sodium per liter?

18. Exactly 1.50 L of 2.00 M HCl was added to a 2.00 L volumetric flask, and the volume brought to 2.00 L with water. What was the molar concentration of HCl in the resulting solution?

19. Give an equation that relates molar concentration, volume of solution, mass of solute, and molar mass of solute.

20. What is the process called when a measured volume of a solution is diluted with water to a specific volume? What remains the same during this process?

21. What are the relationships among the terms *neutral*, *acidic*, *basic*, and *pH*?

22. What does a *saturated* solution have to do with the *solubility* of a solute? What is meant by *supersaturated*?

23. What are the common effects of different temperatures upon the solubilities of (a) solids and (b) gases?

24. Explain how a *standard solution*, *buret*, and *indicator* are used in *titrations*.

25. Exactly 1.00 L of 1.00 M HCl was added to a 2.00 L volumetric flask. Next, a total of 0.25 mol of solid NaOH was added. After the chemical reaction between the HCl and the NaOH was complete, water was added to bring the total volume to 2.00 L. What chemical reaction occurred when the HCl was added? How many moles of HCl were left after this reaction occurred? What was the molar concentration of HCl after the solution was diluted to 2.00 L?

26. A total of 2.50 L of 0.100 M NaCl was mixed with 3.00 L of 0.0800 M NaCl. Assume that the volumes of the solutions add together to give 5.50 L (volumes *may* not be additive with more concentrated solutions). What was the molar concentration of NaCl in the final solution?

27. How is the following equation used in titrimetric calculations?

$$M_{\mathrm{HCl}} \times \text{Volume HCl} = M_{\mathrm{NaOH}} \times \text{Volume NaOH}$$

28. How are *colligative properties* defined and what are three major colligative properties?

29. Why is it desirable to have ethylene glycol mixed with water in an automobile's cooling system even during the summer?

30. Placed in a solution that is about 0.10 M in sodium chloride, blood cells would be somewhat turgid, or "swollen." What property of solutions does this observation illustrate? Explain.

31. What is an *isotonic solution*?

32. What are the nature and uses of *reverse osmosis*?

33. Consider the following generalized reaction, where all species are in solution:

$$aA + bB \rightleftharpoons cC + dD$$

What is the significance of the **double** arrows? Explain *solution equilibrium* in the context of this reaction. Be sure to include a discussion of the *equilibrium constant expression*.

34. Match each reaction on the left with the type of equilibrium that it represents on the right:

(A) $Fe^{3+} + SCN^- \rightleftharpoons FeSCN^{2-}$	1. Oxidation–reduction
(B) $AgCl\ (s) \rightleftharpoons Ag^+ + Cl^-$	2. Solubility
(C) $HAc \rightleftharpoons H^+ + Ac^-$	3. Acid–base
(D) $MnO_4^- + 5Fe^{2+} + 8H^+ \rightleftharpoons$ $Mn^{2+} + 5Fe^{3+} + 4H_2O$	4. Complex ion formation

35. Name, state, and give the mathematical expression for the law that describes the following kind of equilibrium:

$$X\ (g) \rightleftharpoons X\ (aq)$$

36. How do reactions such as the following affect gas solubility?

$$NH_3 + H_2O \rightleftharpoons NH_4^+ + OH^-$$

$$SO_2 + HCO_3^- \text{ (from water alkalinity)} \rightleftharpoons CO_2 + HSO_3^-$$

37. What does the equation

$$P_{O_2} = (1.0000\,\text{atm} - 0.0313\,\text{atm}) \times 0.2095 = 0.2029\,\text{atm}$$

express? What is the "0.0313 atm?"

38. Describe the nature and characteristics of *colloidal particles* that are in colloidal suspension in water. What is it about colloidal particles that causes them to exhibit the *Tyndall effect*?

39. Distinguish among *hydrophilic colloids, hydrophobic colloids,* and *association colloids.* Which consist of *micelles*? What are micelles?

8. CHEMISTRY AND ELECTRICITY

8.1. CHEMISTRY AND ELECTRICITY

Electricity can have a strong effect on chemical systems. For example, passing electricity through a solution of sodium sulfate in water breaks the water down into hydrogen gas and oxygen gas. In Section 6.3, it was discussed how solutions of completely ionized strong electrolytes conduct electricity well, weak electrolytes conduct it poorly, and nonelectrolytes conduct not at all. Electrical current basically involves the activity of electrons, as well as their flow and exchange between chemical species. Since electrons are so important in determining the chemical bonding and behavior of atoms, it is not surprising that electricity is strongly involved with many chemical processes.

The exchange of electrons between chemical species is part of the more general phenomenon of oxidation–reduction, which is defined in more detail in Section 8.2. Oxidation–reduction processes are of particular importance in environmental chemistry. Organic pollutants are degraded and nutrient organic matter utilized for energy and as a carbon source for biomass by oxidation–reduction reactions involving fungi in water and soil. The nature of inorganic species in water and soil depends upon whether the medium is oxidizing (oxygen present, low electron activity) or reducing (oxygen absent, high electron activity). Strong oxidants, such as ozone (O_3) and organic peroxides formed by photochemical processes in polluted atmospheres, are noxious pollutants present in photochemical smog.

The flow of electricity through a chemical system can cause chemical reactions to occur. Similarly, chemical reactions may be used to produce electricity. Such phenomena are called **electrochemical phenomena** and are covered under the category of **electrochemistry**. Examples of electrochemistry abound (Figure 8.1). The transmission of nerve impulses in animals, and even human thought processes, are essentially electrochemical. A dry cell produces electricity from chemical reactions. An automobile storage battery stores electrical energy as chemical energy, and reverses the process when it is discharged. An electrochemical process called electrodialysis can be used for the purification of water. The analytical chemist can use an electrical potential developed at a probe called a glass electrode to measure

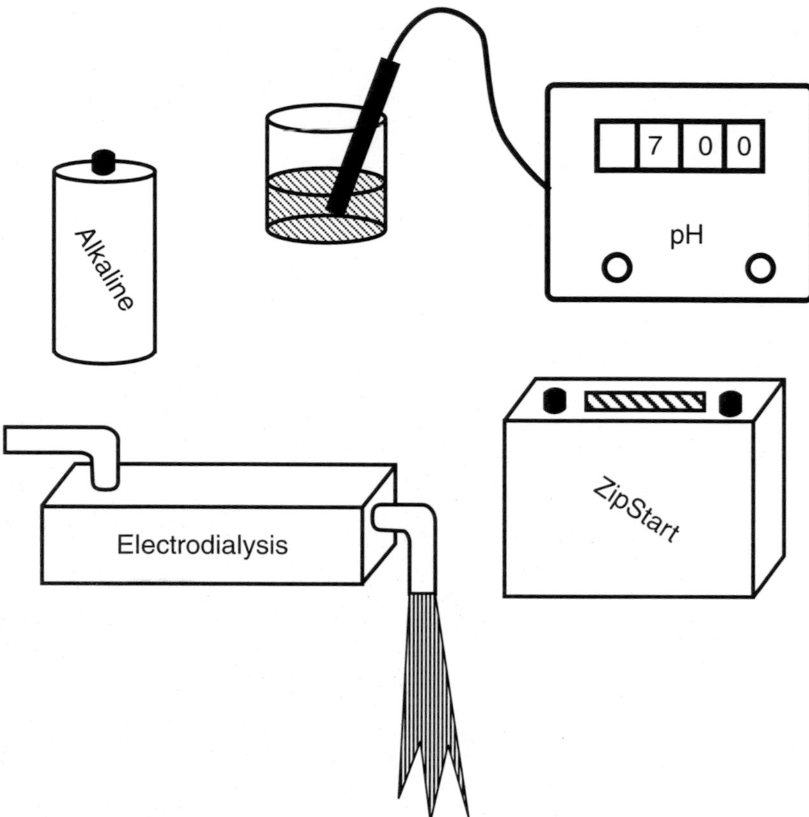

Figure 8.1. The interaction of electricity and chemistry is an important phenomenon in many areas.

the pH of water. The addition of electrons to chemical species (reduction) or removal of electrons (oxidation) is an important aspect of green chemistry in that, in many cases, it can be accomplished without the use of chemical reagents, greatly reducing consumption of chemicals and production of wastes. The electrical aspects of chemistry are of obvious importance, and are covered in detail in this chapter.

8.2. OXIDATION AND REDUCTION

It has been seen that many chemical reactions can be regarded as the transfer of electrically charged electrons from one atom to another. Such a transfer can be arranged to occur through a wire connected to an electrode in contact with atoms that either gain or lose electrons. In that way, electricity can be used to bring about chemical reactions, or chemical reactions can be used to generate electricity. First, however, consider the transfer of electrons between atoms that are in contact with each other.

An oxygen atom, which has a strong appetite for electrons, accepts 2 valence (outer shell) electrons from a calcium atom to form a calcium ion, Ca^{2+}, and an oxide ion, O^{2+} (Figure 8.2). The loss of electrons by the calcium atom is called **oxidation**, and the calcium is said to have been **oxidized**. These terms are derived from the

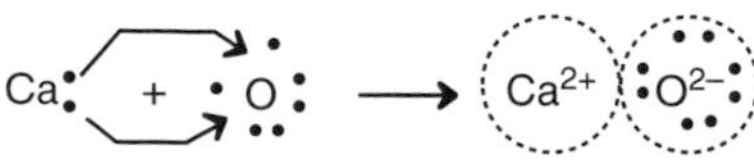

Figure 8.2. Oxygen atoms have a strong tendency to accept electrons, and an atom that has lost electrons is said to be oxidized.

name of oxygen because of its tendency to take electrons from other atoms, such as calcium. The gain of electrons by the oxygen is called **reduction**, and the oxygen atom has been **reduced**; atoms that gain electrons in a transfer such as this are always reduced. The overall reaction is called an **oxidation–reduction reaction**, sometimes abbreviated as a **redox reaction**. Oxidation and reduction always occur together. Whenever something is oxidized, something else is always reduced. In this case, Ca is oxidized when O is reduced.

It is easy to see what is meant by oxidation and reduction when there is a transfer of electrons to form ionic bonds (see Section 4.3). However, the concept can be extended to compounds that are bonded together by covalent bonds. For example, when H combines with O to form H_2O,

$$\text{H}\cdot \; \text{H}\cdot \rightarrow \cdot\ddot{\text{O}}\cdot \longrightarrow \text{H}:\ddot{\text{O}}:\text{H} \qquad (8.2.1)$$

electrons are shared between H and O atoms in the covalent bonds joining them together. However, the sharing is not equal, because the O atom has a much stronger tendency to grab onto electrons. Therefore, the O atom is regarded as "taking" an electron from each of the H atoms. If O really did take both of these electrons completely away from H, it would have a −2 charge. Although the oxygen does not have a charge as such, the oxygen atom in H_2O is assigned an oxidation number with a value of −2 because of its tendency to attract the 2 electrons from the 2 H atoms. The oxidation number is the hypothetical charge that an atom would have if it gained or lost a particular number of electrons when it formed a chemical compound. In virtually all cases encountered in this book, the oxidation number of chemically combined O is µ2. In H_2O, each H atom can be visualized as losing its electron to the O atom because of the unequal sharing. This gives H an oxidation number of +1. In practically all of the compounds encountered in this book, the oxidation number of chemically combined H is +1.

The oxidation number of any element is zero in the elemental form. Therefore, the oxidation number of O in O_2 is zero, and the oxidation number of H in H_2 is zero.

Figure 8.3 gives the oxidation numbers of the chemically combined forms of the first 20 elements in the periodic table. From this table, observe that the chemically combined forms of the elements in the far left column (H together with the

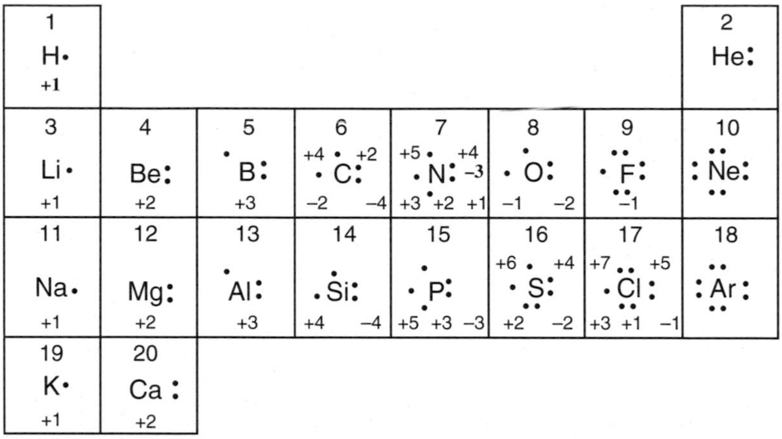

1 H· +1							2 He:
3 Li· +1	4 Be: +2	5 B: +3	6 +4 C: +2 −2 −4	7 +5 N: +4 −3 +3 +2 +1	8 O: −1 −2	9 F: −1	10 :Ne:
11 Na· +1	12 Mg: +2	13 Al: +3	14 Si: +4 −4	15 P: +5 +3 −3	16 +6 S: +4 +2 −2	17 +7 Cl: +5 +3 +1 −1	18 :Ar:
19 K· +1	20 Ca : +2						

Figure 8.3. Common oxidation numbers of the chemically combined forms of the first 20 elements (small bold-faced numbers).

alkali metals: Li, Na, and K) have an oxidation number of +1. Those in the next column over (the alkaline earths: Be, Mg, and Ca) have an oxidation number of +2, and those in the following column (B and Al) have an oxidation number of +3. It has already been mentioned that O almost always has an oxidation number of −2. That of F is always −1. The oxidation number of chemically combined Cl is generally −1. As shown in Figure 8.3, the oxidation numbers of the other elements are variable.

Even though the oxidation numbers of some elements vary, they can usually be determined out by applying the following rules:

- The sum of the oxidation numbers of the elements in a compound equals 0.
- The sum of the oxidation numbers of the elements in an ion equals the charge on the ion.

The application of these rules can be seen in Table 8.1. For each compound in this table, the oxidation number of each element, except for one, is definitely known. For example, in SO_2, each O has an oxidation number of −2. There are 2 Os for a total of −4. The −4 must be balanced with a +4 for S. In SO_3, the total of −6 for 6 O atoms, each with an oxidation number of −2, must be balanced with a +6 for S. In Na_3PO_4, each Na has an oxidation number of +1, and each O has −2. Therefore, the oxidation number of P is calculated as follows:

3 Na times +1 per Na = +3
4 O times −2 per O = −8

That leaves a sum of −5, which must be balanced with a +5:

$$\begin{gathered}(+3 \text{ for } 3 \text{ Na}) + (x \text{ for } 1 \text{ P}) + (-8 \text{ for } 4 \text{ O}) = 0 \\ x = \text{oxidation number of P} = 0 - 3 + 8 = 5\end{gathered} \tag{8.2.2}$$

Table 8.1. Oxidation Numbers of Elements in Some Compounds and Ions

Compound or Ion	Element Whose Oxidation Number Is to Be Calculated	Oxidation Number of Element
SO_2	S	+4
SO_3	S	+6
NH_3	N	−3
CH_4	C	−4
HNO_3	N	+5
Na_3PO_4	P	+5
CO_3^{2-}	C	+4
NO_2^-	N	+3

In the NO_2^- ion, each of the 2 O atoms has an oxidation number of −2, for a total of −4. The whole ion has a net charge of −1, so that the oxidation number of N is +3.

8.3. OXIDATION–REDUCTION IN SOLUTION

Many important oxidation–reduction reactions involve species dissolved in solution. Many years ago, a struggling college student worked during the summer in a petroleum refinery to earn money for the coming academic year. One of the duties assigned to the labor gang at that time involved carrying a solution of copper sulfate, $CuSO_4$, dissolved in water, from one place to another. Castoff steel buckets were used for that purpose. After about the fifth or sixth trip, the bottom would drop out of the bucket, spilling the copper sulfate solution on the ground (Figure 8.4). This unfortunate phenomenon can be explained by an oxidation–reduction reaction. $CuSO_4$ dissolved in water, is present as Cu^{2+} and SO_4^{2-} ions. The Cu^{2+} ion acts as an oxidizing agent and reacts with the iron, Fe, in the bucket as follows:

$$\underbrace{Cu^{2+} + SO_4^{2-}}_{\text{Copper(II) Sulfate}} + Fe \rightarrow \underbrace{Fe^{2+} + SO_4^{2-}}_{\text{Iron(II) Sulfate}} + Cu \qquad (8.3.1)$$

This results in the formation of a solution of iron(II) sulfate and leaves little pieces of copper metal in the bottom of the bucket. A hole is eventually eaten through the side or bottom of the bucket as the iron goes into solution.

The kind of reaction just described is used in a process called **cementation** to purify some industrial wastewaters that contain dissolved metal ions. The water is allowed to flow over iron scraps, and so-called heavy metal ions, such as

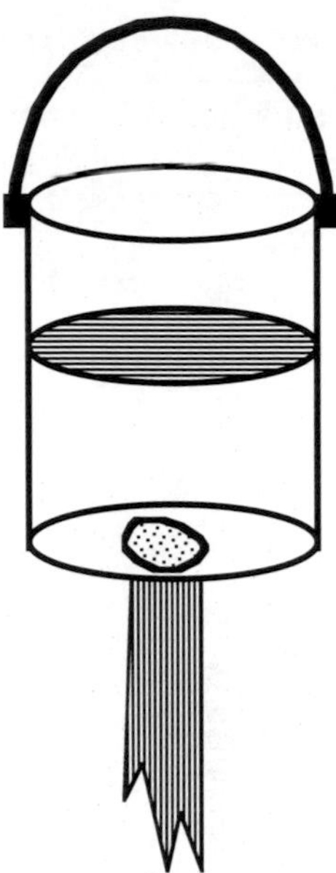

Figure 8.4. A solution of $CuSO_4$ in a steel bucket soon eats its way through the steel, allowing a leak to occur.

Cu^{2+}, Cd^{2+}, and Pb^{2+} precipitate from solution as they are replaced by Fe^{2+}. Cementation results in the replacement of poisonous heavy metal ions in solution by relatively harmless iron.

The reaction of iron and copper sulfate involves the transfer of electrons. These electrons can be forced to go through a wire as electricity and do useful work. To do that, an **electrochemical cell** is set up as shown in Figure 8.5. As can be seen, Cu and a $CuSO_4$ solution are kept in a container that is separated from a bar of Fe

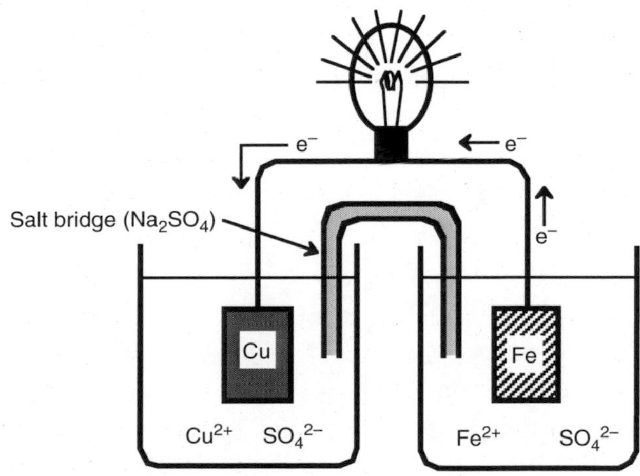

Figure 8.5. An electrochemical cell.

dipping into a solution of $FeSO_4$. These two containers are called **half-cells**. On the left side, Cu^{2+} is reduced to Cu by the **reduction half-reaction**

$$Cu^{2+} + 2e^- \rightarrow Cu \qquad (8.3.2)$$

The electrons required for the reduction of Cu^{2+} are picked up from the bar of copper. On the right side, Fe is oxidized by the **oxidation half-reaction**

$$Fe \rightarrow Fe^{2+} + 2e^- \qquad (8.3.3)$$

In this case, the electrons are left behind on the iron bar, and they go through a wire to the copper bar. In so doing, they may be forced through a light bulb or electric motor and made to do useful work. This is an example of the conversion of chemical energy to electrical energy. The salt bridge shown in Figure 8.5 is just a tube filled with a solution of a salt, such as Na_2SO_4. It completes the circuit by allowing charged ions to move between the two half-cells so that neither has an excess of positive or negative charge. The two half-reactions can be added together to give the total reaction that occurs in the electrochemical cell:

$$\begin{array}{c} Cu^{2+} + 2e^- \rightarrow Cu \\ Fe \rightarrow Fe^{2+} + 2e^- \\ \hline Cu^{2+} + Fe \rightarrow Fe^{2+} + Cu \end{array} \qquad (8.3.4)$$

Recall that this is the same reaction that occurs when $CuSO_4$ solution contacts iron in a steel bucket (when that reaction was written, SO_4^{2-} was shown as a spectator ion—one that does not take part in the reaction).

For an electrical current to flow and light up the bulb shown in Figure 8.5, there must be a **voltage difference** (difference in electrical potential) between the iron and copper bar. These bars are used to transfer electrons between a wire and solution—and in this case, they actually participate in the oxidation–reduction reaction that occurs. The metal bars are called **electrodes**. Because this cell is used to generate a voltage and to extract electricity from a chemical reaction, it is called a **voltaic cell**.

8.4. THE DRY CELL

Dry cells are used as sources of portable electrical energy for flashlights, radios, portable instruments, and many other devices. As shown in Figure 8.6, a dry cell consists of a graphite (carbon) rod in the center of a zinc cylinder, which is filled with a moist paste of manganese dioxide (MnO_2), ammonium chloride (NH_4Cl), and

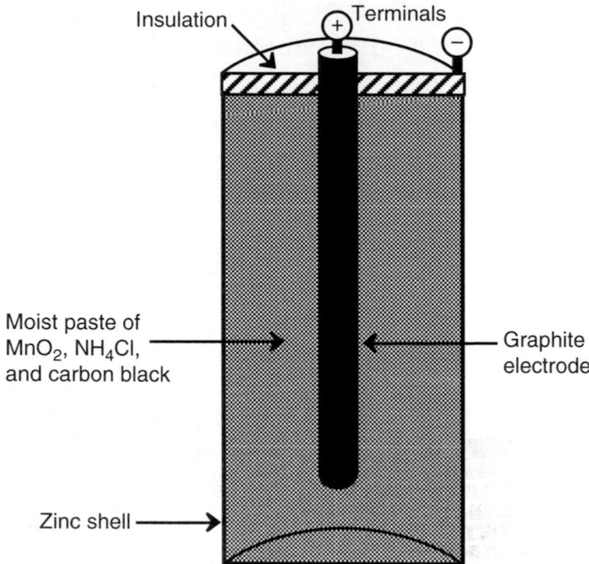

Figure 8.6. The dry cell.

carbon black. The carbon rod and the zinc cylinder make up the two electrodes in the dry cell.

When the two terminals on the electrodes are connected, such as through an electrical circuit powering a portable radio, half-reactions occur at both electrodes. At the graphite electrode, the half-reaction is

$$MnO_2(s) + NH_4^+(aq) + e^- \rightarrow MnO(OH)(s) + NH_3(aq) \qquad (8.4.1)$$

where, as usual, (*s*) indicates a solid materials and (*aq*) one that is dissolved in water. This half-reaction takes negatively charged electrons away from the graphite electrode, leaving it with a positive charge. It is a reduction half-reaction. The electrode at which reduction occurs is always called the **cathode**. Therefore, the graphite rod is the cathode in a dry cell. The half-reaction that occurs at the zinc electrode is

$$Zn(s) \rightarrow Zn^{2+}(aq) + 2e^- \qquad (8.4.2)$$

This reaction leaves a surplus of negatively charged electrons on the zinc electrode, so that it has a negative charge. It is an oxidation half-reaction. The electrode at which oxidation occurs in an electrochemical cell is always called the **anode**. If the two electrodes are connected, electrons will flow from the zinc anode to the carbon cathode. Such a flow of electrons is an electrical current from which useful work can be extracted. Every time that 2 MnO_2 molecules are reduced, as shown in the first half-reaction, 1 Zn is oxidized. Therefore, the overall oxidation–reduction reaction is obtained by multiplying everything in the reduction half-reaction at the cathode by 2 and adding it to the oxidation half-reaction that occurs at the anode. The $2e^-$ cancel on both sides of the equation, yielding the following overall oxidation–reduction reaction:

$$2MnO_2(s) + 2NH_4^+(aq) + Zn(s) \rightarrow Zn^{2+}(aq) + 2MnO(OH)(s) + 2NH_3(aq) \quad (8.4.3)$$

8.5. STORAGE BATTERIES

One of society's generally faithful technological servants is the lead storage battery used in automobiles. This kind of battery provides a steady source of electricity for starting, lights, and other electrical devices on the automobile. Unlike the electrochemical cells already discussed, it can both store electrical energy as chemical energy and convert chemical energy to electrical energy. In the former case, it is being charged; in the latter case, it is being discharged.

A 12 V lead storage battery actually consists of 6 electrochemical cells, each delivering 2 V. The simplest unit of a lead storage battery that can be visualized consists, in the charged state, of two lead grids, one of which is covered by a layer of lead dioxide, PbO_2. These two electrodes are immersed in a sulfuric acid solution, as shown in Figure 8.7. When the battery is discharged, the pure lead electrode acts

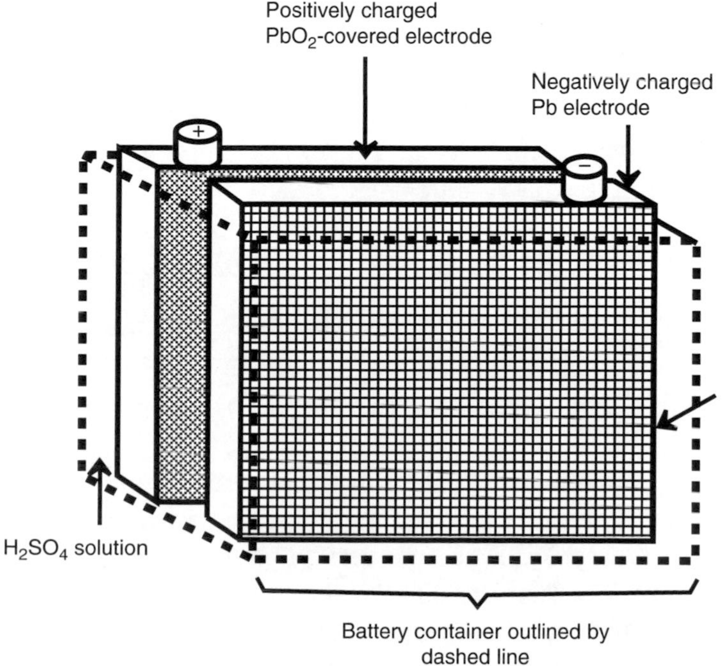

Figure 8.7. Two electrodes from a lead storage battery.

as an anode where the oxidation reaction occurs. In this reaction, a layer of solid lead sulfate, $PbSO_4$, is plated onto the lead electrode:

$$Pb + SO_4^{2-} \rightarrow PbSO_4 + 2e^- \tag{8.5.1}$$

The PbO_2 on the other electrode is reduced as part of the cathode reaction:

$$PbO_2 + 4H^+ + SO_4^{2-} + 2e^- \rightarrow PbSO_4 + 2H_2O \tag{8.5.2}$$

The electrons needed in this half-reaction come from the anode. These electrons are forced to go through the battery leads and through the automobile's electrical system to do useful work. Adding these two half-reactions together gives the overall reaction that occurs as the battery is discharged:

$$Pb + PbO_2 + 4H^+ + 2SO_4^{2-} \rightarrow 2PbSO_4 + 2H_2O \tag{8.5.3}$$

As the battery is discharged, solid $PbSO_4$ is deposited on both electrodes. As the battery is charged, everything is reversed. The half-reactions and overall reaction are

Cathode: $PbSO_4 + 2e^- \rightarrow Pb + SO_4^{2-}$

Anode: $PbSO_4 + 2H_2O \rightarrow PbO_2 + 4H^+ + SO_4^{2-} + 2e^-$

Overall: $2PbSO_4 + 2H_2O \rightarrow Pb + PbO_2 + 4H^+ + 2SO_4^{2-}$ (8.5.4)

Another kind of rechargeable battery is the nickel–cadmium (NiCd, "Nicad") battery formerly widely employed in cordless electric tools, electronic calculators, electronic camera flash attachments, and other applications. During discharge, the reactions that occur in this battery are

Cathode: $NiO_2 + 2H_2O + 2e^- \rightarrow Ni(OH)_2 + 2OH^-$

Anode: $Cd + 2OH^- \rightarrow Cd(OH)_2 + 2e^-$

Overall: $Cd + NiO_2 + 2H_2O \rightarrow Cd(OH)_2 + Ni(OH)_2$ (8.5.5)

As the battery is being charged, both half-reactions and the overall reaction are simply reversed. Larger versions of the Nicad battery have been used in some automobiles and motorcycles. Although very good for storage of electrical energy, Nicad batteries have lost favor because of the toxicity of cadmium. Although the batteries are safe during use, they are often disposed of improperly, and their presence in landfills and incinerators poses pollution problems.

A popular alternative to Nicad batteries for electronic applications is the **lithium ion battery**, the principles of which are illustrated in Figure 8.8. These batteries operate by means of the movement of Li^+ ions between electrodes through an electrolyte and across a separator. One electrode is composed of electrically conducting carbon (graphite) and the other of a mixed oxide of lithium and a transition metal capable of undergoing oxidation and reduction, usually cobalt or manganese. These batteries have become the rechargeable batteries of choice for portable computers and some other electronic devices. They are considered green and consistent with sustainability because they are composed of materials that are relatively nontoxic, abundant, and recyclable. Some of these batteries have caught fire and burned, but that problem now seems to have been solved with improved manufacturing techniques.

High-capacity, readily charged storage batteries are very important to the success of newly developing electrically powered vehicles. These vehicles use a storage

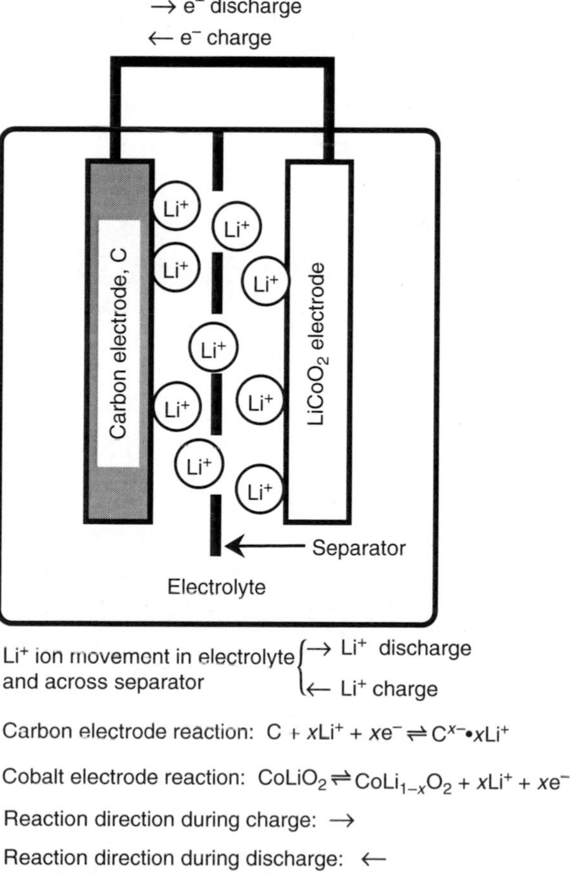

Figure 8.8. Lithium ion rechargeable batteries operate by the movement of lithium ions between electrodes through an electrolyte and across a separator membrane.

battery rechargeable by a small internal combustion engine and by recharge from braking to provide much better overall fuel economy than do conventional vehicles. They are an excellent example of sustainable technology, and significantly reduce emissions of exhaust pollutants and greenhouse-gas carbon dioxide. With current technology, hybrid vehicles use a nickel–metal hydride (NiMH) electrode that takes advantage of the ability of nickel to bond with hydrogen, a characteristic shared by cobalt, manganese, and aluminum. In practice, the nickel is combined as an alloy with the general formula ANi_5, where A represents a mixture of the rare earth metals lanthanum, cerium, neodymium, and praseodymium. The other electrode in this battery consists of a coating of a mixture of NiOOH and $Ni(OH)_2$ on a conducting metal surface. The electrodes are immersed in an electrolyte of an aqueous solution of KOH. Representing the ANi_5 electrode as M, the reaction that it undergoes as the battery is charged is

$$M(s) + H_2O(l) + e^- \rightarrow MH(s) + OH^-(aq) \quad (8.5.6)$$

and during discharge the opposite reaction occurs. As the battery is charged, the reaction at the other electrode is

$$Ni(OH)_2(s) + OH^-(aq) \rightarrow NiOOH(s) + H_2O(l) + e^- \quad (8.5.7)$$

with the reverse occurring during discharge.

8.6. USING ELECTRICITY TO MAKE CHEMICAL REACTIONS OCCUR

Electrolysis of Water: A Green Technology

Electricity is commonly used as an energy source to make chemical reactions occur that do not occur by themselves. Such a reaction is called an **electrolytic reaction**. The cell in which it occurs is an **electrolytic cell** (Figure 8.9). A direct electrical current passing through a solution of a salt in water causes the water to break up and form H_2 and O_2. The process can now be examined in a little more detail. As always, oxidation occurs at the anode. In this case, the external electrical power supply withdraws electrons from the anode, which in turn takes electrons away from water, as shown for the following oxidation half-reaction:

$$2H_2O \rightarrow 4H^+ + O_2(g) + 4e^- \quad (8.6.1)$$

As a result, oxygen gas, O_2, is given off at the anode. The power supply forces electrons onto the cathode, where the following reduction half-reaction occurs:

$$2H_2O + 2e^- \rightarrow H_2(g) + 2OH^- \quad (8.6.2)$$

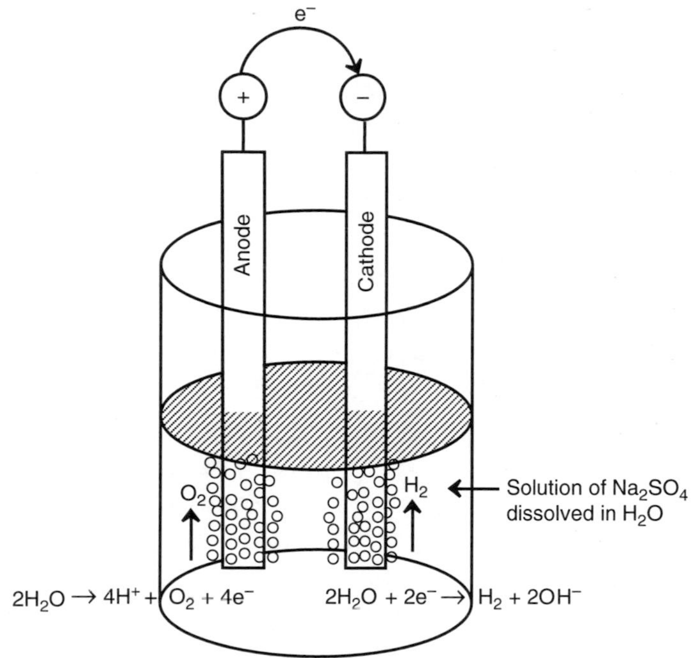

Figure 8.9. The electrolysis of water to form H_2 and O_2.

As a result, hydrogen gas is produced at the cathode. The H^+ ions produced at the anode and the OH^- ions produced at the cathode combine according to the neutralization reaction

$$H^+ + OH^- \rightarrow H_2O \tag{8.6.3}$$

The overall electrolysis reaction is

$$2H_2O + \text{electrical energy} \rightarrow 2H_2(g) + O_2(g) \tag{8.6.4}$$

The electrolysis of water is a potentially sustainable means of generating the greenest of all fuels, elemental hydrogen, which, burned in an engine or oxidized in a fuel cell, provides a pollution-free source of energy that gives water as its only product. Reykjavik, Iceland, has a station for dispensing electrolytically generated hydrogen gas to vehicles, and the city runs three buses powered by hydrogen-fueled fuel cells. Using rooftop pressurized tanks for hydrogen gas storage, each bus can travel about 125 miles before refueling is required. This provides a perfectly acceptable range, especially since the buses operate within a rather close radius to the refueling station.

Electrolytically generated hydrogen gas fuel is perhaps the ultimate green fuel in that the electricity can be generated from renewable sources that may be intermittent in nature. In Iceland, the electricity is generated by abundant renewable hydroelectric and geothermal steam sources. In other areas, solar cells and windpower can be used to produce electricity required to electrolyze water. These sources are inherently variable, and intermittent and electrolytically generated hydrogen, which can be stored at very high pressures underground, provides a means of evening out the power supply from sun or wind. This hydrogen can be used very efficiently in fuel cells to generate power during times when solar or wind sources provide little or no energy output. Additional efficiencies can be gained by using pressure-driven engines to reclaim much of the energy required to store the hydrogen when it is released for fuel cell use. The pure oxygen gas byproduct of electrolytically generated elemental hydrogen has several potential uses, such as in gasification of biomass to produce renewable hydrocarbon fuels. Pure oxygen can be employed as an alternative to oxygen in air for the combustion of hydrogen in gas turbines or its utilization in fuels cells.

Electrolytic Manufacture of Chemicals

Electrolytic processes are widely used to manufacture chemicals. One of the simplest of these is the Downs process for manufacturing liquid sodium. This chemically active metal is used to synthesize organic compounds, such as tetraethyl lead (once widely used as a gasoline additive), and in many other applications. It is made by passing an electrical current through melted sodium chloride, as shown in Figure 8.10. Electrons forced onto the iron cathode by the external source of electricity bring about the reduction half-reaction

$$Na^+ + e^- \rightarrow Na \tag{8.6.5}$$

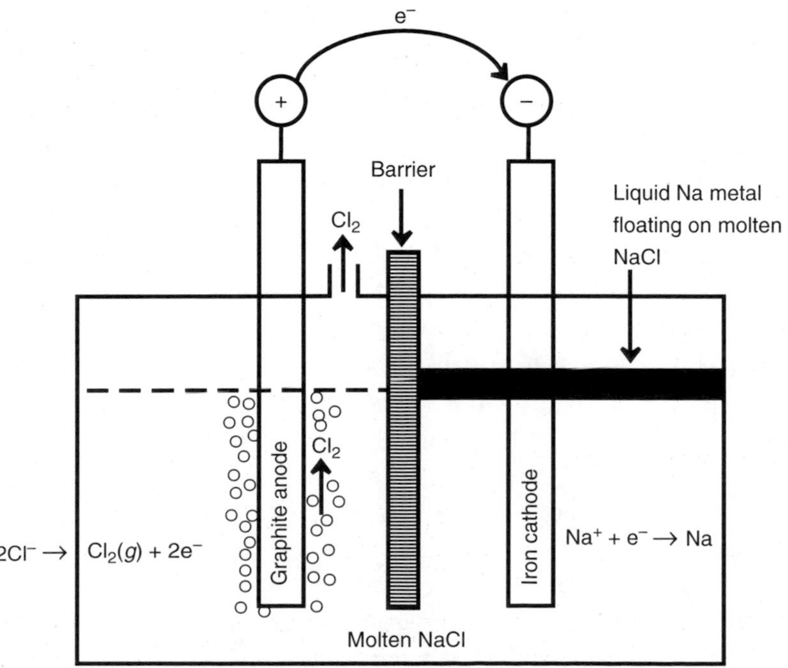

Figure 8.10. Electrolysis of melted sodium chloride to produce liquid sodium metal and chlorine gas.

Withdrawal of electrons from the graphite anode by the external electrical circuit applies the driving force for the oxidation half-reaction

$$2Cl^- \rightarrow Cl_2(g) + 2e^- \tag{8.6.6}$$

The overall oxidation–reduction reaction is,

$$2Na^+ + 2Cl^- \rightarrow 2Na + Cl_2(g) \tag{8.6.7}$$

In industrial practice, a Downs cell is especially designed to prevent contact of the sodium and chlorine, which would, of course, react violently with each other.

The chlorine that is produced during the manufacture of liquid sodium is a valuable byproduct that is used to manufacture chlorinated solvents, pesticides, and in other applications. Far more chlorine is needed than sodium. More economical processes involving the electrolysis of sodium chloride solutions in water are used to manufacture chlorine along with sodium hydroxide.

8.7. ELECTROPLATING

One of the most common uses of electrochemistry is the plating of a thin layer of an expensive or attractive metal onto a base of cheaper metal. The shiny layer of chromium metal that used to be very common on automobile bumpers and other trim before it became too expensive for this application is plated onto steel with an electrochemical process. A thin layer of silver is commonly plated onto "silverware" to make eating utensils look like the "real thing" using an electrochemical process. The use of electrochemistry for plating metals onto surfaces is called **electroplating**.

Many electroplating processes are actually quite complicated. However, an idea of how these processes work can be gained by considering the electroplating of silver metal onto a copper object, as shown in Figure 8.11. The copper object and a piece of silver metal make up the two electrodes. They are dipped into a solution containing dissolved silver nitrate, $AgNO_3$, and other additives that result in a smooth deposit of silver metal. Silver ion in solution is reduced at the negatively charged cathode, leaving a thin layer of silver metal:

$$Ag^+ + e^- \rightarrow Ag \tag{8.7.1}$$

The silver ion is replaced in solution by oxidation of silver metal at the silver metal anode:

$$Ag \rightarrow Ag^+ + e^- \tag{8.7.2}$$

Electroplating is commonly used to prevent corrosion (rusting) of metal. Zinc metal electroplated onto steel prevents rust. The corrosion of "tin cans" is inhibited by a very thin layer of tin plated onto rolled steel.

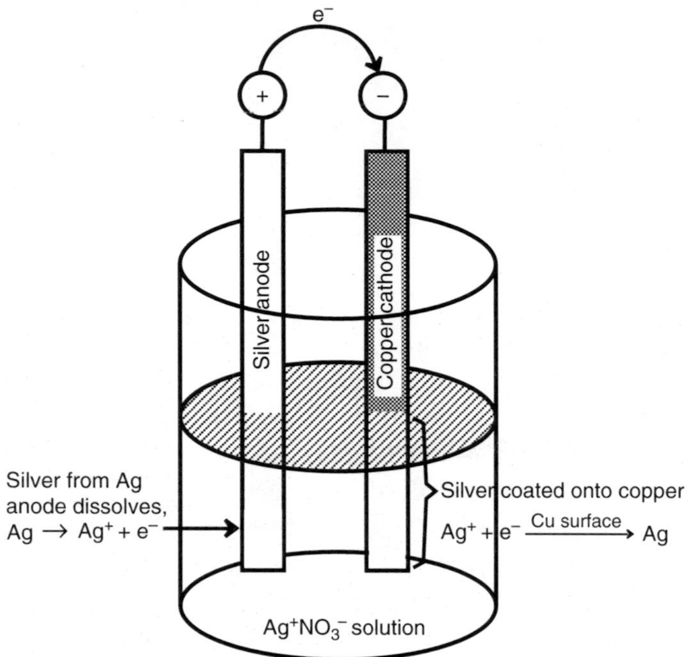

Figure 8.11. Electrochemical cell for electroplating silver onto copper.

8.8. FUEL CELLS

Current methods for the conversion of chemical energy to electrical energy are rather cumbersome and wasteful. Typically, coal is burned in a boiler to generate steam, the steam goes through a turbine, the turbine drives a generator, and the generator produces electricity. This whole process wastes about 60% of the energy originally in the coal. More energy is wasted in transmitting electricity through power lines to users. It is easy to see the desirability of converting chemical energy directly to electricity. This can be done in fuel cells.

A fuel cell that uses the chemical combination of H_2 and O_2 to produce electricity directly from a chemical reaction is shown in Figure 8.12. This device consists of two porous graphite (carbon) electrodes dipping into a solution of potassium hydroxide, KOH. At the anode, H_2 is oxidized, giving up electrons, and at the cathode, O_2 is reduced, taking up electrons. The two half-reactions and the overall chemical reaction are

$$\text{Anode:} \quad 2H_2 + 4OH^- \rightarrow 4H_2O + 4e^-$$

$$\text{Cathode:} \quad O_2 + H_2O + 4e^- \rightarrow 4OH^-$$

$$\text{Overall:} \quad 2H_2 + O_2 \rightarrow 2H_2O \qquad (8.8.1)$$

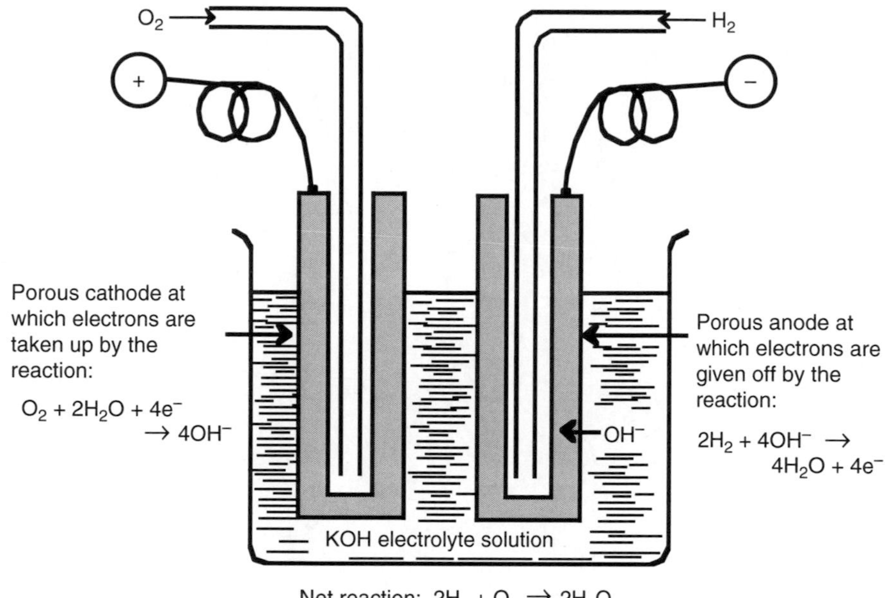

Figure 8.12. Fuel cell for direct production of electricity by oxidation of hydrogen.

This reaction obviously provides a nonpolluting source of energy. Unfortunately, both hydrogen and oxygen are rather expensive to generate.

8.9. SOLAR CELLS

The direct conversion of light energy to electrical energy can be accomplished in **photovoltaic cells**, also called **solar cells**. These consist basically of two layers of silicon, Si, one **doped** with about 1 atom per million of arsenic, As, and the other doped with about 1 atom per million of boron, B. To understand what happens, consider the Lewis symbols of the three elements involved:

$$\cdot\dot{\text{Si}}: \qquad \cdot\dot{\underset{\cdot}{\text{As}}}: \qquad \dot{\text{B}}:$$

As shown by its Lewis symbol above and discussed in Chapter 3, each silicon atom has 4 valence electrons. In a crystal of Si, each of the silicon atoms is covalently bonded to four other Si atoms, as shown in Figure 8.13. If a B atom replaces one of the Si atoms, the 3 valence electrons in B result in a shortage of 1 electron, leaving what is called a **positive hole**. Each As atom has 5 valence electrons, so that replacement of one of the Si atoms by As leaves a surplus of 1 electron, which can move about in the Si crystal. A layer of Si atoms doped with As makes up a **donor** layer because of the "extra" valence electron introduced by each As atom, and Si doped

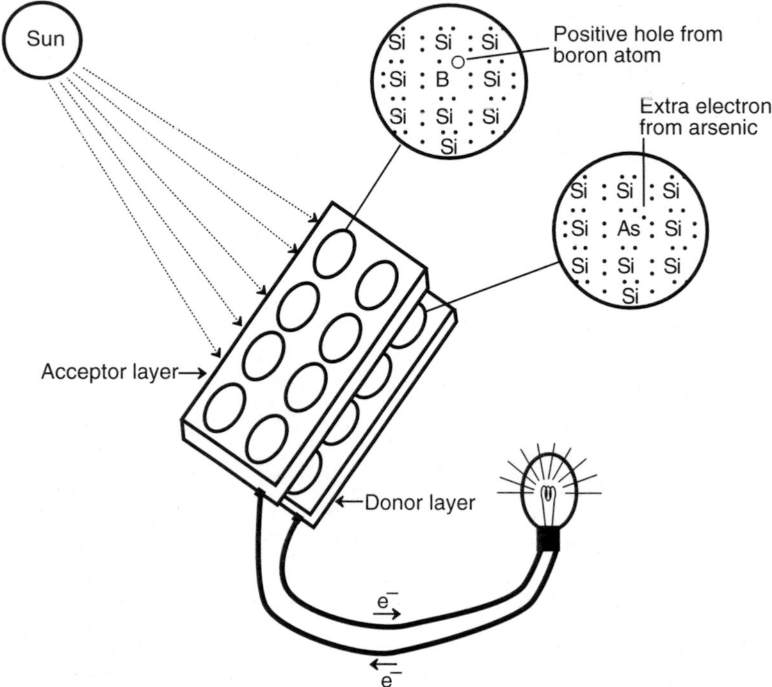

Figure 8.13. A photovoltaic cell. Note the positive hole, ○, at the site of the B atom, and the extra electron around the As atom.

with B makes up an **acceptor** layer because each B atom has 1 less valence electron than does each Si atom. When two such layers are placed in contact, the positive holes near the surface of the acceptor layer become filled with excess electrons from the donor layer. Nothing else happens unless light falls on the solar cell. This light provides the energy required to push electrons back across the boundary. These electrons can be withdrawn by a wire from the donor layer and returned by a wire to the acceptor layer, resulting in a flow of usable electrical current.

8.10. REACTION TENDENCY

It has already been seen that whole oxidation–reduction reactions can be constructed from half-reactions. The direction in which a reaction goes is a function of the relative tendencies of its constituent half-reactions to go to the right or left. These tendencies, in turn, depend upon the concentrations of the half-reaction reactants and products and their relative tendencies to gain or lose electrons. The latter is expressed by a **standard electrode potential** E^0. The tendency of the whole reaction to proceed to the right as written is calculated from the **Nernst equation**, which contains both E^0 and the concentrations of the reaction participants. These concepts are explained further in this section and the following section.

Measurement of E^0

To visualize the measurement of E^0, consider the electrochemical cell shown in Figure 8.14. When the two electrodes are connected by an electrical conductor, a reaction occurs in which hydrogen gas would reduce silver ion to silver metal:

$$2Ag^+ + H_2 \rightleftharpoons 2H^+ + 2Ag \tag{8.10.1}$$

This reaction is composed of the following two half-reactions:

$$\begin{array}{lll} 2Ag^+ + 2e^- & \rightleftharpoons 2Ag & E^0 = 0.7994\ \text{V} \\ -(2H^+ + 2e^- & \rightleftharpoons H_2 & E^0 = 0.000\ \text{V}) \\ \hline 2Ag^+ + H_2 & \rightleftharpoons 2H^+ + 2Ag & E^0 = 0.7994\ \text{V} \end{array}$$

If, in the cell shown in Figure 8.14, the activities of both Ag^+ and H^+ were exactly 1 (approximated by concentrations of 1 mol/L) and the pressure of H_2 exactly 1 atm, the potential registered between the two electrodes by a voltmeter, E, would be 0.7994 V. The platinum electrode, which serves as a conducting surface to exchange electrons, would be negative because of the prevalent tendency for H_2 molecules to leave negatively charged electrons behind on it as they go into solution as H^+ ions. The silver electrode would be positive because of the prevalent tendency for Ag^+ ions to pick up electrons from it and to be deposited as Ag atoms.

The left electrode shown in Figure 8.14 is of particular importance, because it is the standard electrode against which all other electrode potentials are compared.

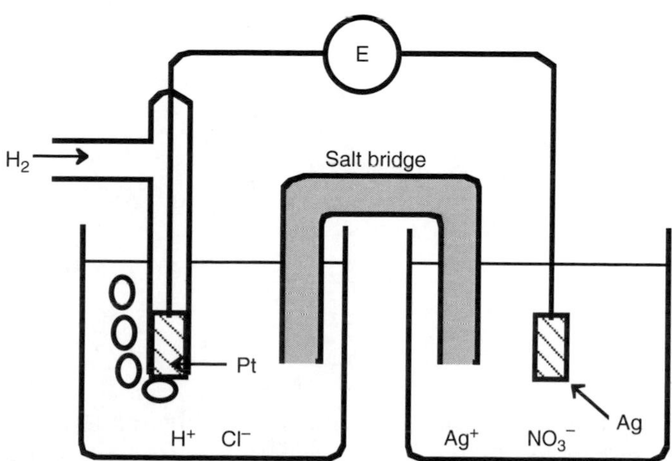

Figure 8.14. Electrochemical cell in which the reaction $2Ag^+ + H_2 \rightleftharpoons 2H^+ + 2Ag$ can be carried out in two half-cells.

It is called the **standard hydrogen electrode (SHE),** and has been assigned a value of exactly 0 V by convention; its half-reaction is written as

$$2H^+ + 2e^- \rightleftharpoons H_2 \qquad E^0 = 0.000\ V \qquad (8.10.2)$$

The measured potential of the right-hand electrode in Figure 8.14 versus the standard hydrogen electrode is called the **electrode potential** E. If Ag^+ and other ions in solution are at unit activity (approximated by a concentration of 1 mol/L), the potential is the **standard electrode potential** E^0. The standard electrode potential for the Ag^+/Ag couple is 0.7994 V, expressed conventionally as

$$Ag^+ + e^- \rightleftharpoons Ag \qquad E^0 = 0.7994\ V \qquad (8.10.3)$$

E^0 Values and Reaction Tendency

Recall that half-reactions can be combined to give whole reactions and that half-reactions can occur in separate half-cells of an electrochemical cell. The inherent tendency of a half-reaction to occur is expressed by a characteristic E^0 value. In favorable cases, E^0 values of half-reactions can be measured directly versus a standard hydrogen electrode by a cell such as that shown in Figure 8.14 or they can be calculated from thermodynamic data. Several half-reactions and their E^0 values are given below:

$$Cl_2 + 2e^- \rightleftharpoons 2Cl^- \qquad E^0 = 1.359\ V \qquad (8.10.4)$$

$$O_2 + 4H^+ + 4e^- \rightleftharpoons 2H_2O \qquad E^0 = 1.229\ V \qquad (8.10.5)$$

$$Ag^+ + e^- \rightleftharpoons Ag \qquad E^0 = 0.7994\ V \qquad (8.10.3)$$

$$Fe^{3+} + e^- \rightleftharpoons Fe^{2+} \qquad E^0 = 0.771\ V \qquad (8.10.6)$$

$$Cu^{2+} + 2e^- \rightleftharpoons Cu \qquad E^0 = 0.337\ V \qquad (8.10.7)$$

$$2H^+ + 2e^- \rightleftharpoons H_2 \qquad E^0 = 0.000\ V \qquad (8.10.2)$$

$$Pb^{2+} + 2e^- \rightleftharpoons Pb \qquad E^0 = -0.126\ V \qquad (8.10.8)$$

$$Zn^{2+} + 2e^- \rightleftharpoons Zn \qquad E^0 = -0.763\ V \qquad (8.10.9)$$

Basically, the E^0 values of these half-reactions express the tendency for the reduction half-reaction to occur when all reactants and products are present at unit activity; the more positive the value of E^0, the greater the tendency of the reduction half-reaction to proceed. (In a simplified sense, the activity of a substance is 1 when its concentration in aqueous solution is 1 mol/L, its pressure as a gas is 1 atm, or it is present as a solid.) With these points in mind, examination of the E^0 values above shows the following:

- The highest value of E^0 shown above is for the reduction of Cl_2 gas to Cl^- ion. This is consistent with the strong oxidizing tendency of Cl_2; chlorine much "prefers" to exist as Cl^- ion rather than highly reactive Cl_2 gas.
- The comparatively high E^0 value of 0.7994 V for the reduction of Ag^+ ion to Ag metal indicates that silver is relatively stable as a metal, which is consistent with its uses in jewelry and other applications where resistance to oxidation is important.
- The value of exactly $E^0 = 0.000$ V is assigned by convention for the half-reaction in which H^+ ion is reduced to H_2 gas; all other E^0 values are relative to this value.
- The lowest (most negative) E^0 value shown above is -0.763 V for the half-reaction $Zn^{2+} + 2e^- \rightleftharpoons Zn$. This reflects the strong tendency for zinc to leave the metallic state and become Zn^{2+} ion; that is, the half reaction tends to lie strongly to the left. Since Zn metal gives up electrons when it is oxidized to Zn^{2+} ion, Zn metal is a good reducing agent.

Half-reactions and their E^0 values can be used to explain observations such as the following: A solution of copper(II) ion flows through a lead pipe, and the lead acquires a layer of copper metal through the reaction

$$Cu^{2+} + Pb \rightarrow Cu + Pb^{2+} \tag{8.10.10}$$

This reaction occurs because the copper(II) ion has a greater tendency to acquire electrons than the lead ion has to retain them. This reaction can be obtained by subtracting the lead half-reaction, Equation 8.10.8, from the copper half-reaction, Equation 8.10.7:

$$\begin{array}{ll} Cu^{2+} + 2e^- \rightleftharpoons Cu & E^0 = 0.337\text{ V} \\ -(Pb^{2+} + 2e^- \rightleftharpoons Pb & E^0 = -0.126\text{ V}) \\ \hline Cu^{2+} + Pb \rightarrow Cu + Pb^{2+} & E^0 = 0.463\text{ V} \end{array}$$

The appropriate mathematical manipulation of the E^0 values of the half-reactions enables calculation of an E^0 for the overall reaction, the positive value of which indicates that Reaction 8.10.10 tends to go to the right as written. This is in fact what occurs when lead metal directly contacts a solution of copper(II) ion. Therefore, if a

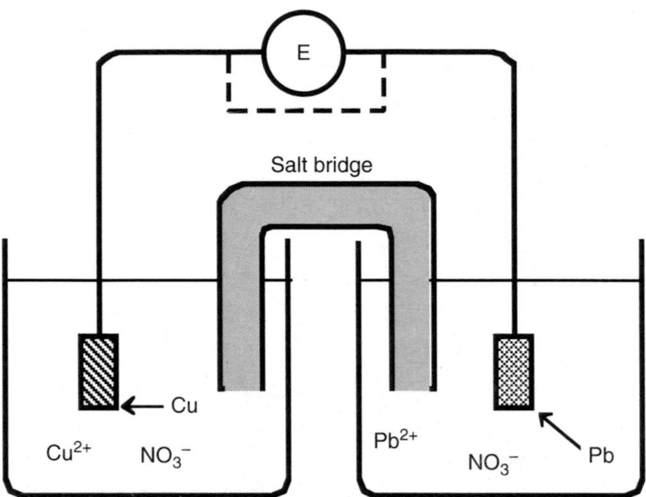

Figure 8.15. Electrochemical cell in which the tendency for the reaction $Cu^{2+} + Pb \rightarrow Cu + Pb^{2+}$ can be measured. In this configuration, the meter "E" has a very high resistance and current cannot flow.

waste solution containing copper(II) ion, a relatively innocuous pollutant, comes into contact with lead in plumbing, toxic lead may go into solution.

Based on the above calculation, if the electrochemical cell shown in Figure 8.15 were set up with activities of both Cu^{2+} and Pb^{2+} at exactly 1 (approximated by concentrations of 1 mol/L), the potential registered by a meter that measures voltage but does not allow any current to flow would be $E = 0.463$ volts. The lead electrode would be negative, because the reaction tendency is for Pb metal to give up negative electrons to the external circuit and go into solution as Pb^{2+} ion, whereas Cu^{2+} ions tend to remove electrons from the copper electrode, giving it a positive charge and coming out of solution as Cu metal. The effects of different concentrations on the potential that would be measured in such a cell are discussed in the following section.

8.11. EFFECT OF CONCENTRATION: NERNST EQUATION

The **Nernst equation** is used to account for the effect of different activities upon electrode potential. Referring to Figure 8.15, if the Cu^{2+} ion concentration $[Cu^{2+}]$ is increased with everything else remaining constant, it can readily be visualized that the potential of the left electrode will become more positive because the higher concentration of electron-deficient Cu^{2+} ions clustered around it tends to draw electrons from the electrode. Decreased $[Cu^{2+}]$ has the opposite effect. If the Pb^{2+} ion concentration $[Pb^{2+}]$ in the right electrode is increased, it is "harder" for Pb atoms to leave the Pb electrode as positively charged ions, therefore, there is less of a tendency for electrons to be left behind on the Pb electrode, and its potential tends to be more positive. At a lower value of $[Pb^{2+}]$ in the right half-cell, the exact opposite is true. Such concentration effects upon E are expressed by the **Nernst equation**. As applied to the reaction in question,

$$Cu^{2+} + Pb \rightarrow Cu + Pb^{2+} \qquad E^0 = 0.463\,V$$

at a temperature of 25°C, the potential E of the cell is given by the Nernst equation

$$E = E^0 + \frac{2.303RT}{nF}\log\frac{[Cu^{2}]}{[Pb^{2+}]} = 0.463 + \frac{0.0591}{2}\log\frac{[Cu^{2}]}{[Pb^{2+}]} \qquad (8.11.1)$$

where R is the molar gas constant, T is the absolute temperature, F is the Faraday constant, n is the number of electrons involved in the half-reaction ($n = 2$ in this case), and the activities are approximated by concentrations. The value of $2.303RT/F$ is 0.0591 at 25°C.

As an example of the application of the Nernst equation, suppose that $[Cu^{2+}] = 3.33 \times 10^{-4}$ mol/L and $[Pb^{2+}] = 0.0137$ mol/L. Substituting into the Nernst equation above gives

$$E = 0.463 + \frac{0.0591}{2}\log\frac{3.33 \times 10^{-4}}{0.0137} = 0.415\ V \qquad (8.11.2)$$

The value of E is still positive and Reaction 8.10.10 still proceeds to the right as written.

8.12. POTENTIOMETRY

Most of this discussion of electrochemistry has dealt with effects involving a flow of electrical current. It has been seen that chemical reactions can be used to produce an electrical current. It has also been seen that the flow of an electrical current through an electrochemical cell can be used to make a chemical reaction occur. Another characteristic of electricity is its voltage, or electrical potential, which was discussed as E and E^0 values above as a kind of "driving force" behind oxidation–reduction reactions.

As noted above, voltage is developed between two electrodes in an electrochemical cell. The voltage depends upon the kinds and concentrations of dissolved chemicals in the solutions contacted by the electrodes. In some cases, this voltage can be used to measure concentrations of some substances in solution. This gives rise to the branch of analytical chemistry known as **potentiometry**. Potentiometry uses **ion-selective electrodes** or **measuring electrodes** whose potentials relative to a **reference electrode** vary with the concentrations of particular ions in solution. The reference electrode that serves as the ultimate standard for potentiometry is the standard hydrogen electrode shown in Figure 8.14. In practice, other electrodes, such as the silver/silver chloride or calomel (mercury metal in contact with Hg_2Cl_2) are used. A reference electrode is hooked to the reference terminal input of a voltmeter, and a measuring electrode is hooked to the measuring terminal; in practice, both electrodes are immersed in the solution being analyzed and the potential of the measuring electrode is read versus the reference electrode.

Two very useful ion-selective electrodes are shown in Figure 8.16. The easier of these to understand is the fluoride ion-selective electrode consisting of a disk of lanthanum fluoride, LaF_3, molded into the end of a plastic body and connected to a wire. When this electrode is placed in solution, its potential relative to a reference electrode (an electrode whose potential does not vary with the composition of the solution) shifts more negative with increasing concentrations of F^- ion in solution. By comparing the potential of the fluoride electrode in a solution of unknown F^- concentration with its potential in a solution of known F^- concentration, it is possible to calculate the value of the unknown concentration.

The potential of the fluoride measuring electrode at 25°C responds according to the Nernst equation in the form

$$E = E_a - 0.0591 \log[F^-] \quad (8.12.1)$$

where E is the measured potential of the fluoride electrode versus a reference electrode. (Strictly speaking, E is a function of *activity* of fluoride ion, which is approximated very well by $[F^-]$ at relatively lower concentrations of F^-.) E_a is not a true E^0, but is treated just like E^0 in the Nernst expression. Equation 8.12.1 shows that for

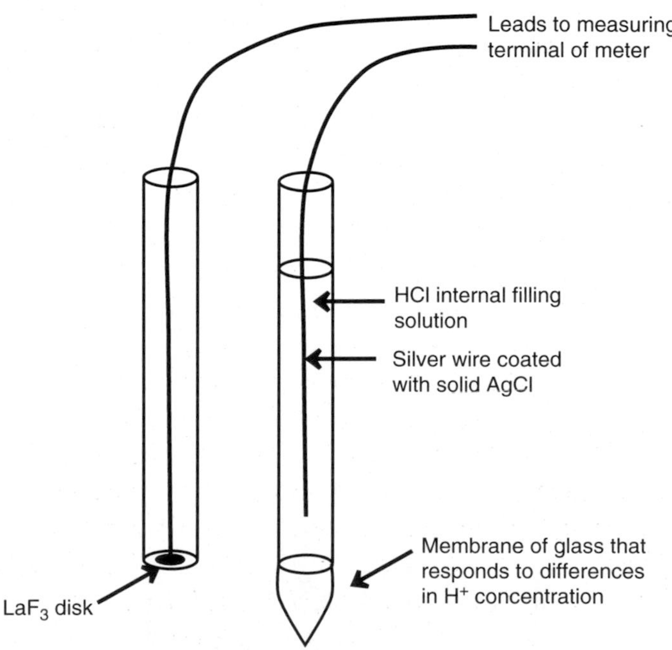

Figure 8.16. A fluoride ion-selective electrode (left) and a glass electrode used to measure pH (right).

every 10-fold increase in [F^-], the potential of the fluoride electrode shifts in a negative direction by 0.0591 V (59.1 mV). Increasing the concentration of a *negatively charged* ion shifts the potential to more *negative* values.

The other electrode shown in Figure 8.16 is the glass electrode used to measure H^+ concentration. Its potential varies with the concentration of H^+ ion in solution. The electrode has a special glass membrane on its end, to which H^+ ions tend to adsorb. The more H^+ ions in the solution in which the electrode is dipped, the more that are adsorbed to the membrane. This adsorption of *positively charged* ions shifts the potential *positive* with increasing H^+. This shift in potential is detected by a special voltmeter with a high resistance, called a **pH meter**. The pH meter makes contact with the glass membrane by way of a silver chloride (AgCl-coated silver wire) dipping into a solution of HCl contained inside the electrode. The potential is, of course, measured relative to a reference electrode, also dipping into the solution. Recall from Section 7.5 that pH is equal to the negative logarithm of the hydrogen ion concentration, so that as the potential of the glass electrode becomes more positive, a lower pH is indicated. The Nernst equation that applies to the glass electrode is

$$E = E_a + 0.0591 \log[H^+] \tag{8.12.2}$$

and, since pH $= -\log[H^+]$, the following applies:

$$E = E_a - 0.0591\,\text{pH} \tag{8.12.3}$$

To measure pH, the pH meter must first be calibrated. This is done by placing the electrodes in a **standard buffer solution** of known pH (a buffer solution is one that resists changes in pH with added acid, base, or water, and a standard buffer solution is one made up to an accurately known pH). After calibration, the pH meter is then adjusted to read that pH. Next, the electrodes are removed from the buffer solution, rinsed to remove any buffer, and placed in the solution of unknown pH. This pH is then read directly from the pH meter.

8.13. CORROSION

Corrosion is defined as *the destructive alteration of metal through interactions with its surroundings*—in short, rust. It is a redox phenomenon resulting from the fact that most metals are unstable in relation to their surroundings. Thus, the steel in cars really prefers to revert back to the iron oxide ore that it came from by reacting with oxygen in the air. The corrosion process is accelerated by exposure to water, which may contain corrosive salt placed on road ice, or acid from acid rain. It is only by the application of anticorrosive coatings and careful maintenance that the process is slowed down.

Corrosion occurs when an electrochemical cell is set up on a metal surface, as shown in Figure 8.17. A layer of moisture serves to dissolve ions and act as a salt

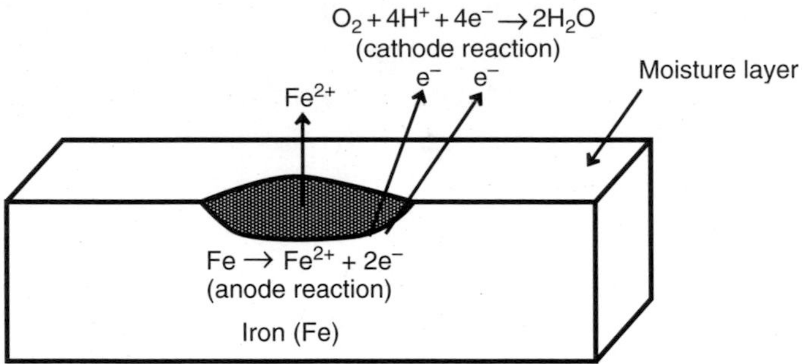

Figure 8.17. Electrochemical cell on a metal surface on which corrosion occurs.

bridge between the anode and cathode. The area in which the metal is oxidized is the anode. Typically, when iron is corroded, the anode reaction is

$$Fe(s) \rightarrow Fe^{2+}(aq) + 2e^- \tag{8.13.1}$$

so that the solid iron goes into solution as Fe^{2+} ion. Several cathode reactions are possible. Usually oxygen is involved. A typical cathode reaction is

$$O_2(g) + 4H^+(aq) + 4e^- \rightarrow 2H_2O \tag{8.13.2}$$

The overall corrosion process is normally very complicated and involves a number of different reactions. Very commonly, bacteria are involved in corrosion. The bacterial cells derive energy by acting as catalysts in the corrosion reactions.

CHAPTER SUMMARY

The chapter summary below is presented in a programmed format to review the main points covered in this chapter. It is used most effectively by filling in the blanks, referring back to the chapter as necessary. The correct answers are given at the end of the summary.

Passing electricity through a solution of sodium sulfate in water [1]______________ ______________________. To a large extent, the nature of inorganic species in water depends upon whether the water solution is [2]_____________ or [3]_____________. The study of the interaction of chemistry and electricity is called [4]_______________. The loss of electrons by an atom is called [5]___________ and the atom is said to have been [6]_____________. A gain of electrons by an atom is called [7]_______________ and the atom is said to have been [8]___________. The hypothetical charge that an atom would have if it gained or lost a particular number of electrons in forming a chemical compound is called its [9]______________. For chemically combined O, the oxidation number is almost always [10]___________, and for chemically combined H, it is almost always [11]__________. The oxidation number of any element in its elemental form is [12]______. The sum of the oxidation numbers of the elements in a compound equals [13]________, and the sum of the oxidation numbers of the elements in an ion equals [14]____________. In chemical compounds, Na, Ca, and F have oxidation numbers of [15]_____, ____, and ___, respectively. When a solution containing Cu^{2+} ions comes into contact with iron metal, the products are [16]______________. As a result of this process, Fe metal is [17]_____________ and Cu^{2+} ion is [18]_____________. A device designed to carry out a reduction half-reaction and an oxidation half-reaction in physically separate locations such that electrical energy may be extracted is called [19]_______________________. Metal or graphite bars that transfer electrons between wires and the solution in an electrochemical cell are called [20]__________. The one of these at which reduction occurs is called the [21]______________ and the one at which oxidation occurs is called the [22]_______________. The function of a salt bridge is to [23]________________________. In a dry cell, the cathode consists of [24]_______________ and the anode is [25]______________________. The overall reaction in a dry cell is [26]_____________________________. A reaction that is made to occur by the passage of electrical current through a solution is called [27]______________ and it takes place in [28]____________________. When an electrical current is passed through a solution of melted NaCl, [29]_____________ is produced at the graphite anode and [30]____________________ is produced at the iron cathode. When an automobile battery is discharged, [31]____________ is oxidized at the anode and [32]_______________ is reduced at the cathode. The device in which the overall reaction $Cd + NiO_2 + 2H_2O \rightarrow Cd(OH)_2 + Ni(OH)_2$ occurs is the [33]____________________________. A device in which H_2 and O_2 are used for the direct production of electrical energy from a chemical reaction is a [34]____________________________. The direct conversion of light energy to electrical energy is accomplished in a [35]_______________. Such a device consists of two layers known as the [36]_____________ and the [37]________________. The process by which electricity is used to plate a layer of metal onto an object made from another metal is called [38]__________. In a fuel cell powered with H_2 and O_2, electrons are produced at the anode by oxidation of [39]_________ and electrons are taken up at the cathode by reduction of [40]______. In a solar cell, doping silicon with boron atoms produces [41]_______________ and doping with arsenic atoms produces [42]_________________. The tendency of a half-reaction to go to the right or to the left as written is expressed by a [43]______________________. Account is taken of the influence of differences in concentration on this tendency by the [44]_________ _________. The standard electrode against which all other electrode potentials are compared is called the [45]______________________ for which the assigned E^0 value is [46]_______ and the half-reaction is [47]_____________. The measured potential of an electrode versus the standard hydrogen electrode when the activities of all the reaction constituents are exactly 1 is called the [48]______ _______________. The fact that the reaction $Cu^{2+} + Pb \rightleftharpoons Cu + Pb^{2+}$ goes to the right as written indicates that E^0 for the half-reaction $Cu^{2+} + 2e^- \rightleftharpoons Cu$

has a 49___________ value than E^0 for the half-reaction $Pb^{2+} + 2e^- \rightleftharpoons Pb$. The Nernst equation applied to the reaction $Cu^{2+} + Pb \rightleftharpoons Cu + Pb^{2+}$ ($E^0 = 0.463$ V) at 25°C is 50___________________. The branch of analytical chemistry in which the voltage developed by an electrode is used to measure the concentration of an ion in solution is called 51__________. The electrodes used for this purpose are called by the general name of 52________________________ and they must always be used with a 53____________________. The voltage developed at a glass membrane is used to measure 54_________________ or 55_________. Before measurement of an unknown pH, a pH meter and its electrode system must be calibrated using a known 56_______________. The destructive alteration of metal through interactions with its surroundings is called 57_______________. In the electrochemical cell typically involved with this phenomenon, the metal is oxidized at the 58________________.

Answers to Chapter Summary

1. breaks the water down into H_2 and O_2
2. oxidizing
3. reducing
4. electrochemistry
5. oxidation
6. oxidized
7. reduction
8. reduced
9. oxidation number
10. −2
11. +1
12. 0
13. 0
14. the charge on the ion
15. +1, +2, −1
16. Cu metal and Fe^{2+} ion in solution
17. oxidized
18. reduced

19. an electrochemical cell
20. electrodes
21 cathode
22. anode
23. allow for transfer of ions between half-cells
24. a graphite rod
25. a zinc cylinder
26. $2MnO_2(s) + 2NH_4^+(aq) + Zn(s) \rightarrow Zn^{2+}(aq) + 2MnO(OH)(s) + 2NH_3(aq)$
27. an electrolytic reaction
28. an electrolytic cell
29. chlorine gas
30. liquid sodium
31. Pb metal
32. PbO_2
33. nickel–cadmium storage battery
34. fuel cell
35. solar cell
36. donor layer
37. acceptor layer
38. electroplating
39. H_2
40. O_2
41. holes
42. a donor layer
43. standard electrode potential E^0
44. Nernst equation
45. standard hydrogen electrode (SHE)
46. 0.000 V

47. $2H^+ + 2e^- \rightleftharpoons H_2$

48. standard electrode potential E^0

49. higher

50. $E = 0.463 + \frac{0.0591}{2} \log \frac{[Cu^2]}{[Pb^{2+}]}$

51. potentiometry

52. ion-selective electrodes

53. reference electrode

54. H^+ ion concentration

55. pH

56. buffer solution

57. corrosion

58. anode

QUESTIONS AND PROBLEMS

1. Which element has such a "strong appetite" for electrons that its name is the basis for the term used to describe loss of electrons?
2. Give the oxidation number of each of the elements followed by an asterisk in the following species: Pb^*O_2, $N^*H_4{}^+$, $Cl^*{}_2$, CaS^*O_4, $Cd^*(OH)_2$.
3. For what purpose is a Downs cell used?
4. How is cementation used to purify water?
5. Until recently, small mercury batteries were often used in cameras, wristwatches, and similar applications. The chemical species involved in a mercury battery are Zn, ZnO, Hg, and HgO, where Zn is the chemical symbol for zinc and Hg is that of mercury. If one were to coat a moist paste of HgO onto Zn, a reaction would occur in which Hg and ZnO are produced. From this information, write the cathode, anode, and overall reactions involved in the discharge of a mercury battery. You may use the oxide ion, O^{2-}, in the half-reactions.
6. What are the major ingredients of the paste used to fill a common dry cell?
7. Pure water does not conduct electricity. However, by passing an electrical current through water, pure H_2 and O_2 may be prepared. What is done to the water to make this possible?

8. In this chapter, it was mentioned that Cl_2 gas could be prepared by electrolysis of a solution of NaCl in water. Elemental Na reacts with water to give NaOH and H_2 gas. Recall that the electrolysis of melted NaCl gives Na metal and Cl_2 gas. From this information, give the cathode, anode, and overall reactions for the electrolysis of a solution of NaCl in water.

9. What is the definition of a cathode? What is the definition of an anode?

10. Give the cathode, anode, and overall reactions when a nickel–cadmium battery is charged.

11. From a knowledge of automobile exhausts and coal-fired power plants, justify the statement that "a fuel-cell is a non-polluting source of energy."

12. What is meant by a positive hole in the acceptor layer of Si in a solar cell? What distinguishes the donor layer?

13. From the E^0 values given in Section 8.10, give the E^0 values of the following reactions:

(A) $2Fe^{3+} + Pb \rightleftharpoons 2Fe^{2+} + Pb^{2+}$

(B) $2Cl_2 + 2H_2O \rightleftharpoons 4Cl^- + O_2 + 4H^+$

(C) $2Ag^+ + Zn \rightleftharpoons 2Ag + Zn^{2+}$

(D) $2Fe^{3+} + Zn \rightleftharpoons 2Fe^{2+} + Zn^{2+}$

14. How can it be deduced that metallic zinc and metallic lead displace hydrogen from strong acid, whereas metallic copper and silver do not?

15. Given that $E^0 = 0.771$ V for $Fe^{3+} + e^- \rightleftharpoons Fe^{2+}$, and that $E^0 = 1.80$ V for $Ce^{4+} + e^- \rightleftharpoons Ce^{3+}$, calculate E^0 for the reaction $Ce^{4+} + Fe^{2+} \rightleftharpoons Ce^{3+} + Fe^{3+}$. Using the Nernst equation, calculate E for this reaction when $[Ce^{4+}] = 9.00 \times 10^{-3}$ mol/L, $[Ce^{3+}] = 1.25 \times 10^{-3}$ mol/L, $[Fe^{3+}] = 8.60 \times 10^{-4}$ mol/L, and $[Fe^{2+}] = 2.00 \times 10^{-2}$ mol/L.

16. In millivolts (mV), the Nernst equation (Equation 8.12.1) applied to the fluoride ion-selective electrode is $E = E_a - 59.1 \log[F^-]$. Suppose that the potential of a fluoride electrode versus a reference electrode is -88.3mV in a solution that is 1.37×10^{-4} mol/L in F^-. What is the potential of this same electrode system in a solution for which $[F^-] = 4.68 \times 10^{-2}$ mol/L?

17. Suppose that the potential of a fluoride electrode versus a reference electrode is -61.2 mV in a solution that is 5.00×10^{-5} mol/L in F^-. What is the concentration of fluoride in a solution in which the fluoride electrode registers a potential of -18.2 V?

18. A glass electrode has a potential of 33.3 mV versus a reference electrode in a medium for which $[H^+] = 1.13 \times 10^{-6}$ mol/L. What is the value of $[H^+]$ in a solution in which the potential of the electrode registers 197.0 mV?

19. A glass electrode has a potential of 127 mV versus a reference electrode in a pH 9.03 buffer. What is the pH of a solution in which the electrode registers a potential of 195 mV?

20. Consider an electrochemical cell in which a standard hydrogen electrode is connected by way of a salt bridge to a lead electrode in contact with a Pb^{2+} solution. How does the potential of the lead electrode change when the concentration of Pb^{2+} ion is increased? Explain.

9. ORGANIC CHEMISTRY

9.1. ORGANIC CHEMISTRY

Most carbon-containing compounds are **organic chemicals** and are addressed by the subject of **organic chemistry**. Organic chemistry is a vast, diverse discipline because of the enormous number of organic compounds that exist as a consequence of the versatile bonding capabilities of carbon. Such diversity is due to the ability of carbon atoms to bond to each other through single bonds (2 shared electrons), double bonds (4 shared electrons), and triple bonds (6 shared electrons) in a limitless variety of straight chains, branched chains, and rings (Figure 9.1).

Among organic chemicals are included the majority of important industrial compounds, synthetic polymers, agricultural chemicals, biological materials, and most substances that are of concern because of their toxicities and other hazards. Pollution of the water, air, and soil environments by organic chemicals is an area of significant concern.

Chemically, most organic compounds can be divided among hydrocarbons, oxygen-containing compounds, nitrogen-containing compounds, sulfur-containing compounds, organohalides, phosphorus-containing compounds, and combinations of these kinds of compounds. Each of these classes of organic compounds is discussed briefly here.

All organic compounds, of course, contain carbon. Virtually all also contain hydrogen and have at least one C—H bond. The simplest organic compounds, and the easiest to understand, are those that contain only hydrogen and carbon. These compounds are called **hydrocarbons** and are addressed first among the organic compounds discussed in this chapter. Hydrocarbons are used here to illustrate some of the most fundamental points of organic chemistry, including organic formulas, structures, and names.

Molecular Geometry in Organic Chemistry

The three-dimensional shape of a molecule, that is, its molecular geometry, is particularly important in organic chemistry. This is because the molecular geometry

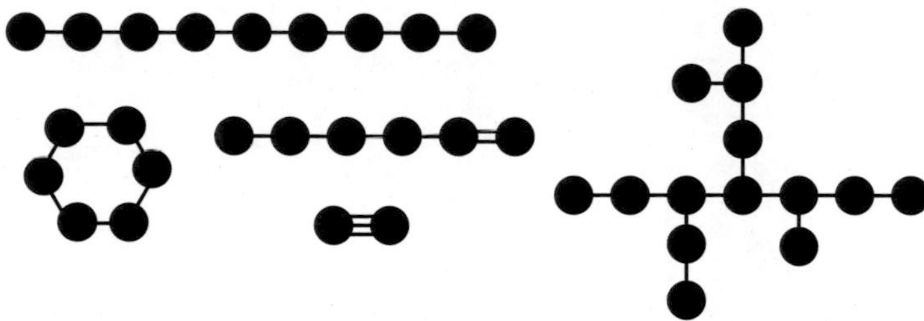

Figure 9.1. Carbon atoms in organic compounds bond with each other in straight chains, branched chains, and rings. In addition to single bonds, carbon atoms may be joined by double and even triple bonds. Because of this remarkable bonding diversity, there are literally millions of known organic compounds.

of an organic molecule determines, in part, its properties, particularly its interactions with biological systems and how it is metabolized by organisms. Shapes of molecules are represented in drawings by lines of normal uniform thickness for bonds in the plane of the paper; broken lines for bonds extending away from the viewer; and heavy lines for bonds extending toward the viewer. These conventions are shown by the example of dichloromethane, CH_2Cl_2, an important organochloride solvent and extractant illustrated in Figure 9.2.

9.2. HYDROCARBONS

As noted above, hydrocarbon compounds contain only carbon and hydrogen. The major types of hydrocarbons are alkanes, alkenes, alkynes, and aromatic (aryl) compounds. Examples of each are shown in Figure 9.3.

Alkanes

Alkanes, also called **paraffins** or **aliphatic hydrocarbons**, are hydrocarbons in which the C atoms are joined by single covalent bonds (sigma bonds) consisting of 2 shared electrons (see Section 4.6). Some examples of alkanes are shown in Figure 9.4.

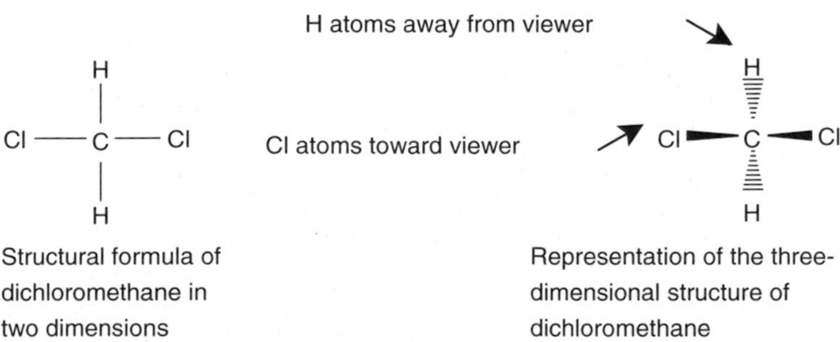

Figure 9.2. Structural formulas of dichloromethane, CH_2Cl_2; the formula on the right provides a three-dimensional representation.

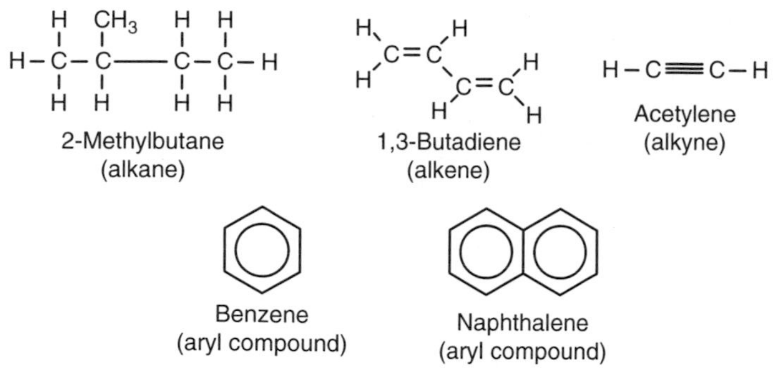

Figure 9.3. Examples of major types of hydrocarbons.

As with other organic compounds, the carbon atoms in alkanes may form straight chains, branched chains, or rings. These three kinds of alkanes are, respectively, **straight-chain alkanes, branched-chain alkanes**, and **cycloalkanes**. As shown in Figure 9.3, a typical branched-chain alkane is 2-methylbutane, a volatile, highly flammable liquid. It is a component of gasoline, which may explain why it is commonly found as an air pollutant in urban air. The general molecular formula for straight- and branched-chain alkanes is C_nH_{2n+2}, and that of cyclic alkanes is C_nH_{2n}. The four hydrocarbon molecules in Figure 9.4 contain 8 carbon atoms each. In one of the molecules, all of the carbon atoms are in a straight chain and in two they are in branched chains, whereas in a fourth, 6 of the carbon atoms are in a ring.

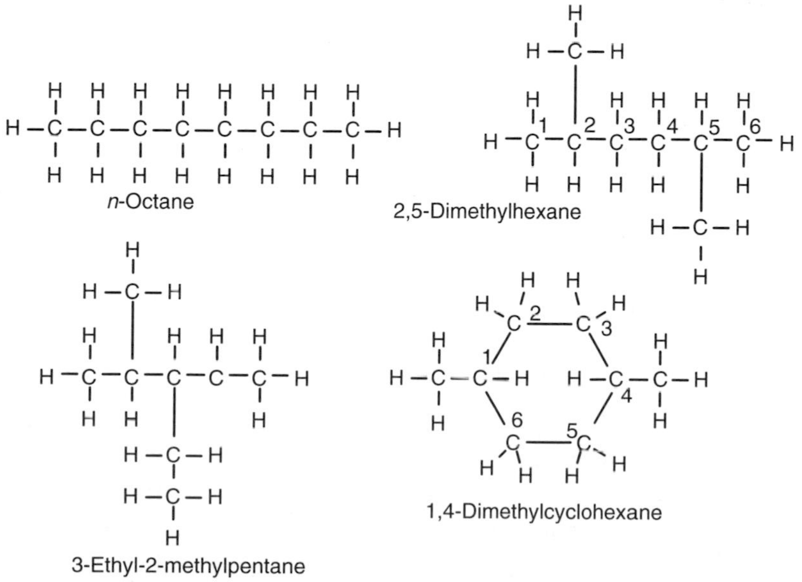

Figure 9.4. Structural formulas of four hydrocarbons, each containing 8 carbon atoms, that illustrate the structural diversity possible with organic compounds. Numbers used to denote locations of atoms for purposes of naming are shown on two of the compounds.

Formulas of Alkanes

Formulas of organic compounds present information at several different levels of sophistication. **Molecular formulas**, such as that of octane (C_8H_{18}), give the number of each kind of atom in a molecule of a compound. As shown in Figure 9.4, however, the molecular formula C_8H_{18} may apply to several alkanes that contain 8 C atoms per molecule, each one of which has unique chemical, physical, and toxicological properties. These different compounds are designated by **structural formulas** showing the order in which the atoms in a molecule are arranged. Compounds that have the same molecular but different structural formulas are called **structural isomers**. Of the compounds shown in Figure 9.4, *n*-octane, 2,5-dimethylhexane, and 3-ethyl-2-methylpentane are structural isomers, all having the formula C_8H_{18}, whereas 1,4-dimethylcyclohexane is not a structural isomer of the other three compounds, because its molecular formula is C_8H_{18}.

Alkanes and Alkyl Groups

Most organic compounds can be derived from alkanes. In addition, many important parts of organic molecules contain one or more alkane groups minus a hydrogen atom bonded as substituents onto the basic organic molecule. As a consequence, the names of many organic compounds are based upon those of alkanes, and it is useful to know the names of some of the more common alkanes and substituent groups derived from them, as shown in Table 9.1.

Names of Alkanes and Organic Nomenclature

Systematic names, from which the structures of organic molecules can be deduced, have been assigned to all known organic compounds. The more common organic compounds also have **common names** that have no structural implications. Although it is not possible to cover organic nomenclature in any detail in this chapter, the basic approach to nomenclature (naming) is presented, along with some pertinent examples. The simplest approach is to begin with names of alkane hydrocarbons.

Consider the alkanes shown in Figure 9.4. The fact that *n*-octane has no side chains is indicated by the prefix "*n*-," that it has 8 carbon atoms is indicated by the prefix "oct," and that it is an alkane is indicated by the suffix "ane." The names of compounds with branched chains or atoms other than H or C attached make use of numbers that stand for positions on the longest continuous chain of carbon atoms in the molecule. This convention is illustrated by the second compound in Figure 9.4. It gets the hexane part of the name from the fact that it is an alkane with 6 carbon atoms in its longest continuous chain ("hex" stands for 6). However, it has a methyl group (CH_3) attached to the second carbon atom of the chain and another to the fifth. Hence, the full systematic name of the compound is 2,5-dimethylhexane, where "di" indicates two methyl groups. In the case of 3-ethyl-2-methylpentane, the longest

Table 9.1 Some Alkanes and Substituent Groups Derived from Them

Alkane	Substituent Groups Derived from Alkane
Methane	Methyl group
Ethane	Ethyl group
Propane	*n*-Propyl group; Isopropyl group
n-Butane	*n*-Butyl group; *sec*-Butyl group; *tert*-Butyl group
n-Pentane	*n*-Pentyl group

```
      H                    H
      |                    |
  H — C — H            H — C —
      |                    |
      H  Methane           H  Methyl group

      H   H                H   H
      |   |                |   |
  H — C — C — H        H — C — C — *
      |   |                |   |
      H   H  Ethane        H   H  Ethyl group

      H   H   H            H   H   H               H   *   H
      |   |   |            |   |   |               |   |   |
  H — C — C — C — H    H — C — C — C — *       H — C — C — C — H
      |   |   |            |   |   |               |   |   |
      H   H   H  Propane   H   H   H               H   H   H
                          n-Propyl group          Isopropyl group

      H   H   H   H        H   H   H   H           H   *   H   H
      |   |   |   |        |   |   |   |           |   |   |   |
  H — C — C — C — C — H H — C — C — C — C — *  H — C — C — C — C — H
      |   |   |   |        |   |   |   |           |   |   |   |
      H   H   H   H        H   H   H   H           H   H   H   H
     n-Butane              n-Butyl group          sec-Butyl group

                                H   *   H
                                |   |   |
                            H — C — C — C — H
                                |   |   |
                                H   |   H
                                    |
                                H — C — H
                                    |
                                    H
                              tert-Butyl group

      H   H   H   H   H        H   H   H   H   H
      |   |   |   |   |        |   |   |   |   |
  H — C — C — C — C — C — H H — C — C — C — C — C — *
      |   |   |   |   |        |   |   |   |   |
      H   H   H   H   H        H   H   H   H   H
        n-Pentane                n-Pentyl group
```

Asterisk denotes point of attachment to molecule

continuous chain of carbon atoms contains 5 carbon atoms, denoted by *pent*ane, an ethyl group, C_2H_5, is attached to the third carbon atom, and a methyl group to the second carbon atom. The last compound shown in the figure has 6 carbon atoms in a ring, indicated by the prefix "cyclo," so it is a cyclo*hex*ane compound. Furthermore, the carbon in the ring to which one of the methyl groups is attached is designated by "1" and another methyl group is attached to the fourth carbon atom around the ring. Therefore, the full name of the compound is 1,4-dimethylcyclohexane.

Summary of Organic Nomenclature as Applied to Alkanes

Naming relatively simple alkanes is a straightforward process. The basic rules to be followed are the following:

1. The name of the compound is based upon the longest continuous chain of carbon atoms. (The structural formula may be drawn such that this chain is not immediately obvious.)
2. The carbon atoms in the longest continuous chain are numbered sequentially from one end. The end of the chain from which the numbering is started is chosen to give the lower numbers for substituent groups in the final name. For example, the compound

```
    H  H  H  H  H
    |  |  |  |  |
 H–C–C–C–C–C–H
    |  |  |  |  |
    H  H  H  |  H
            CH3
```

   could be named 4-methylpentane (numbering the 5-carbon chain from the left), but should be named 2-methylpentane (numbering the 5-carbon chain from the right).
3. All groups attached to the longest continuous chain are designated by the number of the carbon atoms to which they are attached and by the name of the substituent group ("2-methyl" in the example cited in Step 2 above).
4. A prefix is used to denote multiple substitutions by the same kind of group. This is illustrated by 2,2,3-trimethylpentane

```
      H3C CH3
    H  |   |  H  H
    |  |   |  |  |
 H–C–C–C–C–C–H
    |  |   |  |  |
    H  |   H  H  H
      CH3
```

   in which the prefix *tri* is used to show that *three* methyl groups are attached to the pentane chain.
5. The complete name is assigned such that it denotes the longest continuous chain of carbon atoms and the name and location on this chain of each substituent group.

Reactions of Alkanes

Alkanes contain only C—C and C—H bonds, both of which are relatively strong. For that reason, they have little tendency to undergo many kinds of reactions common to some other organic chemicals, such as acid–base reactions or low-temperature oxidation–reduction reactions. However, at elevated temperatures, alkanes readily undergo oxidation, more specifically combustion, with molecular oxygen in air, as shown by the following reaction of propane:

$$C_3H_8 + 5O_2 \rightarrow 3CO_2 + 4H_2O + \text{heat} \tag{9.2.1}$$

Common alkanes are highly flammable, and the more volatile lower-molecular-mass alkanes form explosive mixtures with air. Furthermore, combustion of alkanes in an oxygen-deficient atmosphere or in an automobile engine produces significant quantities of carbon monoxide, CO, a toxic gas that acts as an asphyxiant.

In addition to combustion, alkanes undergo **substitution reactions** in which one or more H atoms on an alkane are replaced by atoms of another element. The most common such reaction is the replacement of H by chlorine, to yield **organochlorine** compounds. For example, methane reacts with chlorine to give chloromethane. This reaction begins with the dissociation of molecular chlorine, usually initiated by ultraviolet electromagnetic radiation:

$$Cl_2 + \text{UV energy} \rightarrow Cl\bullet + Cl\bullet \tag{9.2.2}$$

The Cl• product is a **free radical** species in which the chlorine atom has only 7 outer shell electrons, as shown by the Lewis symbol,

$$:\ddot{\underset{\cdot\cdot}{Cl}}\cdot$$

instead of the favored octet of 8 outer-shell electrons. In gaining the octet required for chemical stability, the chlorine atom is very reactive. It abstracts a hydrogen atom from methane to yield HCl gas and another reactive species with an unpaired electron, $CH_3\bullet$, called the methyl radical:

$$Cl\bullet + CH_4 \rightarrow HCl + CH_3\bullet \tag{9.2.3}$$

The methyl radical attacks molecular chlorine,

$$CH_3\bullet + Cl_2 \rightarrow CH_3Cl + Cl\bullet \tag{9.2.4}$$

to give the chloromethane (CH_3Cl) product and regenerate Cl•, which can attack additional methane as in Reaction 9.2.3. The reactive Cl• and $CH_3\bullet$ species continue to cycle through the two preceding reactions.

The reaction sequence shown above illustrates three important aspects of chemistry that are very important in considering chemical processes in the atmosphere, such as those responsible for photochemical smog formation. The first of these is that a reaction can be initiated by a **photochemical process** in which a photon of "light" (electromagnetic radiation) energy produces a reactive species, in this case the Cl• atom. The second point illustrated is the high chemical reactivity of **free radical species** with unpaired electrons and incomplete octets of valence electrons. The third point illustrated is that of **chain reactions**, which can multiply manyfold the effects of a single reaction-initiating event, such as the photochemical dissociation of Cl_2.

Alkenes and Alkynes

Alkenes or olefins are hydrocarbons that have double bonds consisting of 4 shared electrons. The simplest and most widely manufactured alkene is ethylene (ethane),

$$\begin{array}{c} H \quad\quad H \\ \diagdown\quad\diagup \\ C{=}C \\ \diagup\quad\diagdown \\ H \quad\quad H \end{array}$$

which is used for the production of polyethylene polymer. Another example of an important alkene is 1,3-butadiene (Figure 9.3), which is widely used in the manufacture of polymers, particularly synthetic rubber. The lighter alkenes, including ethylene and 1,3-butadiene, are highly flammable. Like other gaseous hydrocarbons, they form explosive mixtures with air.

Acetylene (Figure 9.3) is an **alkyne**, a class of hydrocarbons characterized by carbon–carbon triple bonds involving 6 shared electrons. Acetylene is a highly flammable gas that forms dangerously explosive mixtures with air. It is used in large quantities as a chemical raw material. Acetylene is the fuel in oxyacetylene torches used for cutting steel and for various kinds of welding applications.

Addition Reactions

The double and triple bonds in alkenes and alkynes have "extra" electrons capable of forming additional bonds. So the carbon atoms attached to these bonds can add atoms without losing any atoms already bonded to them; the multiple bonds are said to be **unsaturated**. Therefore, alkenes and alkynes both undergo **addition reactions** in which pairs of atoms are added across unsaturated bonds, as shown in the reaction of ethylene with hydrogen to give ethane:

$$\begin{array}{c} H \quad\quad H \\ \diagdown\quad\diagup \\ C{=}C \\ \diagup\quad\diagdown \\ H \quad\quad H \end{array} + H{-}H \rightarrow \begin{array}{c} H\;\;H \\ |\;\;\;| \\ H{-}C{-}C{-}H \\ |\;\;\;| \\ H\;\;H \end{array} \tag{9.2.5}$$

This is an example of a **hydrogenation reaction**, a very common reaction in organic synthesis, food processing (manufacture of hydrogenated oils), and petroleum refining. Another example of an addition reaction is that of HCl gas with acetylene to give vinyl chloride:

$$H{-}C{\equiv}C{-}H + H{-}Cl \rightarrow \begin{array}{c} H \quad\quad Cl \\ \diagdown\quad\diagup \\ C{=}C \\ \diagup\quad\diagdown \\ H \quad\quad H \end{array} \tag{9.2.6}$$

This kind of reaction, which is not possible with alkanes, adds to the chemical and metabolic versatility of compounds containing unsaturated bonds, and is a factor contributing to their generally higher toxicities. It makes unsaturated compounds much more reactive, more hazardous to handle in industrial processes, and more active in atmospheric chemical processes such as smog formation.

Alkenes and *Cis–Trans* Isomerism

As shown by the two simple compounds in Figure 9.5, the two carbon atoms connected by a double bond in alkenes cannot rotate relative to each other. For this reason, another kind of isomerism, known as ***cis–trans*** isomerism, is possible for alkenes. *Cis–trans* isomers have different parts of the molecule oriented differently in space, although these parts occur in the same order. Both alkenes illustrated in Figure 9.5 have the molecular formula C_4H_8. In *cis*-2-butene, the two CH_3 (methyl) groups attached to the C=C carbon atoms are on the same side of the double bond, whereas in *trans*-2-butene, they are on opposite sides.

Condensed Structural Formulas

To save space, structural formulas are conveniently abbreviated as **condensed structural formulas** such as $CH_3CH(CH_3)CH(C_2H_5)CH_2CH_3$ for 3-ethyl-2-methylpentane, where the CH_3 (methyl) and C_2H_5 (ethyl) groups are placed in parentheses to show that they are branches attached to the longest continuous chain of carbon atoms, which contains 5 carbon atoms. It is understood that each of the methyl and ethyl groups is attached to the carbon immediately preceding it in the condensed structural formula (methyl attached to the second carbon atom, ethyl to the third).

As illustrated by the examples in Figure 9.6, the structural formulas of organic molecules can be represented in a very compact form by lines and by figures such as hexagons. The ends and intersections of straight line segments in these formulas indicate the locations of carbon atoms. Carbon atoms at the terminal ends of lines are understood to have *3* H atoms attached, C atoms at the intersections of two lines are understood to have 2 H atoms attached to each, *1* H atom is attached to a carbon represented by the intersection of three lines, and *no* hydrogen atoms are bonded to C atoms where four lines intersect. Other atoms or groups of atoms, such as the Cl

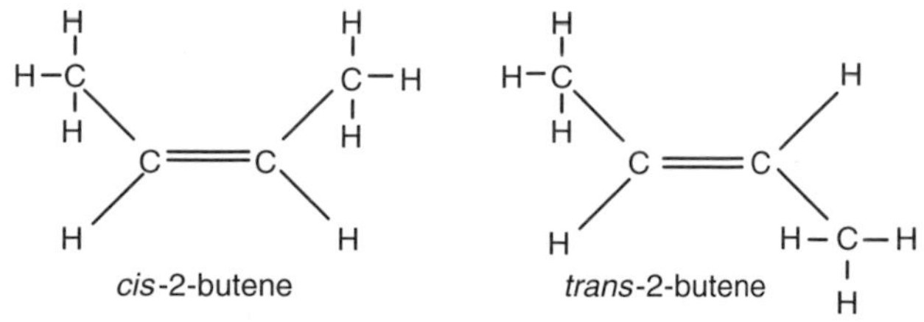

cis-2-butene

trans-2-butene

Figure 9.5. The *cis* and *trans* isomers of the alkene 2-butene.

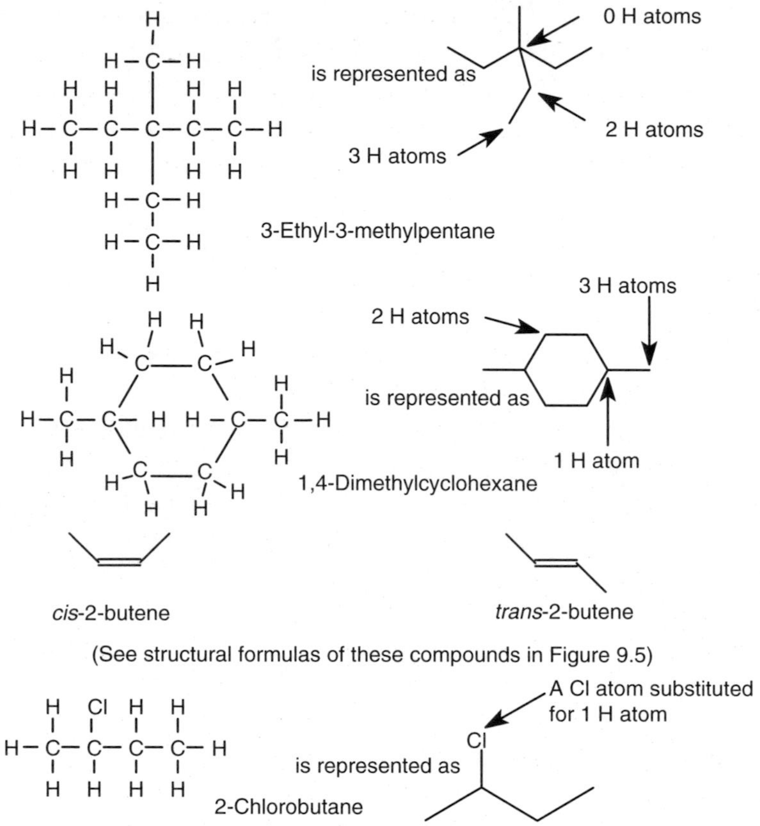

Figure 9.6. Representation of structural formulas with lines. A carbon atom is understood to be at each corner and at the end of each line. The numbers of hydrogen atoms attached to carbons at several specific locations are shown with arrows.

atom or OH group, that are substituted for H atoms are shown by their symbols attached to a C atom with a line.

Aromatic Hydrocarbons

Benzene (Figure 9.7) is the simplest of a large class of **aromatic** or **aryl** hydrocarbons, also known as **arenes**. Many important organic compounds are derived

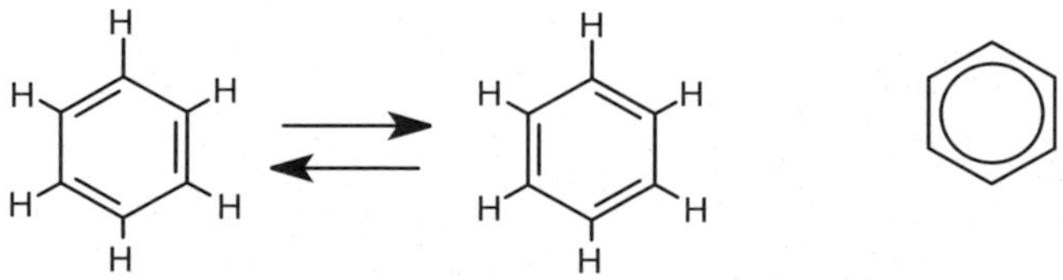

Figure 9.7. Representation of the aromatic benzene molecule with two resonance structures (left) and, more accurately, as a hexagon with a circle in it (right). Unless shown by symbols of other atoms, it is understood that a C atom is at each corner and that 1 H atom is bonded to each C atom.

from aromatic hydrocarbons, with substituent groups containing atoms of elements other than hydrogen and carbon, and are called **aromatic compounds** or **aryl compounds**. Most aromatic compounds discussed in this book contain 6-carbon-atom benzene rings, as shown for benzene, C_6H_6, in Figure 9.7. Aromatic compounds have ring structures and are held together in part by particularly stable bonds that contain delocalized clouds of so-called π (Greek letter pi) electrons.

In an oversimplified sense, the structure of benzene can be visualized as resonating between the two equivalent structures shown on the left in Figure 9.7 by the shifting of electrons in chemical bonds to form a hybrid structure. This structure can be shown more simply and accurately by a hexagon with a circle in it.

Aromatic compounds have special characteristics of **aromaticity**, which include a low hydrogen: carbon atomic ratio; C—C bonds that are quite strong and of intermediate length between such bonds in alkanes and those in alkenes; tendency to undergo substitution reactions rather than the addition reactions characteristic of alkenes; and delocalization of π electrons over several carbon atoms. The last phenomenon adds substantial stability to aromatic compounds, and is known as **resonance stabilization**.

Benzene, toluene, and other lighter aromatic compounds were first isolated from hydrocarbon liquids produced as a byproduct of coal coking, a process in which the coal is heated in the absence of air to produce a carbonaceous residue used in steelmaking. Now recovered during the refining of crude oil, these aromatic liquid compounds have found widespread use as solvents, in chemical synthesis, and as a fuel. Many toxic substances, environmental pollutants, and hazardous waste compounds, such as benzene, naphthalene, and chlorinated phenols, are aromatic compounds (Figure 9.8). As shown in Figure 9.8, some arenes, such as naphthalene and the polycyclic aromatic compound benzo[*a*]pyrene, contain fused rings.

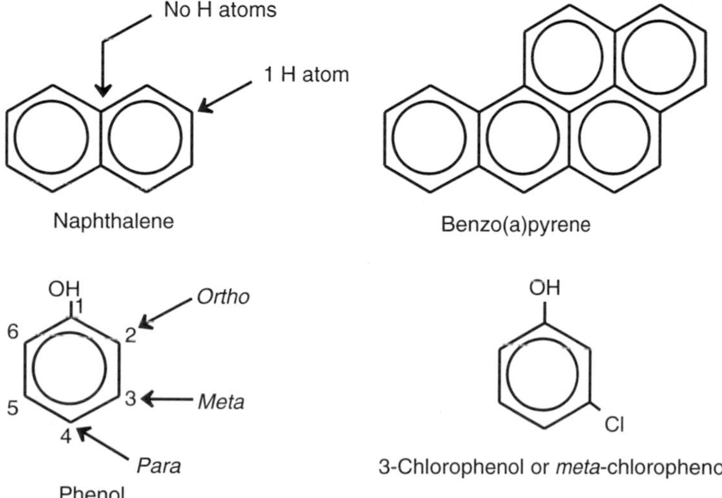

Figure 9.8. Aromatic compounds containing fused rings (top) and showing the numbering of carbon atoms for purposes of nomenclature.

Benzene and Naphthalene

Benzene is a volatile, colorless, highly flammable liquid used to manufacture phenolic and polyester resins, polystyrene plastics, alkylbenzene surfactants, chlorobenzenes, insecticides, and dyes. It is hazardous both for its ignitability and for its toxicity (exposure to benzene causes blood abnormalities that may develop into leukemia). Naphthalene is the simplest member of a large number of polycyclic aromatic hydrocarbons having two or more fused rings. It is a volatile white crystalline solid with a characteristic odor and has been used to make mothballs. The most important of the many chemical derivatives made from naphthalene is phthalic anhydride, from which phthalate ester plasticizers are synthesized.

Polycyclic Aromatic Hydrocarbons

Benzo[*a*]pyrene (Figure 9.8) is one of the most widely studied of the polycyclic aromatic hydrocarbons (PAHs), which are characterized by condensed ring systems ("chicken wire" structures). These compounds are formed by the incomplete combustion of other hydrocarbons, a process that consumes hydrogen in preference to carbon. The carbon residue is left in the thermodynamically favored condensed aromatic ring system of the PAH compounds.

Because there are so many partial combustion and pyrolysis processes that favor production of PAHs, these compounds are encountered abundantly in the atmosphere, soil, and elsewhere in the environment from sources that include engine exhausts, wood stove smoke, cigarette smoke, and charbroiled food. Coal tars and petroleum residues such as road and roofing asphalt have high levels of PAHs. Some PAH compounds, including benzo[*a*]pyrene, are of toxicological concern because they are precursors to cancer-causing metabolites.

9.3. ORGANIC FUNCTIONAL GROUPS AND CLASSES OF ORGANIC COMPOUNDS

The discussion of organic chemistry so far in this chapter has emphasized hydrocarbon compounds, that is, those containing only hydrogen and carbon. It has been shown that hydrocarbons may exist as alkanes, alkenes, alkynes, and arenes, depending upon the kinds of bonds between carbon atoms. The presence of elements other than hydrogen and carbon in organic molecules greatly increases the diversity of their chemical properties. As shown in Table 9.2, **functional groups** consist of specific bonding configurations of atoms in organic molecules. Most functional groups contain at least one element other than carbon or hydrogen, although two carbon atoms joined by a double bond (alkenes) or triple bond (alkynes) are likewise considered to be functional groups. Table 9.2 shows some of the major functional groups that determine the nature of organic compounds.

Table 9.2. Examples of Some Important Functional Groups

Type of Functional Group	Example Compound	Structural Formula of Group[a]
Alkene (olefin)	Propene (propylene)	$H_2C{=}CH{-}CH_3$
Alkyne	Acetylene	$H{-}C{\equiv}C{-}H$
Alcohol (−OH attached to alkyl group)	2-Propanol	$H_3C{-}CH(OH){-}CH_3$
Phenol (−OH attached to aryl group)	Phenol	$C_6H_5{-}OH$
Ketone (when $-\overset{O}{\overset{\|}{C}}-H$ group is on end carbon, compound is an aldehyde)	Acetone	$H_3C{-}C({=}O){-}CH_3$
Amine	Methylamine	$H_3C{-}NH_2$
Nitro compounds	Nitromethane	$H_3C{-}NO_2$
Sulfonic acids	Benzenesulfonic acid	$C_6H_5{-}S({=}O)_2{-}OH$
Organohalides	1,2-Dichloroethane	$Cl{-}CH_2{-}CH_2{-}Cl$

[a] The functional group is outlined by dashed lines.

Organo-Oxygen Compounds

The most common types of compounds with oxygen-containing functional groups are epoxides, alcohols, phenols, ethers, aldehydes, ketones, and carboxylic acids. The functional groups characteristic of these compounds are illustrated by the examples of oxygen-containing compounds shown in Figure 9.9.

Ethylene oxide is a moderately to highly toxic, sweet-smelling, colorless, flammable, explosive gas used as a chemical intermediate, sterilant, and fumigant. It is a

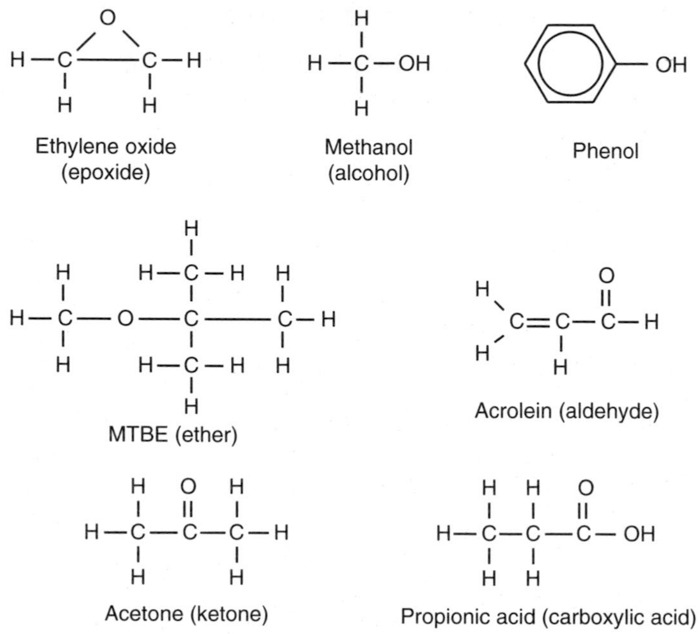

Ethylene oxide (epoxide)

Methanol (alcohol)

Phenol

MTBE (ether)

Acrolein (aldehyde)

Acetone (ketone)

Propionic acid (carboxylic acid)

Figure 9.9. Examples of oxygen-containing organic compounds that may be significant as wastes, toxic substances, or environmental pollutants.

mutagen and a carcinogen to experimental animals. It is classified as hazardous for both its toxicity and ignitability. **Methanol** is a clear, volatile, flammable liquid alcohol used for chemical synthesis, as a solvent, and as a fuel. It is used as a gasoline additive to reduce emissions of carbon monoxide and other air pollutants. Ingestion of methanol can be fatal, and blindness can result from sublethal doses. **Phenol** is a dangerously toxic aromatic hydroxyl compound widely used for chemical synthesis and polymer manufacture. **Methyl tertiarybutyl ether**, MTBE, is an ether that became the octane booster of choice to replace tetraethyllead in gasoline. However, its use has now been phased out, largely because of concern over its water pollution potential.

Acrolein is an alkenic aldehyde and a volatile, flammable, highly reactive chemical. It forms explosive peroxides upon prolonged contact with O_2. An extreme lachrimator and strong irritant, acrolein is quite toxic by all routes of exposure. **Acetone** is the lightest of the ketones. Like all ketones, acetone has a carbonyl (C=O) group that is bonded to *two* carbon atoms (i.e., it is somewhere in the middle of a carbon atom chain). Acetone is a good solvent and is chemically less reactive than the aldehydes, which all have the functional group

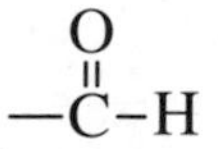

$$-\overset{\overset{\displaystyle O}{\|}}{C}-H \quad \text{Aldehyde group}$$

in which binding of the C=O to H makes the molecule significantly more reactive. **Propionic acid** is a typical organic carboxylic acid. The $-CO_2H$ group of carboxylic acids may be viewed as the most oxidized functional group (other than peroxides) on an oxygenated organic compound, and carboxylic acids can be synthesized by oxidizing alcohols or aldehydes that have an —OH group or C=O group on an end carbon atom.

Organonitrogen Compounds

Figure 9.10 shows examples of three classes of the many kinds of compounds that contain nitrogen (amines, nitrosamines, and nitro compounds). Nitrogen occurs in many functional groups in organic compounds, some of which contain nitrogen in ring structures, or along with oxygen.

Methylamine is a colorless, highly flammable gas with a strong odor. It is a severe irritant, affecting eyes, skin, and mucous membranes. Methylamine is the simplest of the **amine** compounds, which have the general formula

$$R-N\begin{matrix} R' \\ R'' \end{matrix}$$

where the Rs are hydrogen or hydrocarbon groups, at least one of which is the latter.

Dimethylnitrosamine is one of the *N*-nitroso compounds, all characterized by the N—N=O functional group. It was once widely used as an industrial solvent, but caused liver damage and jaundice in exposed workers. Subsequently, numerous other *N*-nitroso compounds, many produced as byproducts of industrial operations and food and alcoholic beverage processing, were found to be carcinogenic.

Solid **2,4,6-trinitrotoluene (TNT)** has been widely used as a military explosive. TNT is moderately to very toxic and has caused toxic hepatitis or aplastic anemia in exposed individuals, a few of whom have died from its toxic effects. It belongs to the

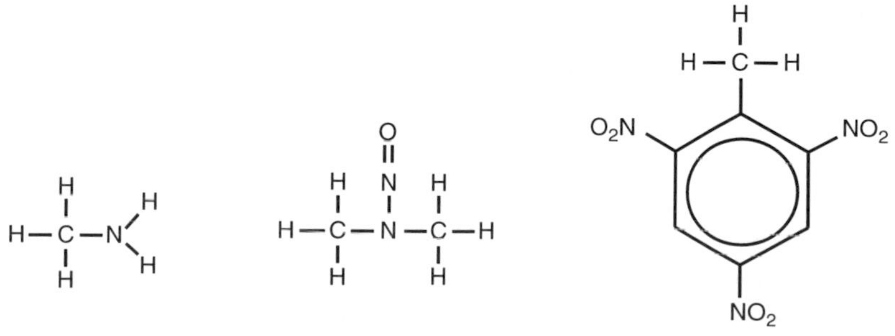

Methylamine   Dimethylnitrosamine (N-nitrosodimethylamine)   2,4,6-trinitrotoluene (TNT)

Figure 9.10. Examples of organonitrogen compounds that may be significant as wastes, toxic substances, or environmental pollutants.

general class of nitro compounds characterized by the presence of $-NO_2$ groups bonded to a hydrocarbon structure.

Some organonitrogen compounds are chelating agents that bind strongly to metal ions and play a role in the solubilization and transport of heavy metal wastes. Prominent among these are salts of the aminocarboxylic acids, which, in their acid form, have $-CH_2CO_2H$ groups bonded to nitrogen atoms. An important example of such a compound is the monohydrate of **trisodium nitrilotriacetate (NTA)**:

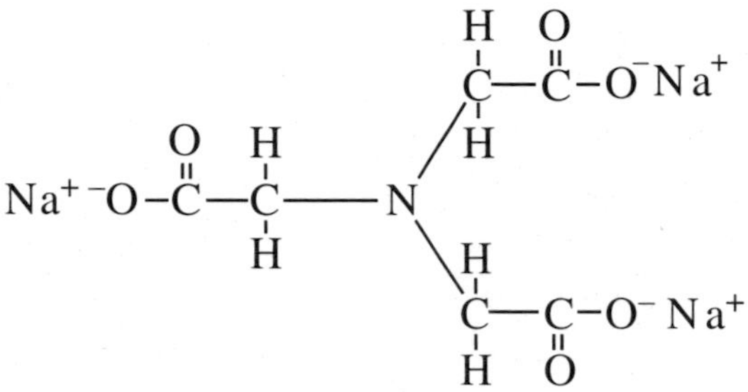

```
                              H  O
                              |  ||
                              C--C-O⁻Na⁺
              O   H          /|
              ||  |         / H
  Na⁺ ⁻O-C--C-------N
                  |          \
                  H           \H
                               |
                               C--C-O⁻Na⁺
                               |  ||
                               H  O
```

This compound can be used as a substitute for detergent phosphates to bind to calcium ion and make the detergent solution basic. NTA is used in metal plating formulations. It is highly water-soluble and quickly eliminated with urine when ingested. It has a low acute toxicity, and no chronic effects have been shown for plausible doses. However, concern does exist over its interaction with heavy metals in waste-treatment processes and in the environment.

Organohalide Compounds

Organohalides (Figure 9.11) exhibit a wide range of physical and chemical properties. These compounds consist of halogen-substituted hydrocarbon molecules, each of which contains at least one atom of F, Cl, Br, or I. They may be saturated (**alkyl halides**), unsaturated (**alkenyl halides**), or aromatic (**aromatic halides**). The most widely manufactured organohalide compounds are chlorinated hydrocarbons, many of which are regarded as environmental pollutants or as hazardous wastes.

Alkyl Halides

Substitution of halogen atoms for one or more hydrogen atoms on alkanes gives **alkyl halides**, example structural formulas of which are given in Figure 9.11. Most of the commercially important alkyl halides are derivatives of alkanes of low molecular mass. A brief discussion of the uses of the compounds listed in Figure 9.11 is given here to provide an idea of the versatility of the alkyl halides.

Dichloromethane is a volatile liquid with excellent solvent properties for nonpolar organic solutes. It has been used as a solvent for the decaffeination of coffee, in paint strippers, as a blowing agent in urethane polymer manufacture, and to depress vapor pressure in aerosol formulations. Once commonly sold as a solvent and stain remover, highly toxic **carbon tetrachloride** is now largely restricted to

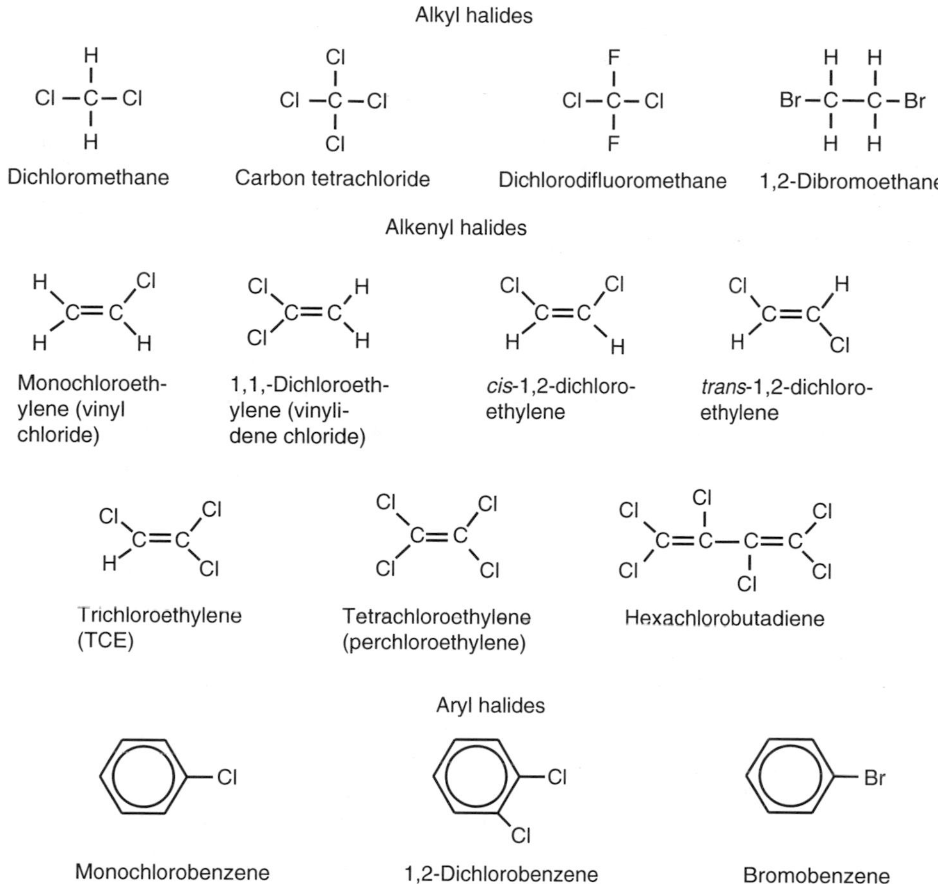

Figure 9.11. Examples of organohalide compounds.

uses as a chemical intermediate under controlled conditions, primarily to manufacture chlorofluorocarbon refrigerant fluid compounds, which are also discussed in this section. Insecticidal **1,2-dibromoethane** has been consumed in large quantities as a lead scavenger in leaded gasoline and to fumigate soil, grain, and fruit (fumigation with this compound has been discontinued because of toxicological concerns). An effective solvent for resins, gums, and waxes, it serves as a chemical intermediate in the synthesis of some pharmaceutical compounds and dyes.

Alkenyl Halides

Viewed as hydrocarbon-substituted derivatives of alkenes, the **alkenyl** or **olefinic organohalides** contain at least one halogen atom and at least one carbon–carbon double bond. The most significant of these, which are materials produced in large quantities for chemical manufacture and other purposes, are the lighter chlorinated compounds, such as those illustrated in Figure 9.11.

Vinyl chloride is consumed in large quantities as a raw material to manufacture pipe, hose, wrapping, and other products fabricated from polyvinylchloride plastic. This highly flammable, volatile, sweet-smelling gas is a known human carcinogen.

As shown in Figure 9.11, there are three possible dichloroethylene compounds, all clear, colorless liquids. Vinylidene chloride forms a copolymer with vinyl chloride used in some kinds of coating materials. The geometrically isomeric 1,2-dichloroethylenes are used as organic synthesis intermediates and as solvents. **Trichloroethylene** is a clear, colorless, nonflammable, volatile liquid. It is an excellent degreasing and drycleaning solvent and has been used as a household solvent and for food extraction (e.g., in decaffeination of coffee). Colorless, nonflammable liquid **tetrachloroethylene** has properties and uses similar to those of trichloroethylene. **Hexachlorobutadiene**, a colorless liquid with an odor somewhat like that of turpentine, is used as a solvent for higher hydrocarbons and elastomers, as a hydraulic fluid, in transformers, and for heat transfer.

Aromatic Halides

Aromatic halide derivatives of benzene and toluene have many uses. They are common intermediates and raw materials in chemical synthesis. In addition, they are used as pesticides and raw materials for pesticides manufacture, as solvents, and for a diverse variety of other applications. These widespread uses over many decades have resulted in substantial human exposure and environmental contamination. Three example aromatic halides are shown in Figure 9.11. Monochlorobenzene is a flammable liquid boiling at 132°C. It is used as a solvent, heat transfer fluid, and synthetic reagent. Used as a solvent, 1,2-dichlorobenzene is employed for degreasing hides and wool. It also serves as a synthetic reagent for dye manufacture. Bromobenzene is a liquid boiling at 156°C that is used as a solvent, motor oil additive, and intermediate for organic synthesis.

Halogenated Naphthalenes and Biphenyls

Two major classes of halogenated aromatic compounds containing two benzene rings are made by the chlorination of naphthalene and biphenyl and have been sold as mixtures with varying degrees of chlorine content. Examples of chlorinated naphthalenes and polychlorinated biphenyls (PCBs discussed later) are shown in Figure 9.12. The less highly chlorinated of these compounds are liquids, and those with higher chlorine contents are solids. Because of their physical and chemical stabilities and other qualities, these compounds have had many uses, including heat transfer fluids, hydraulic fluids, and dielectrics. Polybrominated biphenyls (PBBs) have served as flame retardants. However, because chlorinated naphthalenes, PCBs, and PBBs are extremely persistent, their uses have been severely curtailed.

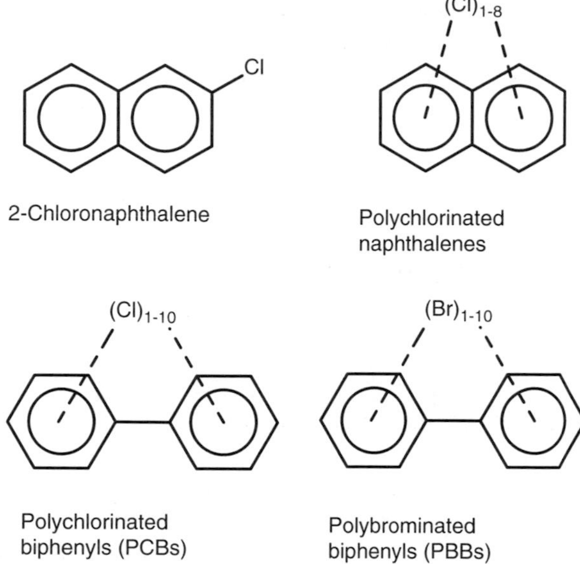

Figure 9.12. Halogenated naphthalenes and biphenyls.

Chlorofluorocarbons, Halons, and Hydrogen-Containing Chlorofluorocarbons

Chlorofluorocarbons (CFCs) are volatile 1- and 2-carbon compounds that contain Cl and F bonded to carbon. Though extremely stable and nontoxic, these compounds have been phased out of use because of their environmental effects. They were once widely used in the fabrication of flexible and rigid foams, and as fluids for refrigeration and air conditioning, but are now banned because of their potential to cause harm to the stratospheric ozone layer. The most widely manufactured of these compounds in the past were CCl_3F (CFC-11), CCl_2F_2 (CFC-12), $C_2Cl_3F_3$ (CFC-113), $C_2Cl_2F_4$ (CFC-114), and C_2ClF_5 (CFC-115). **Halons** are related compounds that contain bromine and are used in fire extinguisher systems. The most commonly produced commercial halons were $CBrClF_2$ (Halon-1211), $CBrF_3$ (Halon-1301), and $C_2Br_2F_4$ (Halon-2402), where the sequence of numbers denotes the number of carbon, fluorine, chlorine, and bromine atoms, respectively, per molecule. Halons have also been implicated as ozone-destroying gases in the stratosphere and are being phased out. However, finding suitable replacements has been difficult, particularly in the crucial area of aircraft fire extinguishers.

Hydrohalocarbons are hydrogen-containing chlorofluorocarbons (HCFCs) and hydrogen-containing fluorocarbons (HFCs) that are now produced as substitutes for chlorofluorocarbons. These compounds include CH_2FCF_3 (HFC-134a, a substitute for CFC-12 in automobile air conditioners and refrigeration equipment), $CHCl_2CF_3$ (HCFC-123, a substitute for CFC-11 in plastic foam-blowing), CH_3CCl_2F (HCFC-141b, a substitute for CFC-11 in plastic foam-blowing), and $CHClF_2$ (HCFC-22, used in air conditioners and the manufacture of plastic foam food containers). Because each molecule of these compounds has at least one H—C bond, which is much more readily broken than C—Cl or C—F bonds, the HCFCs do not persist in the atmosphere and pose essentially no threat to the stratospheric ozone layer.

Chlorinated Phenols

The chlorinated phenols, particularly **pentachlorophenol**

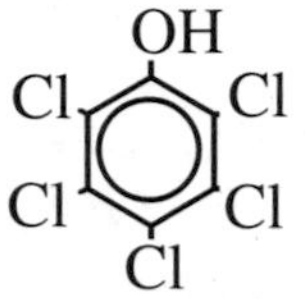

Pentachlorophenol

and the trichlorophenol isomers, are significant hazardous wastes. These compounds are biocides that are used to treat wood to prevent rot by fungi and to prevent termite infestation. They are toxic, causing liver malfunction and dermatitis. However, contaminant polychlorinated dibenzodioxins ("dioxin") may be responsible for some of the observed effects. Pentachlorophenol and other aromatic halides and aromatic hydrocarbons used as wood preservatives are encountered at many hazardous-waste sites in wastewaters and sludges.

Organosulfur Compounds

The chemistry of sulfur is similar to but perhaps more diverse than that of oxygen. Whereas, with the exception of peroxides, most chemically combined organic oxygen is in the −2 oxidation state, sulfur occurs in the −2, +4, and +6 oxidation states. Many organosulfur compounds are noted for their "rotten egg" or garlic odors. A number of example organosulfur compounds are shown in Figure 9.13.

Substitution of alkyl or aromatic hydrocarbon groups such as phenyl and methyl for H on hydrogen sulfide, H_2S, leads to a number of different organosulfur **thiols** (mercaptans, R—SH) and **sulfides**, also called thioethers (R—S—R′). Structural formulas of examples of these compounds are shown in Figure 9.13.

Methanethiol and other lighter alkyl thiols are fairly common air pollutants that have "ultragarlic" odors; both 1- and 2-butanethiol are associated with skunk odor. Gaseous methanethiol is used as an odorant leak-detecting additive for natural gas, propane, and butane; it is also employed as an intermediate in pesticide synthesis. A toxic, irritating volatile liquid with a strong garlic odor, 2-propene-1-thiol (allyl mercaptan) is a typical alkenyl mercaptan. Benzenethiol (phenyl mercaptan) is the simplest of the aromatic thiols. It is a toxic liquid with a severely "repulsive" odor.

Alkyl sulfides or thioethers contain the C—S—C functional group. The lightest of these compounds is dimethyl sulfide, a volatile liquid (bp 38°C) that is moderately toxic by ingestion. It is now known to be a major source of gaseous sulfur entering

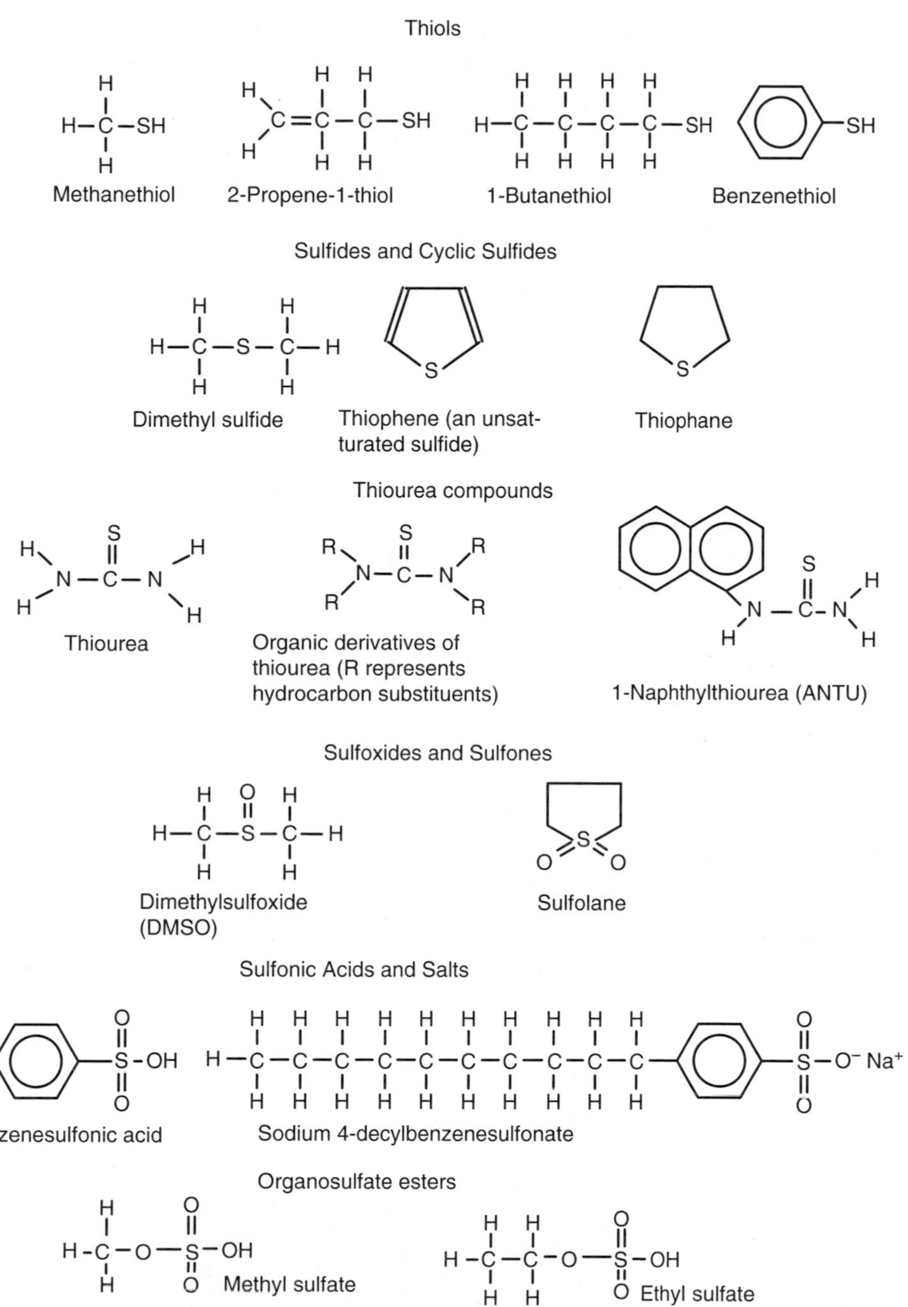

Figure 9.13. Examples of organosulfur compounds.

the atmosphere over the oceans due to its production by marine organisms. Cyclic sulfides contain the C—S—C group in a ring structure. The most common of these compounds is thiophene, a heat-stable liquid (bp 84°C) with a solvent action much like that of benzene, that is used to make pharmaceuticals, dyes, and resins. Its saturated analog is tetrahydrothiophene, or thiophane.

Many important organosulfur compounds also contain nitrogen. One such compound is **thiourea**, the sulfur analog of urea. Its structural formula is shown in Figure 9.13. Thiourea and **phenylthiourea** have been used as rodenticides. Commonly called ANTU, **1-naphthylthiourea** is an excellent rodenticide that is virtually tasteless and has a very high rodent: human toxicity ratio.

Sulfoxides and **sulfones** (Figure 9.13) contain both sulfur and oxygen. **Dimethylsulfoxide (DMSO)** is a liquid used to remove paint and varnish, as a hydraulic fluid, mixed with water as an antifreeze solution, and in pharmaceutical applications as an anti-inflammatory and bacteriostatic agent. **Sulfolane** dissolves both organic and inorganic solutes and is widely used in BTX processing to selectively extract benzene, toluene, and xylene from aliphatic hydrocarbons; as the solvent in the Sulfinol process by which thiols and acidic compounds are removed from natural gas; as a solvent for polymerization reactions; and as a polymer plasticizer.

Sulfonic acids and sulfonate salts contain the $-SO_3H$ and $-SO_3^-$ groups, respectively, attached to a hydrocarbon moiety. The structural formulas of benzenesulfonic acid and sodium 4-decylbenzenesulfonate, a biodegradable detergent surfactant, are shown in Figure 9.13. The common sulfonic acids are water-soluble strong acids that lose virtually all ionizable H^+ in aqueous solution. They are used commercially to hydrolyze fat and oil esters to fatty acids and glycerol.

Replacement of one hydrogen on sulfuric acid, H_2SO_4, with a hydrocarbon group yields an acid ester, and replacement of both yields an ester, examples of which are shown in Figure 9.13. Sulfuric acid esters are used as alkylating agents, which act to attach alkyl groups (such as methyl) to organic molecules in the manufacture of agricultural chemicals, dyes, and drugs. **Methylsulfuric acid** and **ethylsulfuric acid** are oily, water-soluble liquids that are strong irritants to skin, eyes, and mucous tissue.

Organophosphorus Compounds

The first two examples in Figure 9.14 illustrate that the structural formulas of alkyl and aromatic phosphine compounds can be derived by substituting organic groups for the H atoms in phosphine (PH_3), the toxic hydride of phosphorus. **Methylphosphine** is a colorless, reactive gas. Crystalline, solid **triphenylphosphine** has a low reactivity and moderate toxicity when inhaled or ingested.

As shown by the reaction

$$4C_3H_9P + 26O_2 \rightarrow 12CO_2 + 18H_2O + P_4O_{10} \qquad (9.3.1)$$

combustion of aromatic and alkyl phosphines produces tetraphosphorus decoxide, P_4O_{10}, a corrosive, irritant, toxic substance that reacts with moisture in the air to produce droplets of corrosive orthophosphoric acid, H_3PO_4.

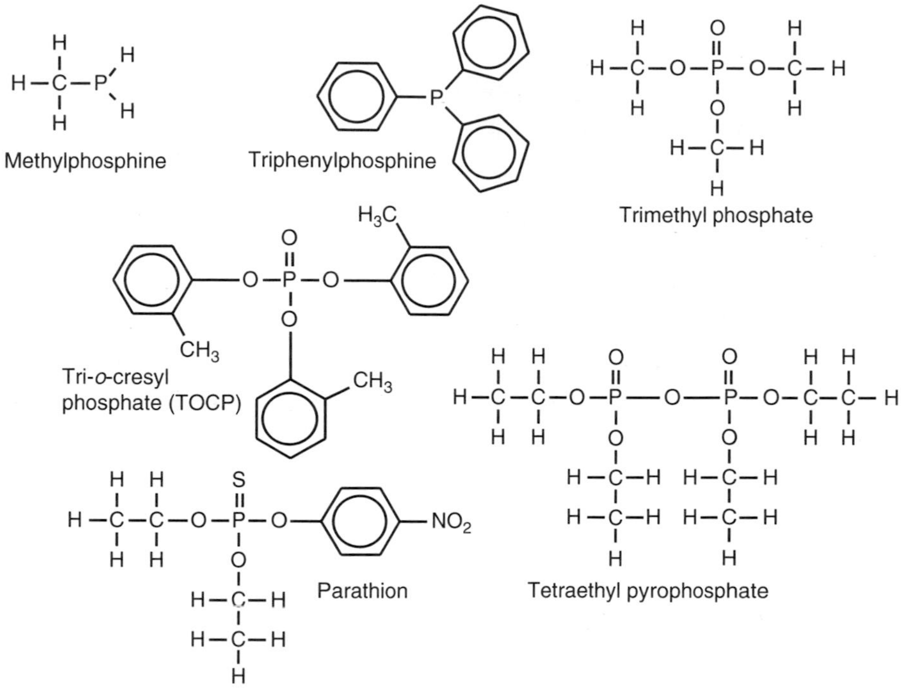

Methylphosphine

Triphenylphosphine

Trimethyl phosphate

Tri-*o*-cresyl phosphate (TOCP)

Parathion

Tetraethyl pyrophosphate

Figure 9.14. Some representative organophosphorus compounds.

The structural formulas of three common esters of orthophosphoric acid and an ester of pyrophosphoric acid, $H_4P_2O_6$, are shown in Figure 9.14. Although **trimethyl phosphate** is considered to be only moderately toxic, **tri-*o*-cresyl phosphate, TOCP,** has a notorious record of poisonings. **Tetraethyl pyrophosphate, TEPP,** was developed in Germany during World War II as a substitute for insecticidal nicotine. Although it is a very effective insecticide, its use in that application was of very short duration because it kills almost everything else, too.

Phosphorothionate Esters

Parathion, shown in Figure 9.14, is an example of a **phosphorothionate** ester. These compounds are used as insecticidal acetylcholinesterase inhibitors. They contain the P=S (thiono) group, which increases their insect:mammal toxicity ratios. Since the first organophosphate insecticides were developed in Germany during the 1930s and 1940s, many insecticidal organophosphate compounds have been synthesized. One of the earliest and most successful of these is **parathion,** *O,O*-diethyl-*O*-*p*-nitrophenylphosphorothionate (banned from use in the U.S. in 1991 because of its acute toxicity to humans). From a long-term environmental standpoint, organophosphate insecticides are superior to the organohalide insecticides that they largely displaced because the organophosphates readily undergo biodegradation and do not bioaccumulate.

9.4. SYNTHETIC POLYMERS

The huge polymer manufacturing industry is significant to the environment both as a source of environmental pollutants and in the manufacture of materials used to alleviate environmental and waste problems. Synthetic **polymers** are produced when small molecules called **monomers** bond together to form a much smaller number of very large molecules. Many natural products are polymers; for example, cellulose produced by trees and other plants, and found in wood, paper, and many other materials, is a polymer of the sugar glucose. Synthetic polymers form the basis of many industries, such as rubber, plastics, and textiles manufacture.

An important example of a polymer is that of polyvinylchloride, shown in Figure 9.15. This polymer is synthesized in large quantities for the manufacture of water and sewer pipe, water-repellant liners, and other plastic materials. Other major polymers include polyethylene (plastic bags and milk cartons), polypropylene (impact-resistant plastics and indoor–outdoor carpeting), polyacrylonitrile (Orlon and carpets), polystyrene (foam insulation), and polytetrafluoroethylene (Teflon coatings and bearings); the monomers from which these substances are made are shown in Figure 9.16.

Many of the hazards from the polymer industry arise from the monomers used as raw materials. Many monomers are reactive and flammable, with a tendency to form explosive vapor mixtures with air. All have a certain degree of toxicity; vinyl chloride is a known human carcinogen. The combustion of many polymers may result in the evolution of toxic gases, such as hydrogen cyanide (HCN) from polyacrylonitrile or hydrogen chloride (HCl) from polyvinylchloride. Another hazard presented by plastics results from the presence of **plasticizers** added to provide essential properties such as flexibility. The most widely used plasticizers are phthalates, which are environmentally persistent, resistant to treatment processes, and prone to undergo bioaccumulation.

Polymers have a number of applications in waste treatment and disposal. Waste disposal landfill liners are made from synthetic polymers, as are the fiber filters that remove particulate pollutants from flue gas in baghouses. Membranes used for

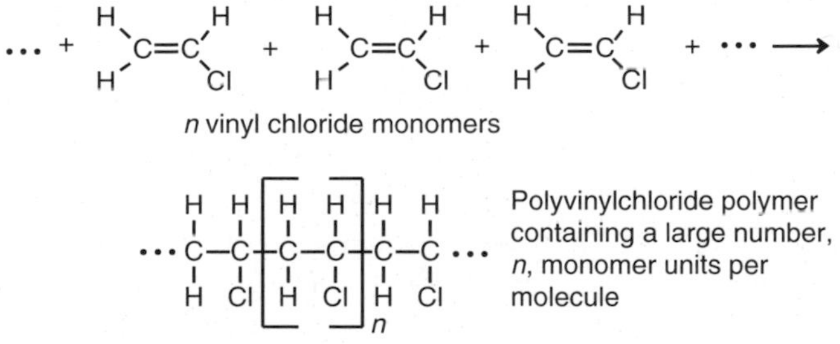

$$\cdots + \mathrm{H_2C{=}CHCl} + \mathrm{H_2C{=}CHCl} + \mathrm{H_2C{=}CHCl} + \cdots \longrightarrow$$

n vinyl chloride monomers

$$\cdots\mathrm{CH_2{-}CHCl}\left[\mathrm{CH_2{-}CHCl}\right]_n\mathrm{CH_2{-}CHCl}\cdots$$

Polyvinylchloride polymer containing a large number, *n*, monomer units per molecule

Figure 9.15. Polyvinylchloride polymer.

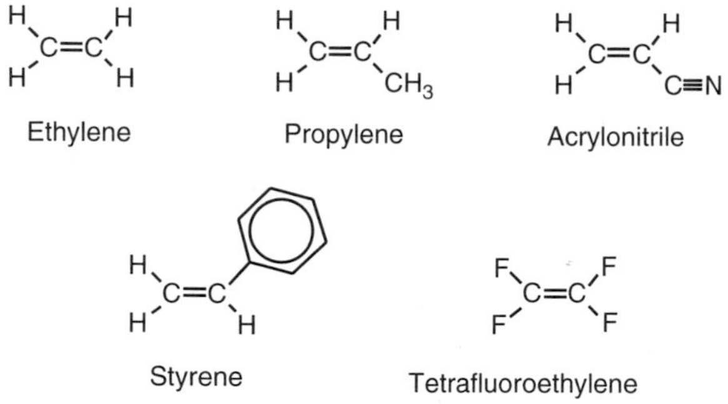

Ethylene Propylene Acrylonitrile

Styrene Tetrafluoroethylene

Figure 9.16. Monomers from which commonly used polymers are synthesized.

ultrafiltration and reverse osmosis treatment of water are composed of very thin sheets of synthetic polymers. Organic solutes can be removed from water by sorption onto hydrophobic (water-repelling) organophilic beads of Amberlite XAD resin. Heavy metal pollutants are removed from wastewater by cation exchange resins made of polymers with anionic functional groups. Typically, these resins exchange harmless sodium ion, Na^+, on the solid resin for toxic heavy metal ions in water. Figure 9.17 shows a segment of the polymeric structure of a cation exchange resin in the sodium form. In the treatment of heavy-metal-containing waste solutions, these resins can exchange toxic heavy metal ions in solution, such as Cd^{2+}, for nontoxic Na^+ ions. Ion exchange resins are used in nuclear reactors to remove traces of metals, some of which may be radioactive, from the water used in the reactor for heat exchange. Ion exchange resins have also been developed in

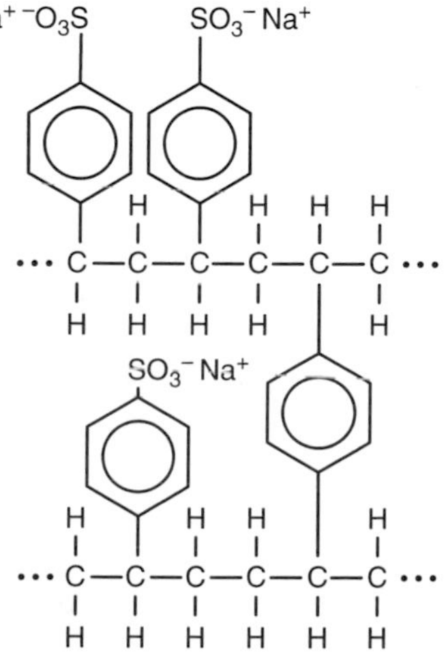

Figure 9.17. Polymeric cation exchanger in the sodium form.

which the ion-exchanging functional group is an iminodiacetate [$-N(CH_2CO_2^-)_2$] group that has a particularly strong affinity for heavy metals.

CHAPTER SUMMARY

The chapter summary below is presented in a programmed format to review the main points covered in this chapter. It is used most effectively by filling in the blanks, referring back to the chapter as necessary. The correct answers are given at the end of the summary.

Organic chemistry is so diverse because of the ability of carbon atoms to bond to each other through [1]__________, ________ and ______, bonds in a huge variety of [2]__________________. All organic compounds contain [3]________ and most contain the element [4]________. Although represented on paper by structures such as,

```
   Cl
   |
H—C—H
   |
   Cl
```

the [5]________________ of organic molecules is particularly important in determining their behavior. The simplest class of organic compounds containing only two elements is the [6]______________, divided into the four general classes of [7]__________ ________________. Hydrocarbons with a general formula such as C_8H_{18} may exist as a variety of different [8]______________________________. The compound,

```
H3C        CH3
    \     /
     C = C
    /     \
   H       H
```

has a "twin" compound with the same molecular formula, and the two compose a pair of [9]__________________. The systematic name of the compound

```
                 H
                 |
              H—C—H
                 |
              H—C—H
                 |
       H   H     |     H   H   H
       |   |     |     |   |   |
    H—C—C—C—C—C—C—H
       |   |   |   |   |   |
       H   H   H   |   H   H
                 H—C—H
                     |
                     H
```

is [10]___________________________ and its condensed structural formula is [11]_________________. The systematic name of an organic compound is based upon the [12]_______________________ chain of carbon atoms. The two most common reactions of alkanes are [13]_________________________________. A free radical is [14]____________

_______________, one of which may be formed initially by a [15]_______________ process. The compound,

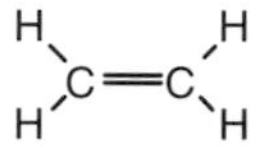

is an example of an [16]_________ hydrocarbon. These compounds undergo [17]___________reactions, which cannot occur with alkanes. Hydrocarbons containing the 6-membered ring C_6H_6 or derivatives of it are examples of [18]_______________ _____. The compound,

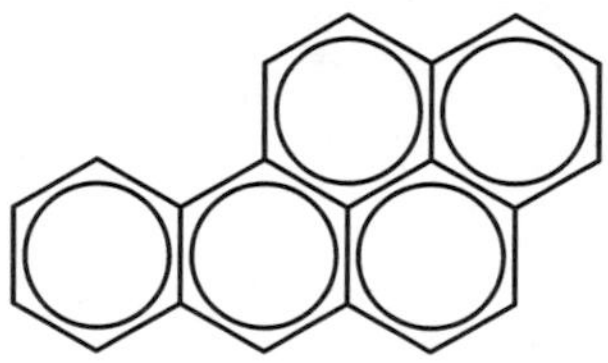

belongs to the class of [19]_______________________, some of which are of concern because [20]__________________. Groups such as those outlined by dashed rectangles below are called [21]___________________:

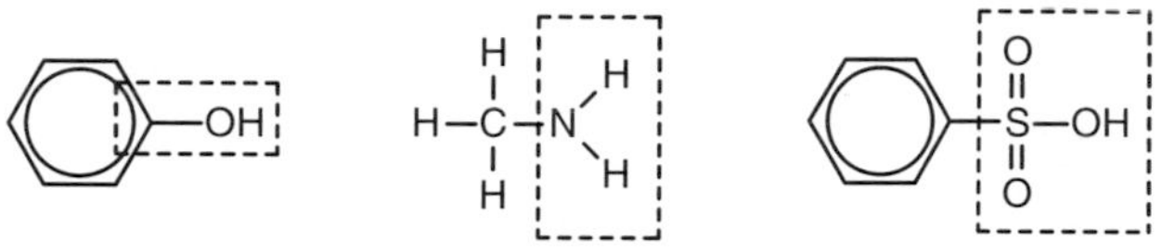

The structural formula of the functional group characteristic of aldehydes is [22]_____ _____________ and that characteristic of carboxylic acids is [23]_______. Amines are characterized by [24]______________________________________. The compound,

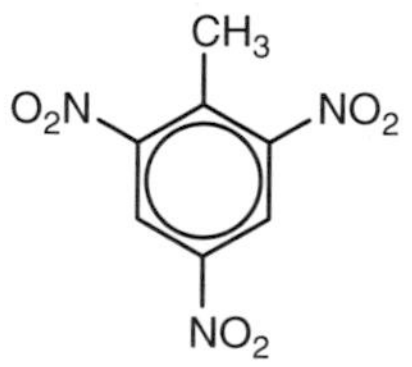

is called [25]__________________ and is used as an [26]______________. Organohalide compounds are those that contain [27]_______________. Of the compounds

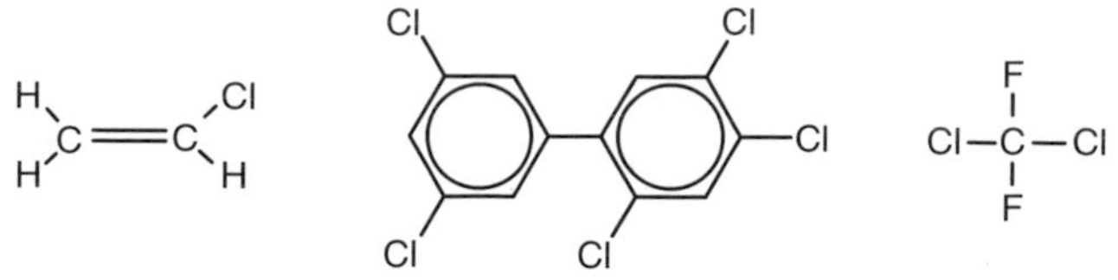

A

B

C

"A" is of concern because it is [28]____________, "B" because it is [29]____________, and "C" because it [30]____________________. The compound

$$H-\overset{\displaystyle H}{\underset{\displaystyle H}{\overset{|}{\underset{|}{C}}}}-SH$$

is an example of a [31]_____________, a class of organosulfur compounds noted for their [32]______________. Sulfonic acids contain the [33]___________ group. Phosphorothionate esters such as parathion, for which the structural formula is [34]______________, are effective as insecticides because they [35]______________. Synthetic polymers are formed when small [36]__________ molecules join together to form [37]_______________________.

Answers to Chapter Summary

1. single, double, triple
2. straight chains, branched chains, and rings
3. carbon
4. hydrogen
5. molecular geometry
6. hydrocarbons
7. alkanes, alkenes, alkynes, and aromatic hydrocarbons
8. structural isomers that have different arrangements of atoms
9. *cis–trans* isomers
10. 3-ethyl-4-methylhexane
11. $CH_3CH_2CH(C_2H_5)CH(CH_3)CH_2CH_3$
12. longest continuous
13. oxidation (combustion) and substitution reactions
14. a species with an unpaired electron, such as $Cl^{\cdot}$
15. photochemical
16. alkene
17. addition

18. aromatic or aryl hydrocarbons
19. polycyclic aromatic hydrocarbons
20. they may cause cancer
21. functional groups
22. $-\overset{\overset{\displaystyle O}{\|}}{C}-H$
23. $-\overset{\overset{\displaystyle O}{\|}}{C}-H$
24. A nitrogen atom bonded to at least one hydrocarbon group and that may be bonded to 1 or 2 H atoms
25. 2,4,6-trinitrotoluene
26. explosive
27. F, Cl, Br, or I
28. a known human carcinogen
29. environmentally persistent
30. depletes stratospheric ozone
31. thiol compound
32. bad odors
33. $-SO_3H$
34. $H_5C_2-O-\overset{\overset{\displaystyle S}{\|}}{\underset{\underset{\displaystyle OC_2H_5}{|}}{P}}-O-C_6H_4-NO_2$
35. inhibit acetylcholinesterase
36. monomer
37. very large molecules

SUPPLEMENTARY REFERENCES

Atkins, Robert C. and Francis A. Carey, *Organic Chemistry: A Brief Course*, 3rd ed., McGraw-Hill, New York, 2002.

Brown, William H. and Christopher S. Foote, *Organic Chemistry*, 4th ed., Thomson Brooks/Cole, Belmont, CA, 2005.

Bruice, Paula Yurkanis, *Organic Chemistry*, 5th ed., Prentice Hall, Upper Saddle River, NJ, 2006.

Carey, Francis A., *Organic Chemistry*, McGraw-Hill, New York, 2007.

Ege, Seyhan N., *Organic Chemistry: Structure and Reactivity*, 5th ed., Houghton Mifflin, Boston, 2004.

Faber, Kurt, *Biotransformations in Organic Chemistry: A Textbook*, 5th ed., Springer Verlag, Berlin, 2004.

McMurry, John, *Organic Chemistry*, 7th ed., Brooks/Cole/Thomson Learning, Pacific Grove, CA, 2007.

McMurry, John and Mary E. Castellion, *Fundamentals of General, Organic and Biological Chemistry*, 5th ed., Prentice Hall, Upper Saddle River, NJ, 2007.

Schwarzenbach, René P., and Phillip M. Gschwend, and Dieter M. Imboden, *Environmental Organic Chemistry*, 2nd ed., John Wiley & Sons, Hoboken, NJ, 2003.

Solomons, T. W. Graham, and Craig B. Fryhle, *Organic Chemistry*, 9th ed., John Wiley & Sons, New York, 2007.

Sorrell, Thomas N., *Organic Chemistry*, 2nd ed., University Science Books, Sausalito, CA, 2006.

Timberlake, Karen C., *Chemistry: An Introduction to General, Organic, and Biological Chemistry*, 10th ed., Pearson Education, Upper Saddle River, NJ, 2009.

Vollhardt, K. Peter C., and Neil E. Schore, *Organic Chemistry: Structure and Function*, 5th ed., W.H. Freeman, New York, 2007.

Wade, L. G., Jr., *Organic Chemistry* , 6th ed., Prentice-Hall, Upper Saddle River, NJ, 2006.

Winter, Arthur, *Organic Chemistry I for Dummies*, Wiley, Hoboken, NJ, 2005.

QUESTIONS AND PROBLEMS

1. Explain the bonding properties of carbon that make organic chemistry so diverse.

2. Distinguish among alkanes, alkenes, alkynes, and aromatic compounds. To which general class of organic compounds do all of these belong?

3. In what sense are alkanes saturated? Why are alkenes more reactive than alkanes?

4. Name the following compound:

```
     H  H    CH3   H  H  H
     |  |    |     |  |  |
   H-C--C----C-----C--C--C-H
     |  |    |     |  |  |
     H  H    CH3   H  H  H
```

5. What is indicated by the prefix "*n*-" in a hydrocarbon name?

6. Discuss the chemical reactivity of alkanes. Why are they chemically reactive or unreactive?

7. Discuss the chemical reactivity of alkenes. Why are they chemically reactive or unreactive?

8. What are the characteristics of aromaticity? What are the chemical reactivity characteristics of aromatic compounds?

9. Describe chain reactions, discussing what is meant by free radicals and photochemical processes.

10. Define, with examples, what is meant by isomerism.

11. Describe how the two forms of 1,2-dichloroethylene can be used to illustrate *cis–trans* isomerism.

12. Give the structural formula corresponding to the condensed structural formula $CH_3CH(C_2H_5)CH(C_2H_5)CH_2CH_3$.

13. Discuss how organic functional groups are used to define classes of organic compounds.

14. Give the functional groups corresponding to (a) alcohols, (b) aldehydes, (c) carboxylic acids, (d) ketones, (e) amines, (f) thiol compounds, and (g) nitro compounds.

15. Give an example compound of each of the following: epoxides, alcohols, phenols, ethers, aldehydes, ketones, and carboxylic acids.

16. Which functional group is characteristic of *N*-nitroso compounds, and why are these compounds toxicologically significant?

17. Give an example of each of the following: alkyl halides, alkenyl halides, and aromatic halides.

18. Give an example compound of a chlorinated naphthalene and of a PCB.

19. What explains the tremendous chemical stability of CFCs? What kinds of compounds are replacing CFCs? Why?
20. How does a thiol compound differ from a thioether?
21. How do sulfoxides differ from sulfones?
22. Which inorganic compound is regarded as the parent compound of alkyl and aromatic phosphines? Give an example of each of these.
23. What are organophosphate esters and what is their toxicological significance?
24. Define what is meant by a polymer and give an example of one.

10. BIOLOGICAL CHEMISTRY

10.1. BIOCHEMISTRY

Many people have had the experience of looking through a microscope at a single cell. It may have been an ameba, alive and oozing about like a blob of jelly on the microscope slide, or a bacterial cell stained with a dye to make it show up more plainly. Or it may have been a beautiful algal cell with its bright green chlorophyll. Even the simplest of these cells is capable of carrying out a thousand or more chemical reactions. These life processes fall under the heading of **biochemistry**, that branch of chemistry dealing with the chemical properties, composition, and biologically mediated processes of complex substances in living systems.

Biochemical phenomena that occur in living organisms are extremely sophisticated. In the human body, complex metabolic processes break down a variety of food materials to simpler chemicals, yielding energy and the raw materials to build body constituents, such as muscle, blood, and brain tissue. Impressive as this may be, consider a humble microscopic cell of a photosynthetic cyanobacterium only about a micrometer in size, which requires only a few simple inorganic chemicals and sunlight for its existence. This cell uses sunlight energy to convert carbon from CO_2, hydrogen and oxygen from H_2O, nitrogen from NO_3^-, sulfur from SO_4^{2-}, and phosphorus from inorganic phosphate into all the proteins, nucleic acids, carbohydrates, and other materials that it requires to exist and reproduce. Such a simple cell accomplishes what could not be done by human endeavors even in a vast chemical factory costing billions of dollars.

Ultimately, most environmental pollutants and hazardous substances are of concern because of their effects upon living organisms. The study of the adverse effects of substances on life processes requires some basic knowledge of biochemistry. This chapter discusses biochemistry, with emphasis upon those aspects that are especially pertinent to environmentally hazardous and toxic substances, including cell membranes, DNA, and enzymes.

Not only are biochemical processes profoundly influenced by chemical species in the environment, but such processes largely determine the nature of these species, their degradation, and even their syntheses, particularly in the aquatic and soil environments. The study of such phenomena forms the basis of **environmental biochemistry**.[1]

Biomolecules

The biomolecules that constitute matter in living organisms are often polymers with molecular masses of the order of a million or even larger. As discussed later in this chapter, these biomolecules may be divided into the categories of carbohydrates, proteins, lipids, and nucleic acids. Proteins and nucleic acids consist of macromolecules, lipids are usually relatively small molecules, and carbohydrates range from relatively small sugar molecules to high-molecular-mass macromolecules such as those in cellulose.

The behavior of a substance in a biological system depends to a large extent upon whether the substance is hydrophilic ("water-loving") or hydrophobic ("water-hating"). Some important toxic substances are hydrophobic, a characteristic that enables them to traverse cell membranes readily. Part of the detoxification process carried on by living organisms is to render such molecules hydrophilic, and therefore water-soluble and readily eliminated from the body.

10.2. BIOCHEMISTRY AND THE CELL

The focal point of biochemistry and biochemical aspects of toxicants is the **cell**, the basic building block of living systems where most life processes are carried out. Bacteria, yeasts, and some algae consist of single cells. However, most living things are made up of many cells. In a more complicated organism, the cells have different functions. Liver cells, muscle cells, brain cells, and skin cells in the human body are quite different from each other and do different things. Cells are divided into two major categories depending upon whether or not they have a nucleus: **eukaryotic** cells have a nucleus and **prokaryotic** cells do not. Prokaryotic cells are found in bacteria. Eukaryotic cells compose organisms other than bacteria.

Major Cell Features

Figure 10.1 shows the major features of the **eukaryotic cell**, which is the basic structure in which biochemical processes occur in multicellular organisms. These features are the following:

- **Cell membrane**. This encloses the cell and regulates the passage of ions, nutrients, lipid-soluble ("fat-soluble") substances, metabolic products, toxicants, and toxicant metabolites into and out of the cell interior because of its varying **permeability** for different substances. The cell membrane protects the contents of the cell from undesirable outside

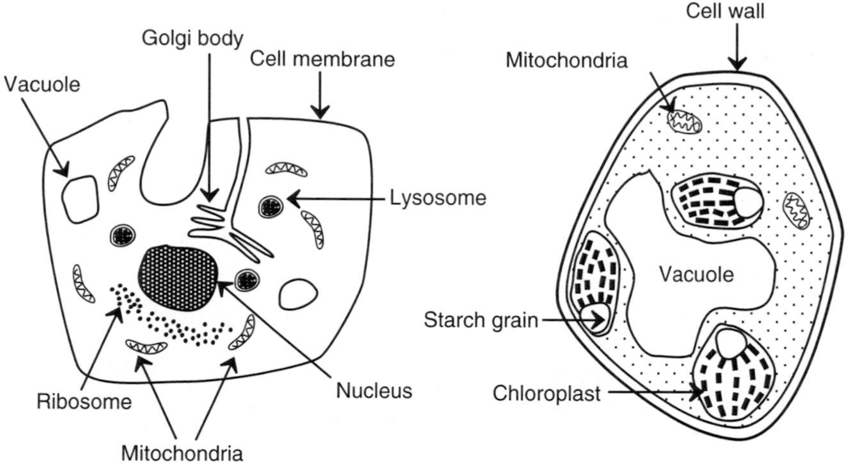

Figure 10.1. Some major features of the eukaryotic cell in animals (left) and plants (right).

influences. Cell membranes are composed in part of phospholipids that are arranged with their hydrophilic heads on the cell membrane surfaces and their hydrophobic tails inside the membrane. Cell membranes contain bodies of proteins that are involved in the transport of some substances through the membrane. One reason the cell membrane is very important in toxicology and environmental biochemistry is because it regulates the passage of toxicants and their products into and out of the cell interior. Furthermore, when its membrane is damaged by toxic substances, a cell may not function properly and the organism may be harmed.

- **Cell nucleus**. This acts as a sort of "control center" of the cell. It contains the genetic directions that the cell needs to reproduce itself. The key substance in the nucleus is **deoxyribonucleic acid (DNA). Chromosomes** in the cell nucleus are made up of combinations of DNA and proteins. Each chromosome stores a separate quantity of genetic information. Human cells contain 46 chromosomes. When DNA in the nucleus is damaged by foreign substances, various toxic effects, including mutations, cancer, birth defects, and defective immune system function may occur.

- **Cytoplasm**. This fills the interior of the cell not occupied by the nucleus. Cytoplasm is further divided into a water-soluble proteinaceous filler called **cytosol**, in which are suspended bodies called **cellular organelles**, such as mitochondria and, in photosynthetic organisms, chloroplasts.

- **Mitochondria**. These are "powerhouses" that mediate energy conversion and utilization in the cell. Mitochondria are sites in which food materials—carbohydrates, proteins, and fats—are broken down to yield carbon dioxide, water, and energy, which is then used by the cell for its energy needs. The best example of this is the oxidation of the sugar glucose, $C_6H_{12}O_6$:

$$C_6H_{12}O_6 + 6O_2 \rightarrow 6CO_2 + 6H_2O + \text{energy}$$

  This kind of process is called **cellular respiration**.

- **Ribosomes**. These participate in protein synthesis.
- **Endoplasmic reticulum**. This is involved in the metabolism of some toxicants by enzymatic processes.
- **Lysosome**. This is a type of organelle that contains potent substances capable of digesting liquid food material. Such material enters the cell through a "dent" in the cell wall, which eventually becomes surrounded by cell material. This surrounded material is called a **food vacuole**. The vacuole merges with a lysosome, and the substances in the lysosome bring about digestion of the food material. The digestion process consists largely of **hydrolysis reactions** in which large, complicated food molecules are broken down into smaller units by the addition of water.
- **Golgi bodies**. These occur in some types of cells. They are flattened bodies of material that serve to hold and release substances produced by the cells.
- **Cell walls** (of plant cells). These are strong structures that provide stiffness and strength. Cell walls are composed mostly of cellulose, which will be discussed later in this chapter.
- **Vacuoles** (inside plant cells). These often contain materials dissolved in water.
- **Chloroplasts** (in plant cells). These are involved in photosynthesis (the chemical process that uses energy from sunlight to convert carbon dioxide and water to organic matter). Photosynthesis occurs in these bodies. Food produced by photosynthesis is stored in the chloroplasts in the form of **starch grains**.

10.3. PROTEINS

Proteins are nitrogen-containing organic compounds that are the basic units of living systems. Cytoplasm, the jelly-like liquid filling the interior of cells, is made up largely of proteins. Enzymes, which act as catalysts of biological reactions, are proteins; they are discussed later in the chapter. Proteins are made up of amino acids joined together in huge chains. Amino acids are organic compounds that contain the carboxylic acid group, $-CO_2H$, and the amino group, $-NH_2$; that is, they are both carboxylic acids and amines (see Chapter 9). Proteins are polymers or macromolecules of amino acids containing from approximately forty to several thousand amino acid groups joined by peptide linkages. Smaller-molecule amino acid polymers, containing only about ten to about forty amino acids per molecule, are called **polypeptides**. A portion of the amino acid left after the elimination of water during polymerization is called a **residue**. The amino acid sequence of these residues is designated by a series of three-letter abbreviations for each amino acid.

The amino acids that occur naturally in proteins and polypeptides all have the following chemical group:

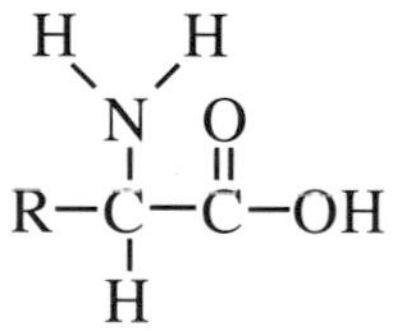

$$R-CH(NH_2)-C(=O)-OH$$

In this structure the $-NH_2$ group is always bonded to the carbon next to the $-CO_2H$ group. This is called the "alpha" (α) location, so natural amino acids are α-amino acids. Other groups, designated as "R," are attached to the basic α-amino acid structure. The R group may be as simple as an atom of H found in glycine or it may be as complicated as the structure of the R group in tryptophan:

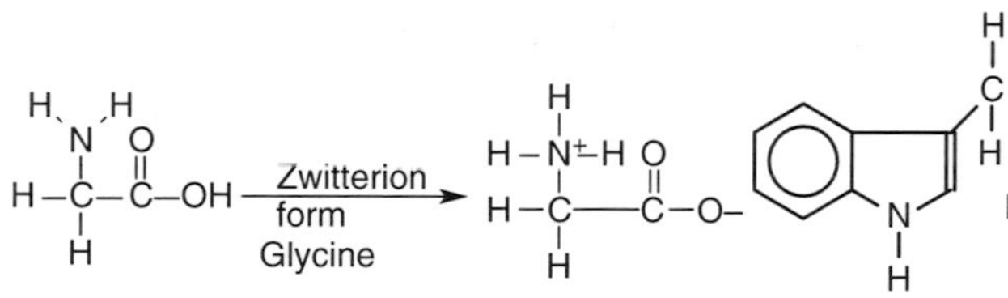

$$H-CH(NH_2)-C(=O)-OH \xrightarrow[\text{Glycine}]{\text{Zwitterion form}} H-CH(NH_3^+)-C(=O)-O^-$$

R group found in tryptophan

As shown in Figure 10.2, there are 20 common amino acids in proteins. These are shown with uncharged $-NH_2$ and $-CO_2H$ groups. Actually, these functional groups exist in the charged zwitterion form as shown for glycine above.

Amino acids in proteins are joined together in a specific way. These bonds constitute the **peptide linkage**. The formation of peptide linkages is a condensation process involving the loss of water. Consider as an example the condensation of alanine, leucine, and tyrosine shown in Figure 10.3. When these three amino acids join together, two water molecules are eliminated. The product is a tripeptide, since there are three amino acids involved. The amino acids in proteins are linked as shown for this tripeptide, except that many more monomeric amino acid groups are involved.

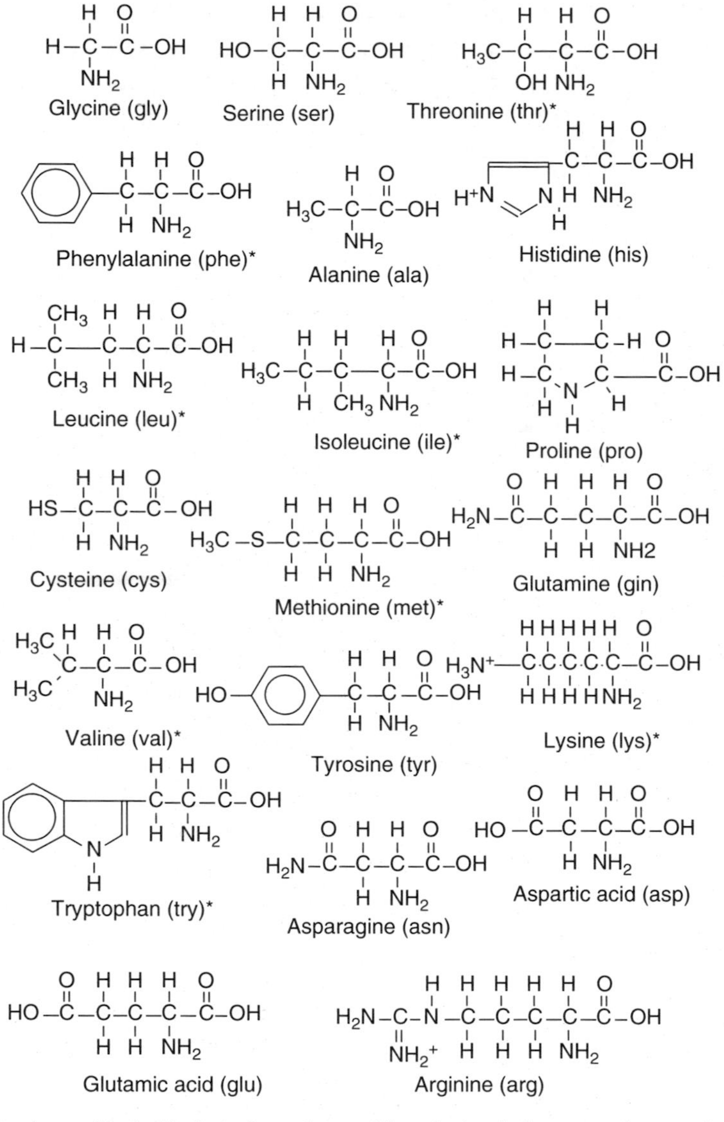

Glycine (gly) Serine (ser) Threonine (thr)*

Phenylalanine (phe)* Alanine (ala) Histidine (his)

Leucine (leu)* Isoleucine (ile)* Proline (pro)

Cysteine (cys) Methionine (met)* Glutamine (gin)

Valine (val)* Tyrosine (tyr) Lysine (lys)*

Tryptophan (try)* Asparagine (asn) Aspartic acid (asp)

Glutamic acid (glu) Arginine (arg)

Figure 10.2. Amino acids (with their three-letter abbreviations) that occur in proteins. Those marked with an asterisk cannot be synthesized by the human body and must come from dietary sources.

Proteins may be divided into several major types that have widely varying functions. These are listed in Table 10.1.

Protein Structure

The order of amino acids in protein molecules, and the resulting three-dimensional structures that result, provide an enormous variety of possibilities for **protein structure**. This is what makes life so diverse. Proteins have primary, secondary, tertiary, and quaternary structures. The structures of protein molecules

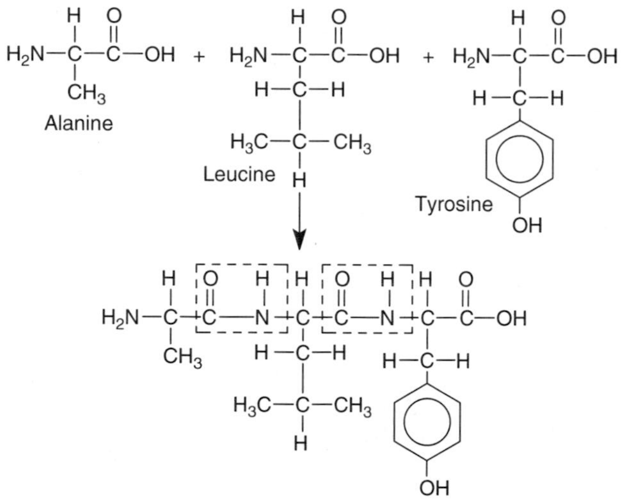

Figure 10.3. Condensation of alanine, leucine, and tyrosine to form a tripeptide consisting of three amino acids joined by peptide linkages (outlined by dashed lines).

determine the behavior of proteins in crucial areas such as the processes by which the body's immune system recognizes substances that are foreign to the body. The very specific functions of enzymes depend upon their protein structures.

The order of amino acids in the protein molecule determines its **primary structure**. The **secondary structure** of a protein results in part from the folding

Table 10.1. Major Types of Proteins

Type of Protein	Example	Function and Characteristics
Nutrient	Casein (milk protein)	Food source. Organisms must have an adequate supply of nutrient protein with the right balance of amino acids for adequate nutrition.
Storage	Ferritin	Storage of iron in animal tissues
Structural	Collagen (tendons), keratin (hair)	Structural and protective components in organisms
Contractile	Actin, myosin	Strong, fibrous proteins that can contract muscle tissue and cause movement to occur
Transport	Hemoglobin	Transport inorganic and organic species across cell membranes, in blood, between organs
Defense	—	Antibodies against foreign agents (e.g., viruses), produced by the immune system
Regulatory	Insulin, human growth hormone	Regulate biochemical processes such as sugar metabolism and growth by binding to sites inside cells or on cell membranes
Enzymes	Acetylcholinesterase	Catalysts of biochemical reactions (see Section 10.6)

of polypeptide chains to produce the maximum number of hydrogen bonds between peptide linkages:

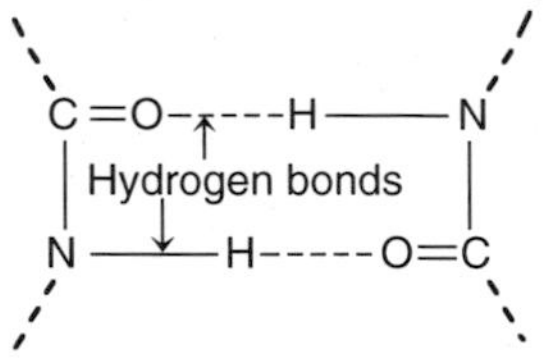

```
 \                         /
  \                       /
   C=O----H———N
   |   ↑              |
   | Hydrogen bonds   |
   |   ↓              |
   N———H-----O=C
  /                       \
 /                         \
```

Illustration of hydrogen bonds between N and O atoms in peptide linkages, which constitute protein secondary structures

The secondary structure is also determined by interactions, both attractive and repulsive, between amino acid side chains ("R"): for example, small R groups enable protein molecules to be hydrogen-bonded together in a parallel arrangement, whereas larger R groups produce a spiral form known as an **α**-helix.

Tertiary structures are formed by the twisting of α-helices into specific shapes. They are produced and held in place by the interactions of amino side chains on the amino acid residues constituting the protein macromolecules. Tertiary protein structure is very important in the processes by which enzymes identify specific proteins and other molecules upon which they act. It is also involved in the action of antibodies, which recognize foreign proteins by their shape and react to them. This is what happens in the phenomenon of disease immunity, where antibodies recognize specific proteins from viruses or bacteria and initiate a process leading to the destruction of these pathogens.

Two or more protein molecules consisting of separate polypeptide chains may be further attracted to each other to produce a **quaternary structure**.

Some proteins are **fibrous proteins**, which occur in skin, hair, wool, feathers, silk, and tendons. The molecules in these proteins are long and threadlike and are laid out in parallel in bundles. Fibrous proteins are quite tough, and they do not dissolve in water.

An interesting fibrous protein is keratin, which is found in hair. The cross-linking bonds between protein molecules in keratin are –S—S– bonds formed from two HS– groups in two molecules of the amino acid cysteine. These bonds largely hold hair in place, thus keeping it curly or straight. A "permanent" consists of breaking the bonds chemically, setting the hair as desired, and then reforming the cross-links to hold the desired shape.

Aside from fibrous protein, the other major type of protein form is the **globular protein**. These proteins are in the shape of balls and oblongs. Globular proteins are relatively soluble in water. A typical globular protein is hemoglobin, the oxygen-carrying protein in red blood cells. Enzymes are generally globular proteins.

Denaturation of Proteins

Secondary, tertiary, and quaternary protein structures are easily changed by a process called **denaturation**. These changes can be quite damaging. Heating, exposure to acids or bases, and even violent physical action can cause denaturation to occur. The albumin protein in egg white is denatured by heating so that it forms a semisolid mass. Almost the same thing is accomplished by the violent physical action of an egg beater in the preparation of meringue. Heavy metal poisons such as lead and cadmium change the structures of proteins by binding to functional groups on the protein surface.

10.4. CARBOHYDRATES

Carbohydrates have the approximate simple formula CH_2O and include a diverse range of substances composed of simple sugars such as glucose:

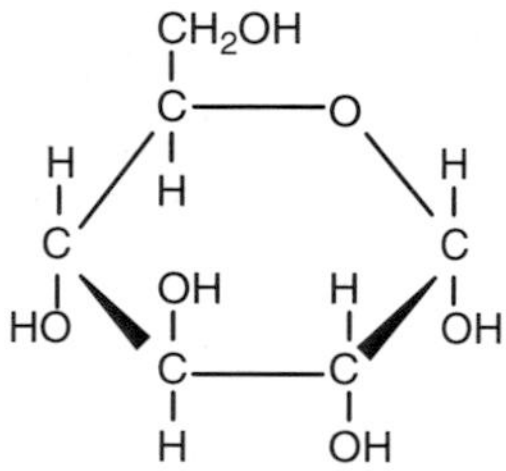

Glucose molecule

High-molecular-mass **polysaccharides**, such as starch and glycogen ("animal starch"), are biopolymers of simple sugars.

Photosynthesis in a plant cell converts the energy from sunlight to chemical energy in a carbohydrate, $C_6H_{12}O_6$. This carbohydrate may be transferred to some other part of the plant for use as an energy source. It may be converted to a water-insoluble carbohydrate for storage until it is needed for energy. Or it may be transformed to cell wall material and become part of the structure of the plant. If the plant is eaten by an animal, the carbohydrate is used for energy by the animal.

The simplest carbohydrates are the **monosaccharides**. These are also called **simple sugars**. Because they have 6 carbon atoms, simple sugars are sometimes called *hex*oses. Glucose (formula shown above) is the most common simple sugar involved in cell processes. Other simple sugars with the same formula but different structures are fructose, mannose, and galactose. These must be changed to glucose before they can be used in a cell. Because of its use for energy in body processes, glucose is found in the blood. Normal levels are from 65 to 110 mg glucose per 100 mL of blood. Higher levels may indicate diabetes.

Units of two monosaccharides make up several very important sugars known as **disaccharides**. When two monosaccharide molecules (not necessarily of the same monosaccharide) join together to form a disaccharide, a molecule of water is lost:

$$C_6H_{12}O_6 + C_6H_{12}O_6 \rightarrow C_{12}H_{22}O_{11} + H_2O \qquad (10.4.1)$$

Recall that proteins are also formed from smaller amino acid molecules by condensation reactions involving the loss of water molecules. Disaccharides include sucrose (cane sugar, used as a sweetener), lactose (milk sugar), and maltose (a product of the breakdown of starch).

Polysaccharides consist of many simple sugar units hooked together. One of the most important polysaccharides is **starch**, which is produced by plants for food storage. Animals produce a related material called **glycogen**. The chemical formula of starch is $(C_6H_{10}O_5)_n$, where n may represent a number as high as several hundreds. What this means is that the very large starch molecule consists of many units of $C_6H_{10}O_5$ joined together. For example, if $n = 100$, there are 6×100 carbon atoms, 10×100 hydrogen atoms, and 5×100 oxygen atoms in the molecule. Its chemical formula is $C_{600}H_{1000}O_{500}$.The atoms in a starch molecule are actually present as linked rings represented by the structure shown in Figure 10.4. Starch occurs in many foods, such as bread and cereals. It is readily digested by animals, including humans.

Cellulose is a polysaccharide that is also made up of $C_6H_{10}O_5$ units. Molecules of cellulose are huge, with molecular masses of around 400 000. The cellulose structure (Figure 10.5) is similar to that of starch. Cellulose is produced by plants, and forms the structural material of plant cell walls. Wood is about 60% cellulose, and cotton contains over 90% of this material. Fibers of cellulose are extracted from wood and pressed together to make paper.

Humans and most other animals cannot digest cellulose. Ruminant animals (cattle, sheep, goats, and moose) have bacteria in their stomachs that break down cellulose into products that can be used by the animal. Chemical processes are available to convert cellulose to simple sugars by the reaction

$$\underset{\text{cellulose}}{(C_6H_{10}O_5)_n} + nH_2O \rightarrow \underset{\text{glucose}}{nC_6H_{12}O_6} \tag{10.4.2}$$

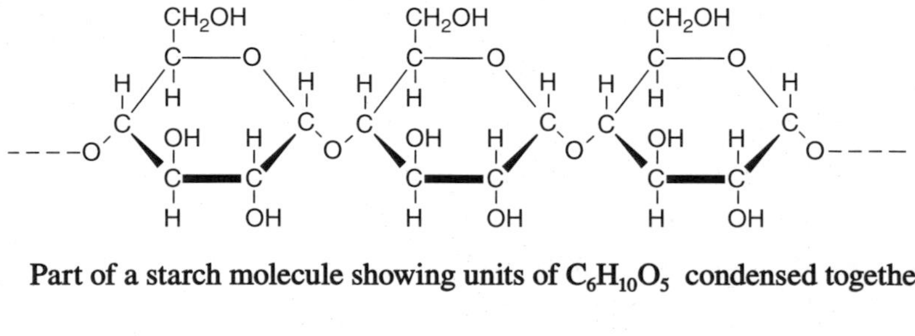

Figure 10.4. Part of a starch molecule showing units of $C_6H_{10}O_5$ condensed together.

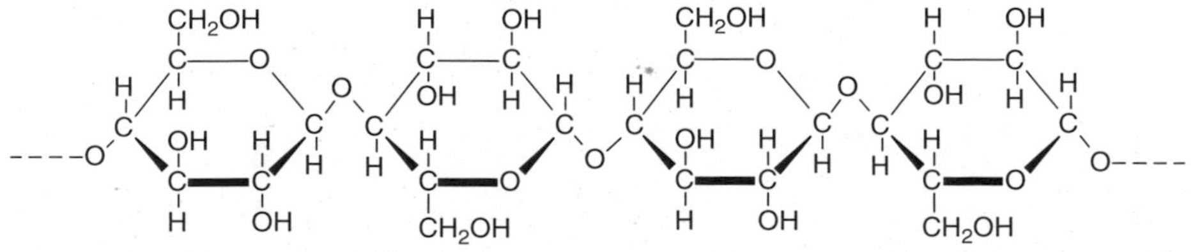

Figure 10.5. Part of the structure of cellulose.

where n may be 2000–3000. This involves breaking the linkages between units of $C_6H_{10}O_5$ by adding a molecule of H_2O at each linkage, in a hydrolysis reaction. Large amounts of cellulose from wood, sugar cane, and agricultural products go to waste each year. The hydrolysis of cellulose enables these products to be converted to sugars, which can be fed to animals.

Carbohydrate groups are attached to protein molecules in a special class of materials called **glycoproteins**. Collagen is a crucial glycoprotein that provides structural integrity to body parts. It is a major constituent of skin, bones, tendons, and cartilage.

10.5. LIPIDS

Lipids are substances that can be extracted from plant or animal matter by organic solvents, such as chloroform, diethyl ether, and toluene (Figure 10.6). Whereas

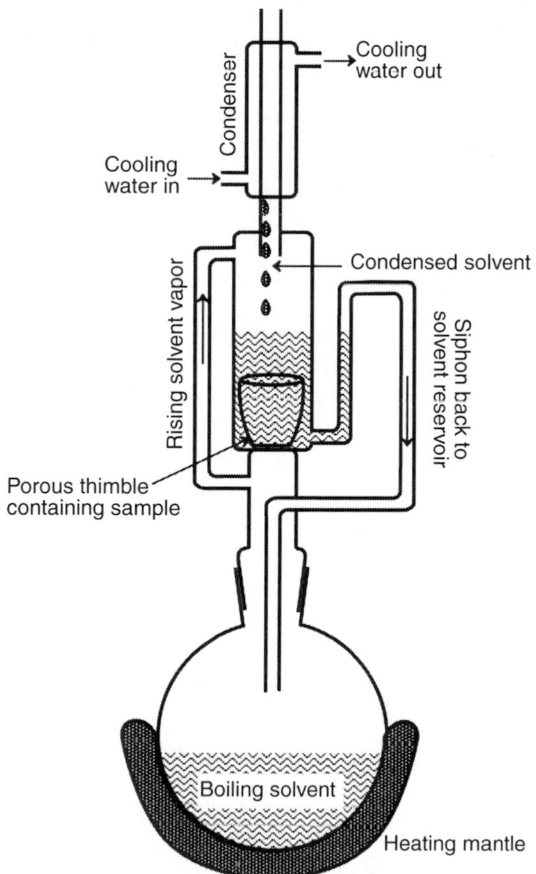

Figure 10.6. Lipids are extracted from some biological materials with a Soxhlet extractor (above). The solvent is vaporized in the distillation flask by the heating mantle, rises through one of the exterior tubes to the condenser, and is cooled to form a liquid. The liquid drops onto the porous thimble containing the sample. Siphon action periodically drains the solvent back into the distillation flask. The extracted lipid collects as a solution in the solvent in the flask.

carbohydrates and proteins are characterized predominately by the monomers (monosaccharides and amino acids) from which they are composed, lipids are defined essentially by their physical characteristic of organophilicity. The most common lipids are fats and oils composed of **triglycerides** (or **triacylglycerols**) formed from the alcohol glycerol, $CH_2(OH)CH(OH)CH_2(OH)$ and long-chain fatty acids such as stearic acid, $CH_3(CH_2)_{16}C(O)OH$ (Figure 10.7). Numerous other biological materials, including waxes, cholesterol, and some vitamins and hormones, are classified as lipids. Common foods, such as butter and salad oils, are lipids. The longer-chain fatty acids, such as stearic acid, are also organic-soluble and are classified as lipids.

Lipids are toxicologically important for several reasons. Some toxic substances interfere with lipid metabolism, leading to detrimental accumulation of lipids. Many toxic organic compounds are poorly soluble in water, but are lipid-soluble, so that bodies of lipids in organisms serve to dissolve and store toxicants.

An important class of lipids consists of **phosphoglycerides** (or **glycerophosphatides**). These compounds may be regarded as triglycerides in which one of the acids bonded to glycerol is orthophosphoric acid. These lipids are especially important because they are essential constituents of cell membranes. These membranes consist of bilayers in which the hydrophilic phosphate ends of the molecules are on the outside of the membrane and their hydrophobic "tails" are on the inside.

Waxes are also esters of fatty acids. However, the alcohol in a wax is not glycerol. It is often a very long-chain alcohol. For example, one of the main compounds in beeswax is myricyl palmitate,

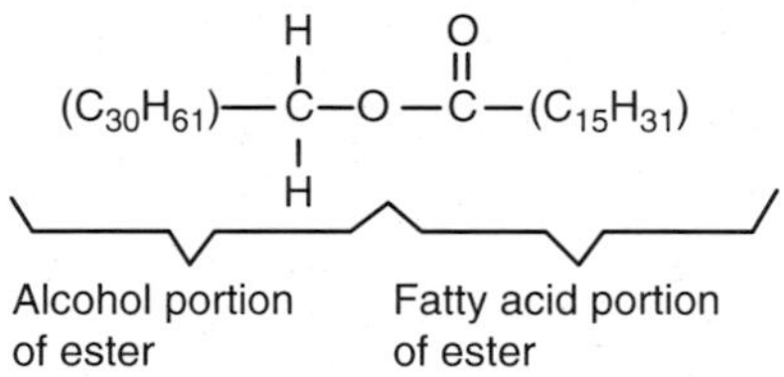

$$\underbrace{(C_{30}H_{61})-\overset{H}{\underset{H}{C}}-O}_{\text{Alcohol portion of ester}}\underbrace{-\overset{O}{\overset{\|}{C}}-(C_{15}H_{31})}_{\text{Fatty acid portion of ester}}$$

in which the alcohol portion of the ester has a very large hydrocarbon chain. Waxes are produced by both plants and animals, largely as protective coatings. Waxes are found in a number of common products. Lanolin is one of these. It is the "grease" in sheep's wool. When mixed with oils and water, it forms stable colloidal emulsions

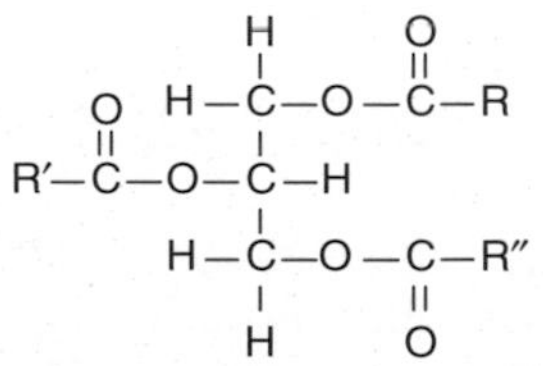

$$\begin{array}{l} \qquad\qquad\quad H-\overset{H}{C}-O-\overset{O}{\overset{\|}{C}}-R \\ R'-\overset{O}{\overset{\|}{C}}-O-C-H \\ \qquad\qquad\quad H-\underset{H}{C}-O-\underset{\underset{O}{\|}}{C}-R'' \end{array}$$

Figure 10.7. General formula of triglycerides, which make up fats and oils. The R, R′, and R″ groups are from fatty acids and are hydrocarbon chains, such as $-(CH_2)_{16}CH_3$.

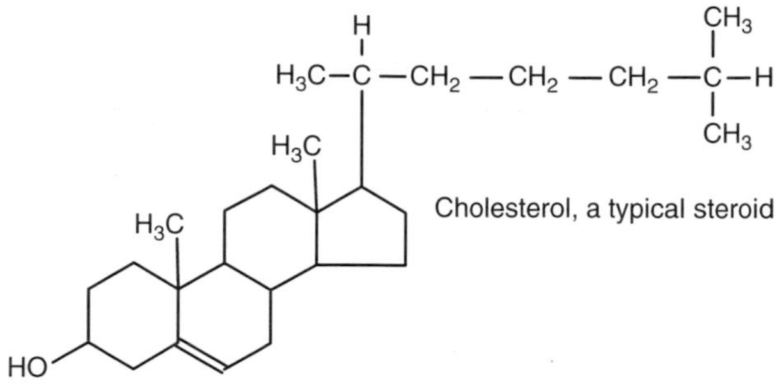

Cholesterol, a typical steroid

Figure 10.8. Steroids are characterized by the ring structure shown here for cholesterol.

consisting of extremely small oil droplets suspended in water. This makes lanolin useful for skin creams and pharmaceutical ointments. Carnauba wax occurs as a coating on the leaves of some Brazilian palm trees. Spermaceti wax is composed largely of cetyl palmitate,

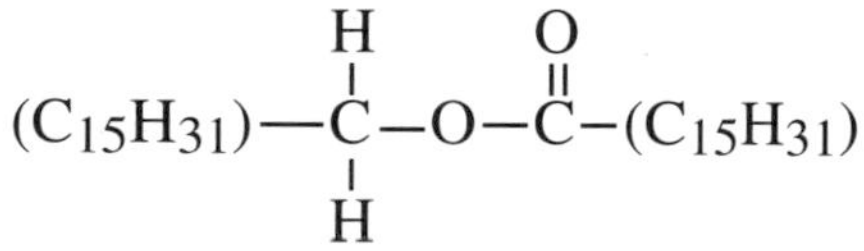

$$(C_{15}H_{31})-\underset{\displaystyle H}{\overset{\displaystyle H}{\underset{|}{\overset{|}{C}}}}-O-\overset{\displaystyle O}{\overset{\|}{C}}-(C_{15}H_{31})$$

extracted from the blubber of the sperm whale. It is very useful in some cosmetics and pharmaceutical preparations.

Steroids are lipids found in living systems, all of which are based on the ring system shown in Figure 10.8 for cholesterol. Steroids occur in bile salts, which are produced by the liver and then secreted into the intestines. Bile salts act upon fats in the intestine. They suspend very tiny fat droplets in the form of colloidal emulsions. This enables the fats to be broken down chemically and digested.

Some steroids are **hormones**. Hormones act as "messengers" from one part of the body to another. As such, they start and stop a number of body functions. Male and female sex hormones are examples of steroid hormones. Hormones are produced by glands in the body called **endocrine glands**. The locations of the important endocrine glands are shown in Figure 10.9.

10.6. ENZYMES

Catalysts are substances that speed up a chemical reaction without themselves being consumed in the reaction. The most sophisticated catalysts of all are those found in living systems. They bring about reactions that could not be performed at all, or only with great difficulty, outside a living organism. These catalysts are called **enzymes**. In addition to speeding up reactions by as much as 10–100 million-fold, enzymes are extremely selective in the reactions that they promote.

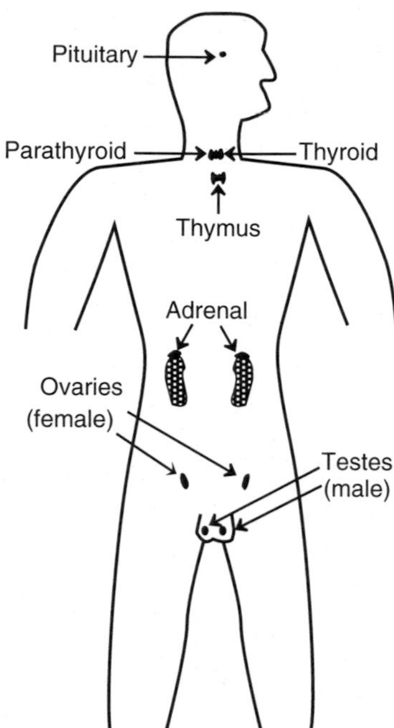

Figure 10.9. Locations of important endocrine glands.

Enzymes are proteins with highly specific structures that interact with particular substances or classes of substances called **substrates**. Enzymes act as catalysts to enable biochemical reactions to occur, after which they are regenerated intact to take part in additional reactions. The extremely high specificity with which enzymes interact with substrates results from their "lock-and-key" action based upon the unique shapes of enzymes, as illustrated in Figure 10.10: an enzyme "recognizes" a particular substrate by its molecular structure and binds to it to produce an **enzyme–substrate complex**. This complex then breaks apart to form one or more products different from the original substrate, regenerating the unchanged enzyme, which is then available to catalyze additional reactions. The basic process for an enzyme reaction is, thus,

$$\begin{aligned} \text{enzyme} + \text{substrate} &\rightleftharpoons \text{enzyme–substrate complex} \\ &\rightleftharpoons \text{enzyme} + \text{product} \end{aligned} \quad (10.6.1)$$

Several important things should be noted about this reaction. As shown in Figure 10.10, an enzyme acts on a specific substrate to form an enzyme–substrate complex because of the fit between their structures. As a result, something happens to the substrate molecule. For example, it might be split in two at a particular location. Then the enzyme–substrate complex comes apart, yielding the enzyme and products. The enzyme is not changed in the reaction and is now free to react again. Note that the arrows in the formula for enzyme reaction point both ways. This means

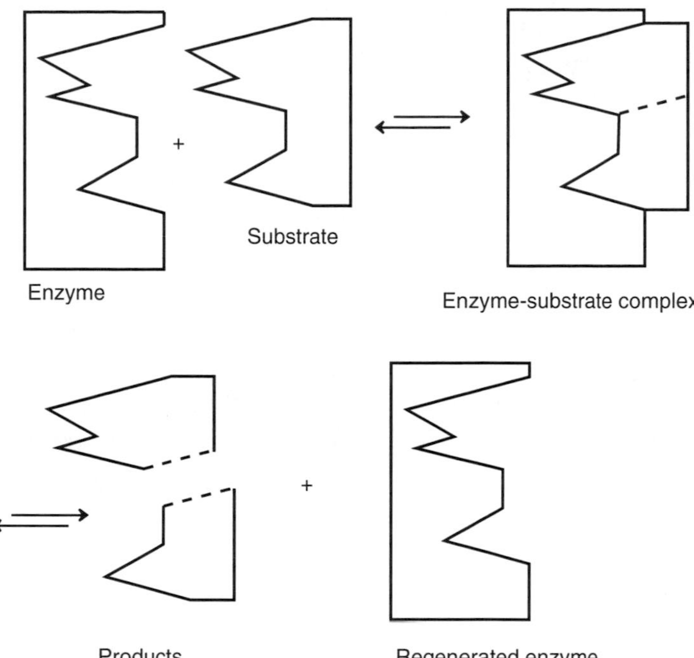

Figure 10.10. Representation of the "lock-and-key" mode of enzyme action that enables the very high specificity of enzyme-catalyzed reactions.

that the reaction is **reversible**. An enzyme–substrate complex can simply go back to the enzyme and the substrate. The products of an enzymatic reaction can react with the enzyme to form the enzyme–substrate complex again. It, in turn, may again form the enzyme and the substrate. Therefore, the same enzyme may act to cause a reaction to go either way.

Some enzymes cannot function by themselves. In order to work, they must first be attached to **coenzymes**. Coenzymes normally are not proteins. Some of the vitamins are important coenzymes.

Enzymes, and classes of enzymes, are generally named for what they do. For example, the stomach produces an enzyme, pepsin, that splits proteins as part of the digestion process; this enzyme is a *gastric proteinase*. The "gastric" part of the name refers to the enzyme's origin in the stomach; the "proteinase" denotes that it splits up protein molecules. Similarly, the enzyme produced by the pancreas that breaks down fats (lipids) is a *pancreatic lipase*; it is commonly called steapsin. In general, lipases cause lipid triglycerides to dissociate and form glycerol and fatty acids.

The enzymes mentioned above are **hydrolyzing enzymes** (or **hydrolases**), which bring about the breakdown of high-molecular-mass biological compounds by the addition of water. This is one of the most important types of reaction involved in digestion. The three main classes of energy-yielding foods that animals eat are carbohydrates, proteins, and fats. Recall that the higher carbohydrates that humans eat are largely disaccharides (sucrose, or table sugar) and polysaccharides (starch).

These are formed by the joining together of units of simple sugars, $C_6H_{12}O_6$, with the elimination of an H_2O molecule at the linkage where they join. Proteins are formed by the condensation of amino acids, again with the elimination of a water molecule at each linkage. Fats are esters that are produced when glycerol and fatty acids link together. A water molecule is lost for each of these linkages when a protein, fat, or carbohydrate is synthesized. In order for these substances to be used as a food source, the reverse process must occur to break down large, complicated molecules of protein, fat, or carbohydrate to simple, soluble substances that can penetrate a cell membrane and take part in chemical processes in the cell. This reverse process is accomplished by hydrolyzing enzymes.

Biological compounds with long chains of carbon atoms are broken down into molecules with shorter chains by the breaking of carbon–carbon bonds. This commonly occurs by the elimination of $-CO_2H$ groups from carboxylic acids. For example, the enzyme *pyruvic decarboxylase* acts upon pyruvic acid to split off CO_2 and produce a compound with one less carbon:

$$\underset{\text{Pyruvic acid}}{H_3C{-}C(=O){-}C(=O){-}OH} \xrightarrow[\text{Decarboxylase}]{\text{Pyruvate}} \underset{\text{Acetaldehyde}}{H_3C{-}C(=O){-}H} + CO_2 \qquad (10.6.2)$$

It is by such carbon–carbon breakdown reactions that long-chain compounds are eventually degraded to CO_2 in the body, and long-chain hydrocarbons undergo biodegradation by bacteria.

Oxidation and reduction are the major reactions for the exchange of energy in living systems. Cellular respiration, discussed in Section 10.2, is an oxidation reaction in which a carbohydrate, $C_6H_{12}O_6$, is broken down to carbon dioxide and water with the release of energy:

$$C_6H_{12}O_6 + 6O_2 \rightarrow 6CO_2 + 6H_2O + \text{energy} \qquad (10.6.3)$$

Actually, this overall reaction occurs in living systems by a complicated series of individual steps. Some of these steps involve oxidation. The enzymes that bring about oxidation in the presence of free O_2 are called **oxidases.** In general, biological oxidation–reduction reactions are catalyzed by **oxidoreductases**.

In addition to the types of enzymes discussed above, there are many enzymes that perform miscellaneous duties in living systems. Typical of these are **isomerases**, which form isomers of particular compounds. For example, there are several simple sugars with the formula $C_6H_{12}O_6$. However, only glucose can be used directly for cellular processes. The other isomers are converted to glucose by the action of isomerases. **Transferases** move chemical groups from one molecule to another, **lyases** remove chemical groups without hydrolysis and participate in the formation of C=C bonds or addition of species to such bonds, and **ligases** work in conjunction with ATP (adenosine triphosphate, a high-energy molecule that plays a crucial role in energy-yielding, glucose-oxidizing metabolic processes) to link molecules together with the formation of bonds such as carbon–carbon or carbon–sulfur bonds.

Enzyme action may be affected by many different things. Enzymes require a certain hydrogen ion concentration to function best. For example, pepsin requires the acid environment of the stomach to work well. When it passes into the much less acidic intestines, it stops working. This prevents damage to the intestinal walls, which would occur if the enzyme tried to digest them. Temperature is critical. Not surprisingly, the enzymes in the human body work best at around 98.6°F (37°C), which is the normal body temperature. Heating these enzymes to around 140°F permanently destroys them. Some bacteria that thrive in hot springs have enzymes that work best at relatively high temperatures. Other, "cold-seeking," bacteria have enzymes adapted to near the freezing point of water.

One of the greatest concerns regarding the effects of surroundings upon enzymes is the influence of toxic substances. A major mechanism of toxicity is the alteration or destruction of enzymes by agents such as cyanide, heavy metals, and organic compounds, such as the insecticide parathion. An enzyme that has been destroyed obviously cannot perform its designated function, whereas one that has been altered may either not function at all or may act improperly. Toxicants can affect enzymes in several ways. Parathion, for example, bonds covalently to the nerve enzyme acetylcholinesterase, which can then no longer serve to stop nerve impulses. Heavy metals tend to bind to sulfur atoms in enzymes (such as sulfur from the amino acid cysteine shown in Figure 10.2), thereby altering the shape and function of the enzyme. An enzymes can be denatured by some poisons, causing it to "unravel" so that it no longer has its crucial specific shape.

10.7. NUCLEIC ACIDS

The "essence of life" is contained in **deoxyribonucleic acid** (**DNA**, which stays in the cell nucleus) and **ribonucleic acid** (**RNA**, which functions in the cytoplasm). These substances, which are known collectively as **nucleic acids**, store and pass on essential genetic information that controls reproduction and protein synthesis.

The structural formulas of the monomeric constituents of nucleic acids are given in Figure 10.11. These are pyrimidine or purine nitrogen-containing bases, two sugars, and phosphate. DNA molecules are made up of the bases **adenine**, **guanine**, **cytosine**, and **thymine**; phosphoric acid (H_3PO_4); and the simple sugar 2-deoxy-β-D-ribofuranose (commonly called **deoxyribose**). RNA molecules are composed of the bases adenine, guanine, cytosine, and **uracil**; H_3PO_4; and the simple sugar β-D-ribofuranose (**ribose**).

The formation of nucleic acid polymers from their monomeric constituents may be viewed as the following steps:

- Monosaccharide (simple sugar) + cyclic nitrogenous base yields a **nucleoside**:

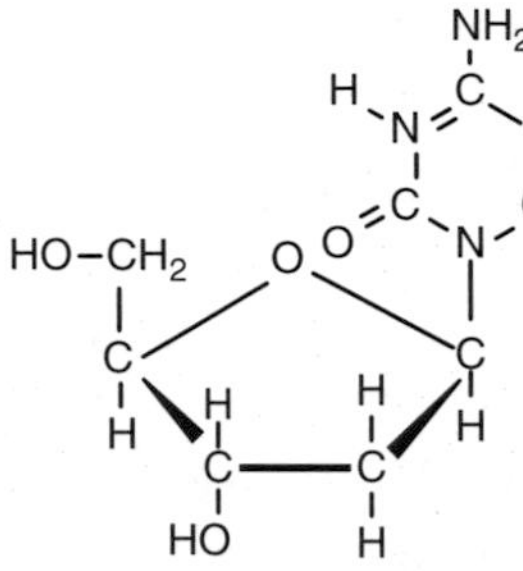

Deoxycytidine formed by the dimerization of cytosine and deoxyribose with the elimination of a molecule of H_2O

- Nucleoside + phosphate yields a phosphate ester, a **nucleotide**:

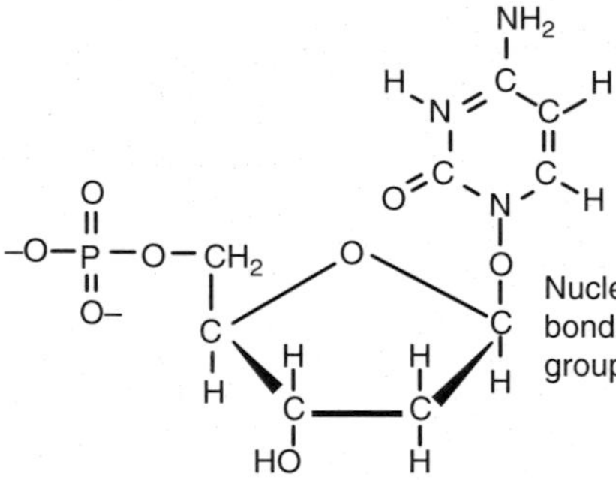

Nucleotide formed by the bonding of a phosphate group to deoxycytidine

- Polymerized nucleotide yields a **nucleic acid**:

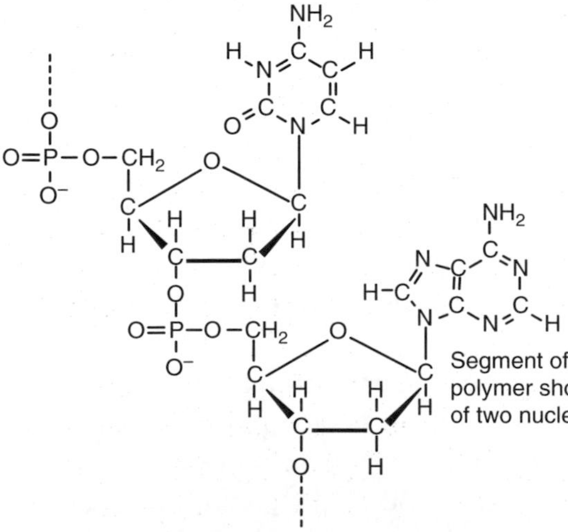

Segment of the DNA polymer showing linkage of two nucleotides

In the nucleic acid, the phosphate negative charges are neutralized by metal cations (such as Mg^{2+}) or positively charged proteins (**histones**).

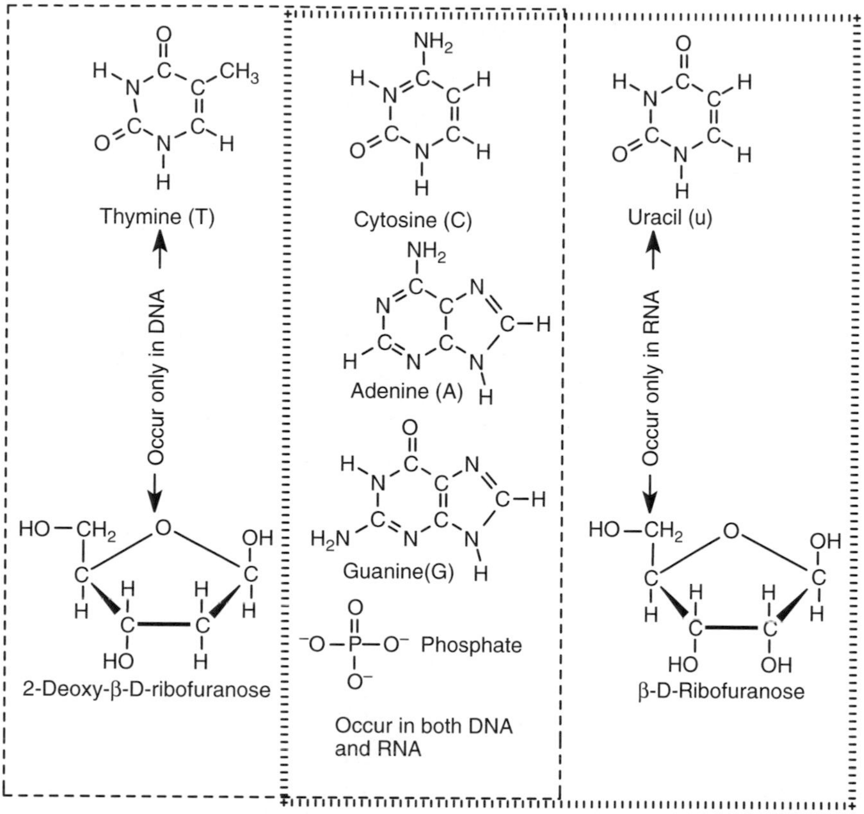

Figure 10.11. Constituents of DNA (enclosed by ----) and of RNA (enclosed by ıııı).

Molecules of DNA are huge, with molecular masses greater than one billion. Molecules of RNA are also quite large. The structure of DNA is that of the famed "double helix" (Figure 10.12). It was figured out in 1953 by an American scientist, James D. Watson, and a British scientist, Francis Crick. They received the Nobel Prize for this scientific milestone in 1962. This model visualizes DNA as a so-called double α-helix structure of oppositely wound polymeric strands held together by hydrogen bonds between opposing pyrimidine and purine groups. As a result, DNA has both a primary and a secondary structure; the former is due to the sequence of nucleotides in the individual strands of DNA and the latter results from the α-helix interaction of the two strands. In the secondary structure of DNA, only cytosine can be opposite guanine and only thymine can be opposite adenine, and vice versa. Basically, the structure of DNA is that of two spiral ribbons "counter-wound" around each other as illustrated in Figure 10.12. The two strands of DNA are **complementary**. This means that a particular portion of one strand fits like a key in a lock with the corresponding portion of another strand. If the two strands are pulled apart, each manufactures a new complementary strand, so that two copies of the original double helix result. This occurs during cell reproduction.

The molecule of DNA acts like a coded message. This "message," the genetic information contained in and transmitted by nucleic acids, depends upon the sequence of bases from which they are composed. It is somewhat like a message sent

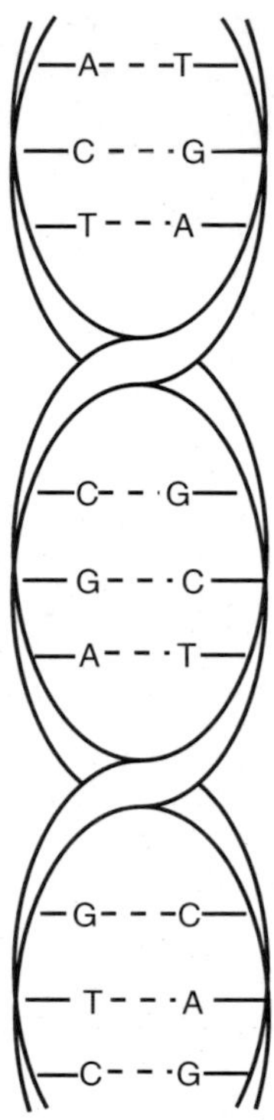

Figure 10.12. Representation of the double-helix structure of DNA showing the allowed base pairs held together by hydrogen bonding between the phosphate/sugar polymer "backbones" of the two strands of DNA. The letters stand for adenine (A), cytosine (C), guanine (G), and thymine (T). The dashed lines, ---, represent hydrogen bonds.

by telegraph, which consists only of dots, dashes, and spaces in between. The key aspect of DNA structure that enables storage and replication of this information is the famed double helix mentioned above.

Portions of the DNA double helix may unravel, and one of the strands of DNA may produce a strand of RNA. This substance then goes from the cell nucleus out into the cell and regulates the synthesis of new protein. In this way, DNA regulates the function of the cell and acts to control life processes.

Nucleic Acids in Protein Synthesis

Whenever a new cell is formed, the DNA in its nucleus must be accurately reproduced from the parent cell. Life processes are absolutely dependent upon accurate protein synthesis as regulated by cellular DNA. The DNA in a single cell must be capable of directing the synthesis of up to 3000 or even more different proteins. The directions for the synthesis of a single protein are contained in a segment of DNA called a **gene**. The process of transmitting information from DNA to a newly synthesized protein involves the following steps:

- The DNA undergoes **replication**. This process involves separation of a segment of the double helix into separate single strands, which then replicate such that guanine is opposite cytosine (and vice versa) and adenine is opposite thymine (and vice versa). This process continues until a complete copy of the DNA molecule has been produced.

- The newly replicated DNA produces **messenger RNA (mRNA)**, a complement of the single strand of DNA, by a process called **transcription**.
- A new protein is synthesized using mRNA as a template to determine the order of amino acids in a process called **translation**.

Modified DNA

DNA molecules may be modified by the unintentional addition or deletion of nucleotides or by substituting one nucleotide for another. The result is a **mutation** that is transmittable to offspring. Mutations can be induced by chemical substances. This is a major concern from a toxicological viewpoint because of the detrimental effects of many mutations and because substances that cause mutations often cause cancer as well. DNA malfunction may result in birth defects, and the failure to control cell reproduction results in cancer. Radiation from X-rays and radioactivity also disrupts DNA, and may cause mutation.

10.8. RECOMBINANT DNA AND GENETIC ENGINEERING

As noted above, segments of DNA contain information for the specific syntheses of particular proteins. Within the last two decades, it has become possible to transfer this information between organisms by means of **recombinant DNA technology**, which has resulted in a new industry based on **genetic engineering**. Most often, the recipient organisms are bacteria, which can be reproduced (cloned) over many orders of magnitude from a cell that has acquired the desired qualities. Therefore, to synthesize a particular substance, such as human insulin or growth hormone, the required genetic information can be transferred from a human source to bacterial cells, which then produce the substance as part of their metabolic processes.

The first step in recombinant DNA gene manipulation is to lyze ("open up") a donor cell to remove needed DNA material by using enzyme action to cut the sought-after genes from the donor DNA chain. These are next spliced into small DNA molecules. These molecules, called **cloning vehicles**, are capable of penetrating the host cell and becoming incorporated into its genetic material. The modified host cell is then reproduced many times and carries out the desired biosynthesis.

Early concerns about the potential of genetic engineering to produce "monster organisms" or new and horrible diseases have been largely allayed, although caution is still required with this technology. In the environmental area, genetic engineering offers some hope for the production of bacteria engineered to safely destroy troublesome wastes and to produce biological substitutes for environmentally damaging synthetic pesticides.

10.9. METABOLIC PROCESSES

Biochemical processes that involve the alteration of biomolecules fall under the category of **metabolism**. Metabolic processes may be divided into the two major categories of **anabolism** (synthesis) and **catabolism** (degradation of substances). An organism may use metabolic processes to yield energy or to modify the constituents of biomolecules.

Energy-Yielding Processes

Organisms can gain energy by the following three processes:

- **Respiration**, in which organic compounds undergo catabolism that requires molecular oxygen (**aerobic or oxic respiration**) or that occurs in the absence of molecular oxygen (**anaerobic or anoxic respiration**). Aerobic respiration uses the **Krebs cycle** to obtain energy from the reaction

$$C_6H_{12}O_6 + 6O_2 \rightarrow 6CO_2 + 6H_2O + \text{energy}$$

About half of the energy released is converted to short-term stored chemical energy, particularly through the synthesis of the nucleoside **adenosine triphosphate** (**ATP**). For longer-term energy storage, glycogen or starch polysaccharides are synthesized, and for still longer-term energy storage, lipids (fats) are generated and retained by the organism.

- **Fermentation**, which differs from respiration in not having an electron transport chain. Yeasts produce ethanol from sugars by fermentation:

$$C_6H_{12}O_6 \rightarrow 2CO_2 + 2C_2H_5OH$$

- **Photosynthesis**, in which light energy captured by plant and algal chloroplasts is used to synthesize sugars from carbon dioxide and water:

$$6CO_2 + 6H_2O + h\nu \rightarrow C_6H_{12}O_6 + 6O_2$$

Plants cannot always get the energy that they need from sunlight. During the dark, they must use stored food. Plant cells, like animal cells, contain mitochondria in which stored food is converted to energy by cellular respiration.

Plant cells, which use sunlight for energy and CO_2 for carbon, are said to be **autotrophic**. In contrast, animal cells must depend upon organic material manufactured by plants for their food. These are called **heterotrophic** cells. They act as "middlemen" in the chemical reaction between oxygen and food material, using the energy from the reaction to carry out their life processes.

CHAPTER SUMMARY

The chapter summary below is presented in a programmed format to review the main points covered in this chapter. It is used most effectively by filling in the blanks, referring back to the chapter as necessary. The correct answers are given at the end of the summary.

The biomolecules that constitute matter in living organisms are often high-molecular-mass [1]______, divided into the categories of [2]_____________. In respect to affinity for water versus lipids, the behavior of a substance in a biological system depends to a large extent upon whether the substance is [3]______ or [4]_______. The two major kinds of cells are [5]______ cells that have a defined nucleus and [6]______ cells that do not. Several major features of the eukaryotic cell are [7]_________. Three other features specifically characteristic of plant cells are [8]_______. Proteins are macromolecular biomolecules composed of [9]_______. A generalized structure for amino acids is [10]_______, where "R" represents a group characteristic of each amino acid. Amino acids are bonded together in proteins by a specific arrangement of atoms and bonds called a [11]________. Casein is a protein that functions as a [12]_________, hemoglobin is a [13]________ protein, and collagen is a [14]______ protein. Proteins exhibit [15]______ different orders of structure, the first of which is determined by the [16]_____. Protein denaturation consists of disruption of [17]________. Carbohydrates have the approximate simple formula [18]________. The molecular formula of glucose is [19]_______. Polysaccharides are carbohydrates that consist of many [20]_________. An important polysaccharide produced by plants for food storage is [21]________. Cellulose is a polysaccharide produced by [22]________ that forms [23]__________. Substances that can be extracted from plant or animal matter by organic solvents such as chloroform are called [24]__________, the most common of which are triglycerides formed from [25]_________. Two reasons that lipids are toxicologically important are that [26]_________________. A special class of lipids with a characteristic ring system, such as that in cholesterol, consists of [27]_____________. Some steroids are [28]____________ that act as "messengers" from one part of the body to another. Enzymes are proteins with highly specific structures that function as biochemical [29]________ and interact with particular substances or classes of substances called [30]___________. Enzymes are very specific in the kinds of substance that they interact with because of their unique [31]____________. In order to work, some enzymes must first be attached to [32]________. Enzymes are named for [33]_______________. For example, the gastric proteinase pepsin comes from the [34]_________ and acts to [35]_______. It is an example of a general class of enzymes called [36]________. As another example, biological oxidation–reduction reactions are catalyzed by [37]_________ Deoxyribonucleic acid (DNA) and ribonucleic acid (RNA) are the two major kinds of biomolecules classified as [38] ________. The three kinds of monomers from which they are composed are [39]__________. Cellular DNA regulates synthesis of [40]__________. Biochemical processes that involve the alteration of biomolecules are termed [41]_______. The three processes by which organisms gain energy are [42]_______.

Answers to Chapter Summary

1. polymers
2. carbohydrates, proteins, lipids, and nucleic acids
3. hydrophilic
4. hydrophobic
5. eukaryotic
6. prokaryotic
7. cell membrane, cell nucleus, cytoplasm, mitochondria, ribosomes, endoplasmic reticulum, lysosome, Golgi bodies
8. cell walls, vacuoles, and chloroplasts
9. amino acids
10.
```
     H   H
      \ /
       N  O
       |  ||
   R—C—C—OH
       |
       H
```
11. peptide linkage
12. nutrient
13. transport
14. structural
15. four
16. order of amino acids
17. secondary, tertiary, and quaternary protein structures
18. CH_2O
19. $C_6H_{12}O_6$
20. simple sugar units linked together
21. starch

22. plants
23. the structural material of plant cell walls
24. lipids
25. glycerol and long-chain fatty acids
26. some toxic substances interfere with lipid metabolism, and many toxic organic compounds are lipid-soluble
27. steroids
28. hormones
29. catalysts
30. substrates
31. shapes or structures
32. coenzymes
33. what they do
34. stomach
35. split apart protein molecules
36. hydrolyzing enzymes
37. oxidoreductase enzymes
38. nucleic acids
39. nitrogen-containing bases, phosphoric acid, and simple sugars
40. proteins
41. metabolism
42. respiration, fermentation, and photosynthesis

LITERATURE CITED

1. Stanley E. Manahan. *Toxicological Chemistry and Biochemistry*, 3rd ed. Taylor & Francis/CRC Press, Boca Raton, Florida, 2003.
2. Horton, Robert, Laurence A. Moran, Gray Scrimgeour, Marc Perry, and David Rawn. *Principles of Biochemistry*, 4th ed. Prentice Hall, Upper Saddle River, NJ, 2005.

SUPPLEMENTARY REFERENCES

Berg, Jeremy M., John L. Tymoczko, and Lubert Stryer. *Biochemistry*, 6th ed. W. H. Freeman, New York, 2006.

Bettelheim, Frederick A., William H. Brown, and Jerry March. *Introduction to Organic and Biochemistry*, 5th ed. Brooks Cole, Pacific Grove, CA, 2003.

Boyer, Rodney F. *Concepts in Biochemistry*, 3rd ed. Wiley, Hoboken, NJ, 2006.

Campbell, Mary K. and Shawn O. Farrell. *Biochemistry*, 6th ed. Brooks Cole, Pacific Grove, CA, 2007.

Champe, Pamela C., Richard A. Harvey, and Denise R. Ferrier. *Biochemistry*. Lippincott Williams & Wilkins, Philadelphia, 2005.

Denniston, Katherine J., Joseph J. Topping, and Robert L. Caret. *General, Organic, and Biochemistry*. McGraw-Hill, Boston, 2007.

Elliott, William H. and Dephne C. Elliott. *Biochemistry and Molecular Biology*, 3rd ed. Oxford University Press, New York, 2005.

Garrett, Reginald H. and Charles M. Grisham. *Biochemistry*, 3rd ed. Thomson Brooks/Cole, Belmont, CA, 2007.

Hames, B. D., and Nigel Hooper. *Biochemistry*, 3rd ed. Taylor & Francis, New York, 2005.

Hein, Morris. *Introduction to General, Organic, and Biochemistry*. Wiley, Hoboken, NJ, 2005.

Nelson, David, and Michael M. Cox, Eds. *Lehninger Principles of Biochemistry*, 4th ed. W. H. Freeman, New York, 2006.

Raymond, Kenneth W., *General, Organic, and Biological Chemistry: An Integrated Approach*. Wiley, Hoboken, NJ, 2006.

Swanson, Todd A., Sandra I. Kim, and Marc J. Glucksman. *Biochemistry and Molecular Biology*, 4th ed. Lippincott Williams & Wilkins, Baltimore, 2007.

Voet, Donald and Judith G. Voet. *Fundamentals of Biochemistry*, 2nd ed. Wiley, Hoboken, NJ, 2005.

QUESTIONS AND PROBLEMS

1. What is the toxicological importance of lipids? How do lipids relate to hydrophobic ("water-disliking") pollutants and toxicants?

2. What is the function of a hydrolase?

3. Match the cell structure on the left with its function on the right:

A. Mitochondria	1. Toxicant metabolism
B. Endoplasmic reticulum	2. Fills the cell
C. Cell membrane	3. Deoxyribonucleic acid
D. Cytoplasm	4. Mediate energy conversion and utilization
E. Cell nucleus	5. Encloses the cell and regulates the passage of materials into and out of the cell interior

4. The formula of simple sugars is $C_6H_{12}O_6$. The simple formula of higher carbohydrates is $C_6H_{10}O_5$. Of course, many of these units are required to make a molecule of starch or cellulose. If higher carbohydrates are formed by joining together molecules of simple sugars, why is there a difference in the ratios of C, H, and O atoms in the higher carbohydrates as compared with the simple sugars?

5. Why does wood contain so much cellulose?

6. What would be the chemical formula of a *tri*saccharide made by the bonding together of three simple sugar molecules?

7. The general formula of cellulose may be represented as $(C_6H_{10}O_5)_x$. If the molecular weight of a molecule of cellulose is 400 000, what is the estimated value of x?

8. During one month, a factory for the production of simple sugars, $C_6H_{12}O_6$, by the hydrolysis of cellulose processes 10^6 kg of cellulose. The percentage of cellulose that undergoes the hydrolysis reaction is 40%. How many kilograms of water are consumed in the hydrolysis of cellulose each month?

9. What is the structure of the largest group of atoms common to all amino acid molecules?

10. Glycine and phenylalanine can join together to form two different dipeptides. What are the structures of these two dipeptides?

11. One of the ways in which two parallel protein chains are joined together, or cross-linked, is by way of an –S–S– link. What amino acid to you think might be most likely to be involved in such a link? Explain your choice.

12. Fungi, which break down wood, straw, and other plant material, have what are called "exoenzymes." Fungi have no teeth and cannot break up plant

material physically by force. Knowing this, what do you suppose an exoenzyme is? Explain how you think it might operate in the process by which fungi break down something as tough as wood.

13. Many fatty acids of lower molecular weight have a bad odor. Speculate as to the reasons that rancid butter has a bad odor. What chemical compound is produced that has this bad odor? What sort of chemical reaction is involved in its production?

14. The long-chain alcohol with 10 carbons is called decanol. What do you think would be the formula of decyl stearate? To what class of compounds would it belong?

15. Write an equation for the chemical reaction between sodium hydroxide and cetyl stearate. What are the products?

16. What are two endocrine glands that are found only in females? Which of these glands is found only in males?

17. The action of bile salts is a little like that of soap. What function do bile salts perform in the intestine? Look up the action of soaps, and explain how you think bile salts may function somewhat like soap.

18. If the structure of an enzyme is illustrated as

how might the structure of its substrate be represented?

19. Look up the structures of ribose and deoxyribose. Explain where the “deoxy” came from in the name, deoxyribose.

20. In what respect are an enzyme and its substrate like two opposite strands of DNA?

21. For what discovery are Watson and Crick noted?

22. Why does an enzyme no longer work if it is denatured?

INDEX

C

D

E

I

K

L

T

U

V

W